ROLEX

PHILOSOPHY

ISBN: 978 1 78884 239 6

First published in Italian under the title Rolex Philosophy in 2023 by White Star

WHITE STAR PUBLISHERS

WS whitestar™ is a trademark property of White Star s.r.l.

Piazzale Luigi Cadorna, 6
20123 Milan, Italy
www.whitestar.it

This translation published in 2023 by ACC Art Books by agreement with White Star s.r.l.
Reprinted 2025, 2026

The author and publisher gratefully acknowledge the permission granted to reproduce the copyright material in this book. Every effort has been made to trace copyright holders and to obtain their permission for the use of copyright material. Any errors or omissions are entirely unintentional and should be addressed to the publisher.

A CIP catalogue record for this book is available from the British Library

English Edition
Translator: Megan Bredeson
Editors: Phillip Gaskill, White Star; Sue Bennett, ACC Art Books
Production: Steve Farrow, ACC Art Books

EU GPSR Authorised Representative:
Easy Access System Europe Oü, 16879218
Address: Mustamäe tee 50, 10621 Tallinn, Estonia
Email: gpsr@easproject.com Tel: +358 40 500 3575

FSC® www.fsc.org MIX Paper from responsible sources FSC® C178000

Printed in China
for ACC Art Books Ltd, Woodbridge, Suffolk, UK
www.accartbooks.com

ACC ART BOOKS

MARA CAPPELLETTI

ROLEX

PHILOSOPHY

ACC ART BOOKS

CONTENTS

A GLIMPSE AT THE LEGEND

....

'Luxury causes one to dream. The concept of luxury envelops products in a chic veil of mystery, products that without this mysterious aura would remain in the shadows upon shelves and appear to us all some sort of incomprehensible whim, or a social code of identification reserved for wealthy people who are all related to one another.'

Thierry Paquot, *Elogio del lusso. Ovvero l'utilità dell'inutile*
(In praise of luxury. Or the utility of the useless)

According to one anecdote connected to Rolex, a friend of André Heiniger, the company's chairman in the seventies, once asked him 'How's the watch business?', and Heiniger replied, 'I have no idea.' To his friend's surprised expression, he elaborated: 'Rolex is not in the watch business. We are in the luxury business.'

To speak of Rolex inevitably means to speak of luxury. Here the term should be understood, however, not only as a form of consumption connected to money, but one of exclusivity founded on conscious, successful lifestyle decisions. While a company's success can be measured in concrete terms, an encounter with its intangible facets, reinforced by the ineffability of legend, is inescapable.

Rolex is a company holding many records. According to data from analysts, it has the highest sales turnover of any company in the watch industry, a record it has successfully held for years. Included on lists of companies with the highest financial value in Europe and the world, Rolex is one of Switzerland's top companies – one of the world's most sought-after brands and one of the luxury labels that generates the most desire. In addition to growing a significant volume of business, the brand has reached the highest Media Impact Value (MIV) online and on social media in recent years. It is also the most imitated watch brand in the world. Again, the watch brand that bears a crown is recognised as having the best, most clearly established identity, one to which authenticity is attributed in both its product and its message. The Geneva-based watchmaker has also won first place for the best brand reputation for several years, topping rankings drawn up by RepTrak, the Boston-based company that each year compiles a report bringing together the hundred best-known brands in the world.

Its success story is one that endures over

time, a textbook case. But how did all this come about, and to what is this popularity due? Could the desire to 'have the time' possibly be so strong as to sell millions of watches – and not just any watches, but timepieces that also create status? Analysts argue that a brand represents a set of intangible attributes and values and can be defined by way of its name, packaging, price, history, reputation and advertising. In the luxury-goods industry, to which Rolex belongs, a brand is synonymous with the name of the company's goods and services; all these components complement one another. If the product has physical properties that characterise it, the brand, defined through images and associations, concepts and ideas, instead belongs to the realm of the senses, and has the capacity to arouse desire.

Concretely, Rolex is recognised for its high degree of professionalism, thanks to its high-tech laboratories, manual production and customer service; but the value of the brand goes further than this, evoking positive emotional connections that are skillfully nurtured. The marketing strategies it has carried out over the years, the involvement of testimonials and ambassadors, and all communication that has been developed have indeed ensured that Rolex timepieces are perceived as a luxury item par excellence, not only within the watchmaking sector, but in a more general context as well.

Rolex, for example, is one of the few luxury brands – if not the only one – that implements communication aimed at the mass market, even if production is limited and watches are available in small quantities. Deprivation marketing, a tool to fuel desire, customer loyalty, sharing of values, and communication: truly, many extremely interesting features characterise Rolex branding. These will be explored in this book through the analysis of experts in psychology, sociology consultants and specialists in communication; each is in a position to offer a stimulating, in-depth reading into one element in the Geneva-based house's complex universe. Equally valuable are the thoughts of major international specialists from the world's most prestigious auction houses, who share their views on the phenomenon of collecting as it pertains to Rolex and their experience surrounding record sales. This set of conversations has birthed a study that investigates the company's philosophy, but which is also itself philosophical, analysing a series of material principles while simultaneously observing and seeking to better understand the 'principle' of an indeterminate reality.

The emblem of a crown refers to something precious and regal, to a prize, to a form of gratification. This mythological component is a strong one in the case of Rolex, understood as both a highly idealised object and something able to draw aspirations and generate desires. And the myth, by its nature, has secret, non-explicit facets: it evokes rather than defining, it narrates rather than speaking – it beckons in a highly symbolic manner, instead of referring to this or that model. It is, after all, about the creation and nurturing of a dream: the dream of Hans Wilsdorf, the illustrious and enlightened founder of Rolex, has certainly come true and continues on through time.

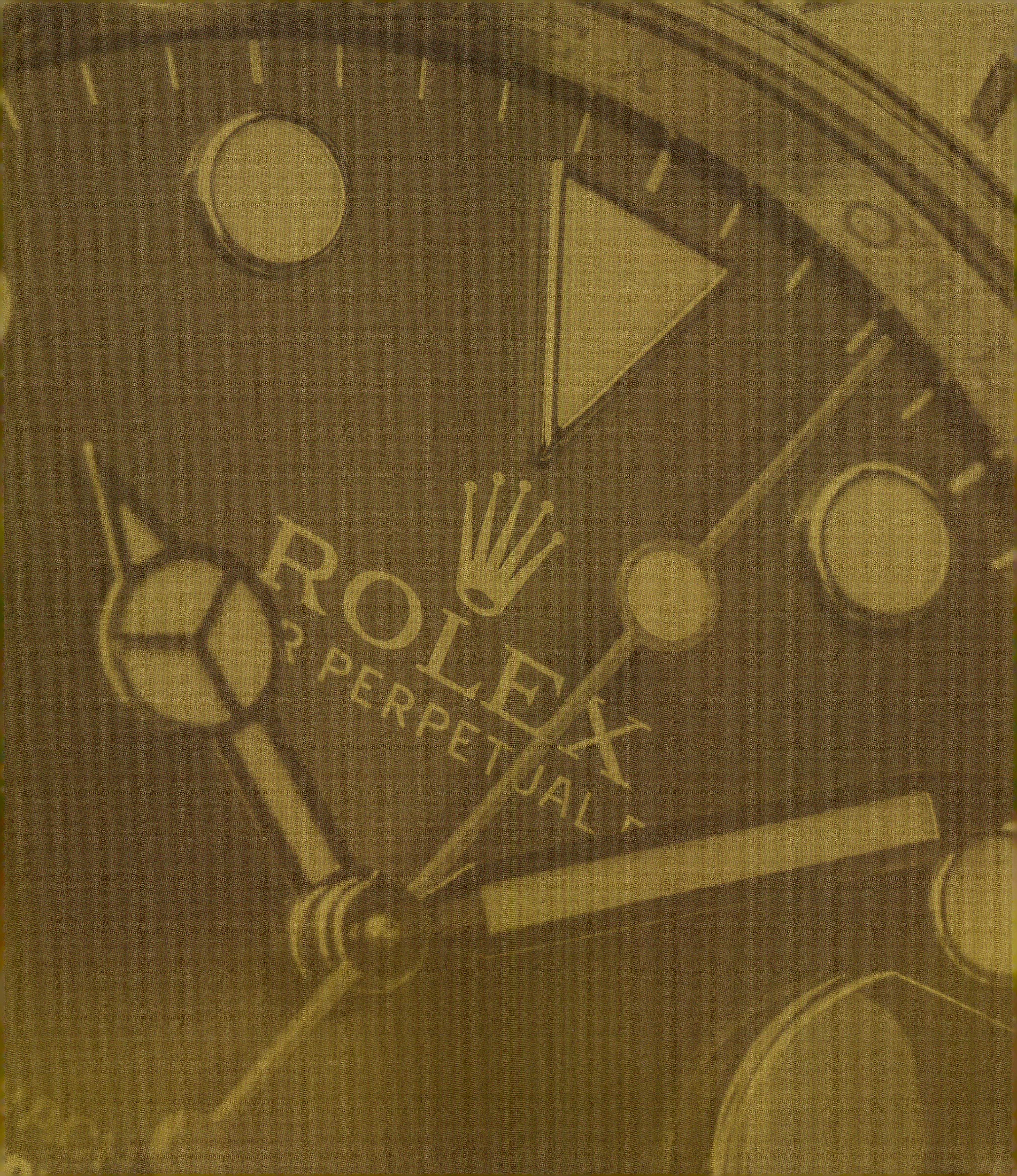
ROLEX
PERPETUAL

THE HISTORY OF THE BRAND

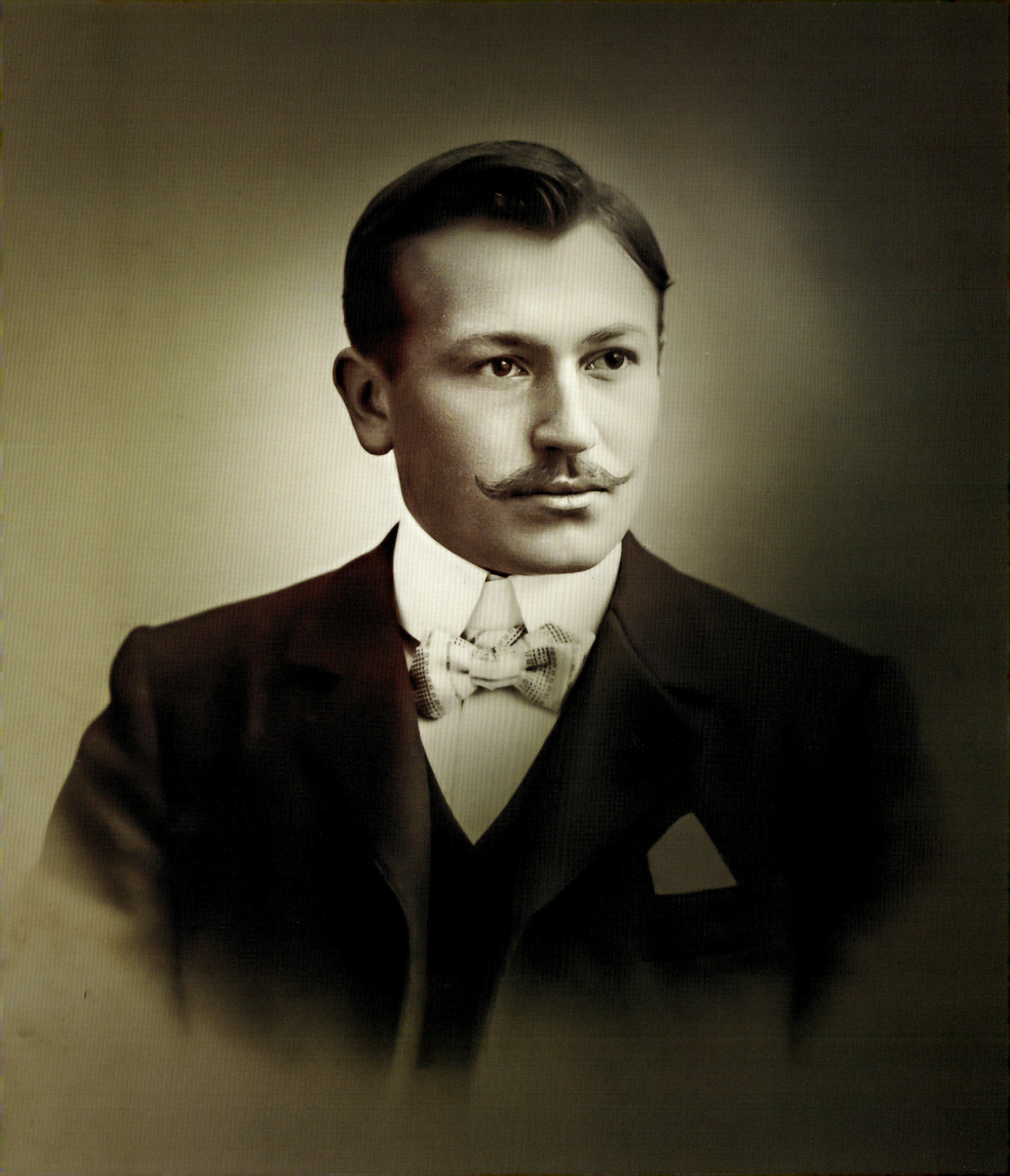

ROLEX: INFINITELY MODERN

....

There are those who have the capacity to see beyond the present day, beyond the limitations imposed by customs, society, economy or logic. These people understand how to make observations and develop a concept – and then to transform that idea into a tangible and complete plan. They see the situation surrounding them objectively and are able to imagine the future. They offer a vision not just as a concept, but as part of the evolution of past eras as their products are continuously, inevitably surpassed.

These individuals are destined to live in a time that is infinite and modern – to create objects taking part in the progress, continually renewing itself within certain aesthetic canons.

Rolex founder Hans Wilsdorf's story falls within the canon of a 19th-century narrative. His origins are fairly mysterious: Wilsdorf was born in Kulmbach, Germany in 1881, and official sources report that, alongside a brother and a sister, the young boy would become an orphan. This was a fate common to many youth in the late 19th century – Coco Chanel, for example, another illustrious personality who would permanently alter women's fashion, had grown up in a convent as little Gabrielle Bonheur. Orphanhood, however, does not at all appear to have limited Hans or Gabrielle. Perhaps, in their determination to overcome life's difficulties, they actually both found the tools that led to their successes.

What is certain is that Hans, together with his brothers, had been taken into the custody of an uncle, who took care to guarantee him a good scholastic education so he might have some job prospects. At the boarding school he attended, Wilsdorf showed a particular predilection for mathematics and languages, which would later facilitate business contacts abroad.

Portrait of Hans Wilsdorf as a young man.

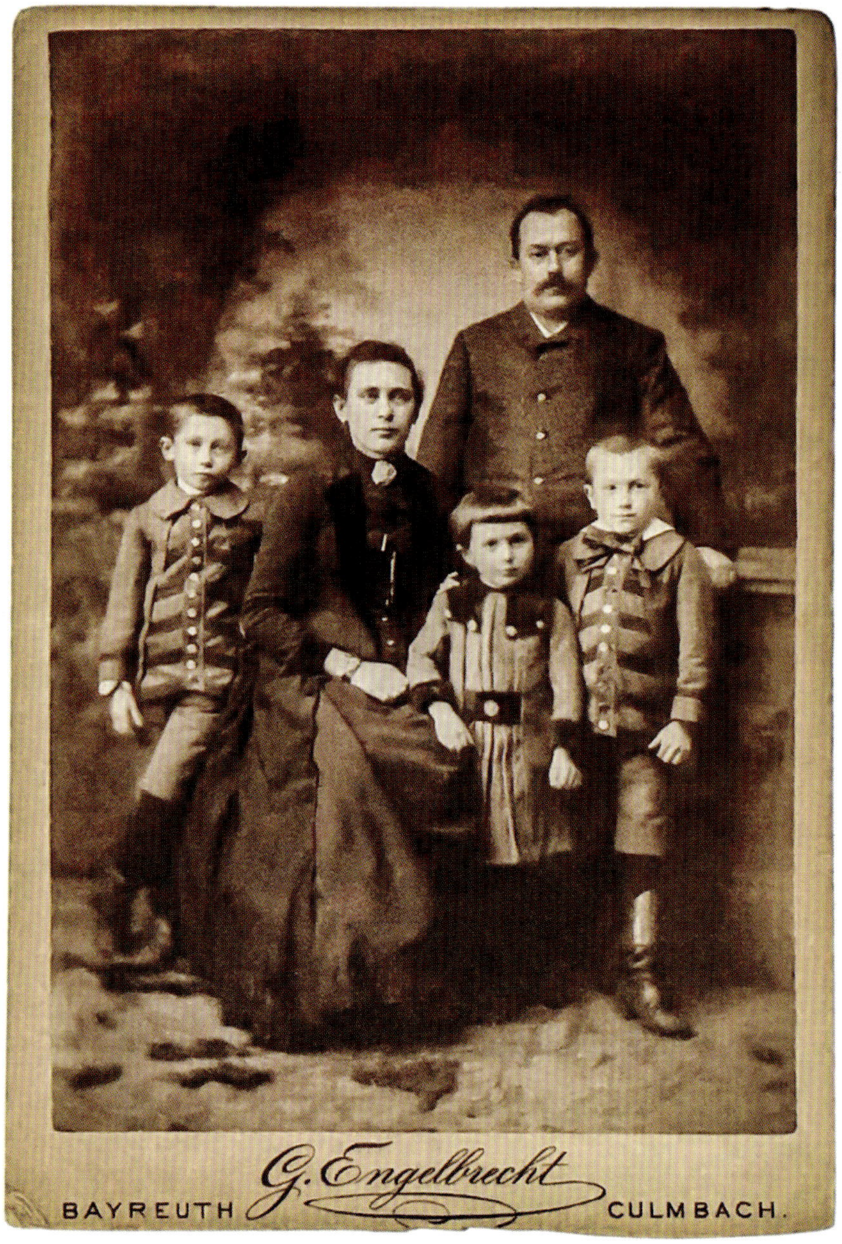

Hans Wilsdorf (front row right) stands in front of his father, with his mother sitting between his sister and brother.

His first work experience was an apprenticeship at a pearl export company. The company's sales organisation covered the entire globe, and in this work he was able to employ the languages he had learned. The skill would prove valuable for the course of his career.

In 1900, at the age of 19, Rolex's founder-to-be arrived in La Chaux-de-Fonds, Switzerland, a predominant centre of watchmaking, by way of a school friend.

His first job was with Cuno Korten, a watch-export company where Wilsdorf conducted correspondence with English-speaking customers. It was here that he began to take an interest in movements and their mechanics.

Fascinated by the precision and accuracy of the timepieces, with a strong belief in the opportunity this industry offered, he moved to London. In 1905, together with businessman and investor Alfred James Davis, he founded the firm Wilsdorf & Davis based at 83 Hatton Garden, London's jewelry district. In 1908, Davis married Wilsdorf's sister Anna and became his brother-in-law, but he does not seem to have ever been actively involved in the company's management.

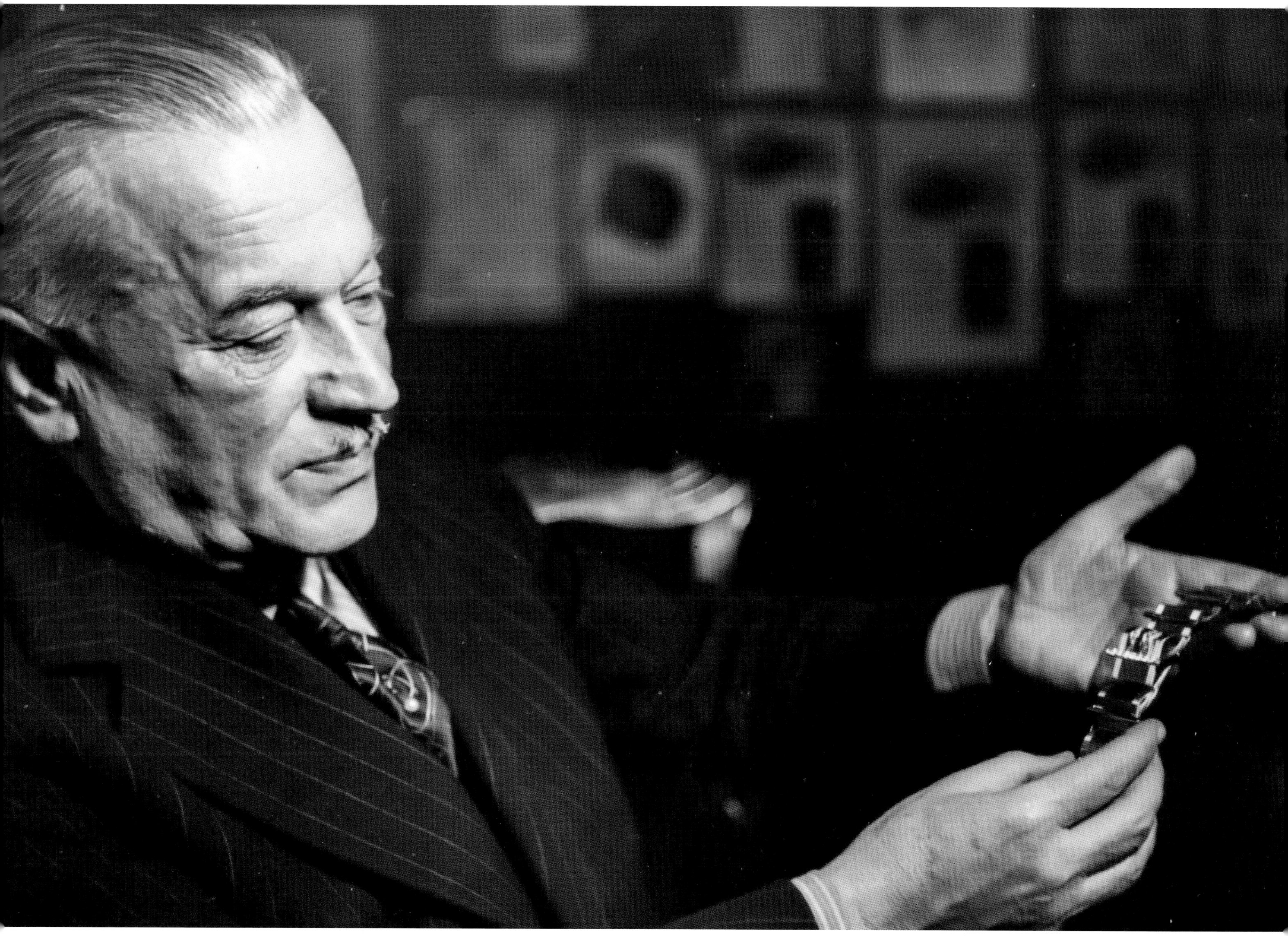

Hans Wilsdorf admires a watch from the *maison*.

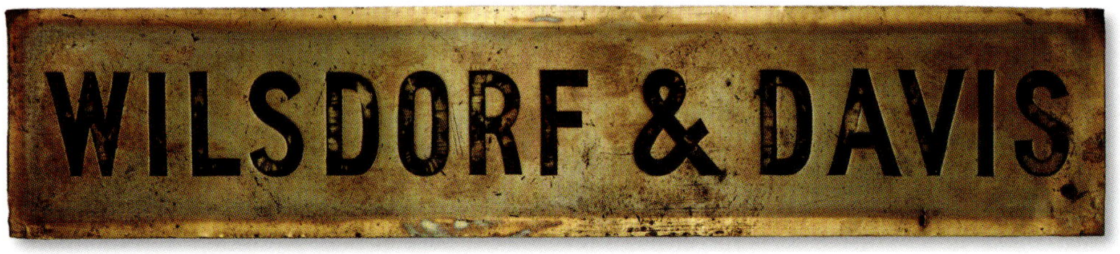

A metal Wilsdorf & Davis company name plate.

First Steps in Entrepreneurship

In its early days, Wilsdorf & Davis imported a wide variety of objects at various price points: one of the first items to gain success was a travel watch in a leather case. Although then the general trend toward pocket watches still prevailed, Wilsdorf sensed that they would soon be replaced by wrist timepieces. In the *Vade Mecum*, a collection of booklets that would be published in 1945 telling the story of Rolex's founding, Wilsdorf recalls doubts that were raised about the success of wrist timepieces:

'Watchmakers all over the world remained skeptical as to its possibilities and believed that this newfangled object was bound to prove a failure.

'Their arguments against the wristlet watch were, inter alia, the following: [...] the mechanism required by this type of watch must of necessity be small and delicate and could therefore never withstand the violent gestures of hand and arm[;] dust and damp would rapidly spoil the mechanism, even if it were well constructed [; and] accuracy and regularity of working could never be attained with so small a movement.'

Wilsdorf writes that he was instead convinced that the wristwatch was the way of the future. He thought there would be a better market in wristwatches than pocket watches, in part because they were more prone to damage (needing replacements) and because, unlike a pocket watch having been passed from generation to generation, wristwatches were to be fashion items customised for the wearer, and customers would choose a series of models to combine with various types of clothing. He therefore decided that this was the type of watch his company would produce; it was now a matter of finding the right technical partner.

While in La Chaux-de-Fonds, Wilsdorf had encountered the Aegler watchmaking company in Bienne, which specialised in the production of small movements and had an excellent reputation for its precision. The manufacturer also had a good supply of pieces and spare parts at affordable prices, thanks to the mechanised production methods it employed.

Wilsdorf sensed that this could be his ideal partner. In 1905, he travelled to Bienne to meet the owner, Hermann Aegler, and secured the provision of an enormous number of wristwatches, then the largest ever ordered. Wilsdorf also states in the *Vade Mecum* that the first timepieces produced under this agreement were men's and women's models with silver cases and leather straps, and that their immediate success after launch prompted him to broaden the range, introducing a line with gold cases.

At the time, Wilsdorf & Davis was not a watch-manufacturing company: it purchased watches from Swiss manufacturers and sold them on the British market through a network of dealers, thanks to which it was also able to reach other international markets, in Asia, for instance, in areas historically connected to the British Empire. While initially no hallmark was required, British law changed in 1907, and new rules required that all gold and silver watch cases imported onto English soil be marked by the Goldsmiths' Company Assay Office prior to sale in Great Britain. This legal change prompted Wilsdorf & Davis to register the 'W&D' trademark to be placed on the watches they marketed. The W&D mark, more specifically, was a responsibility mark, called a 'sponsor's mark' in Britain, identifying the company promoting the item rather than its manufacturer.

This need to adopt a mark for identification likely served as a stimulus to find a name that better represented the young company. Wilsdorf realised that it would be more profitable to build his own brand than to simply act as an anonymous importer, and he began to consider a name. Soon, the brand destined to become one of the most famous in the world would be born.

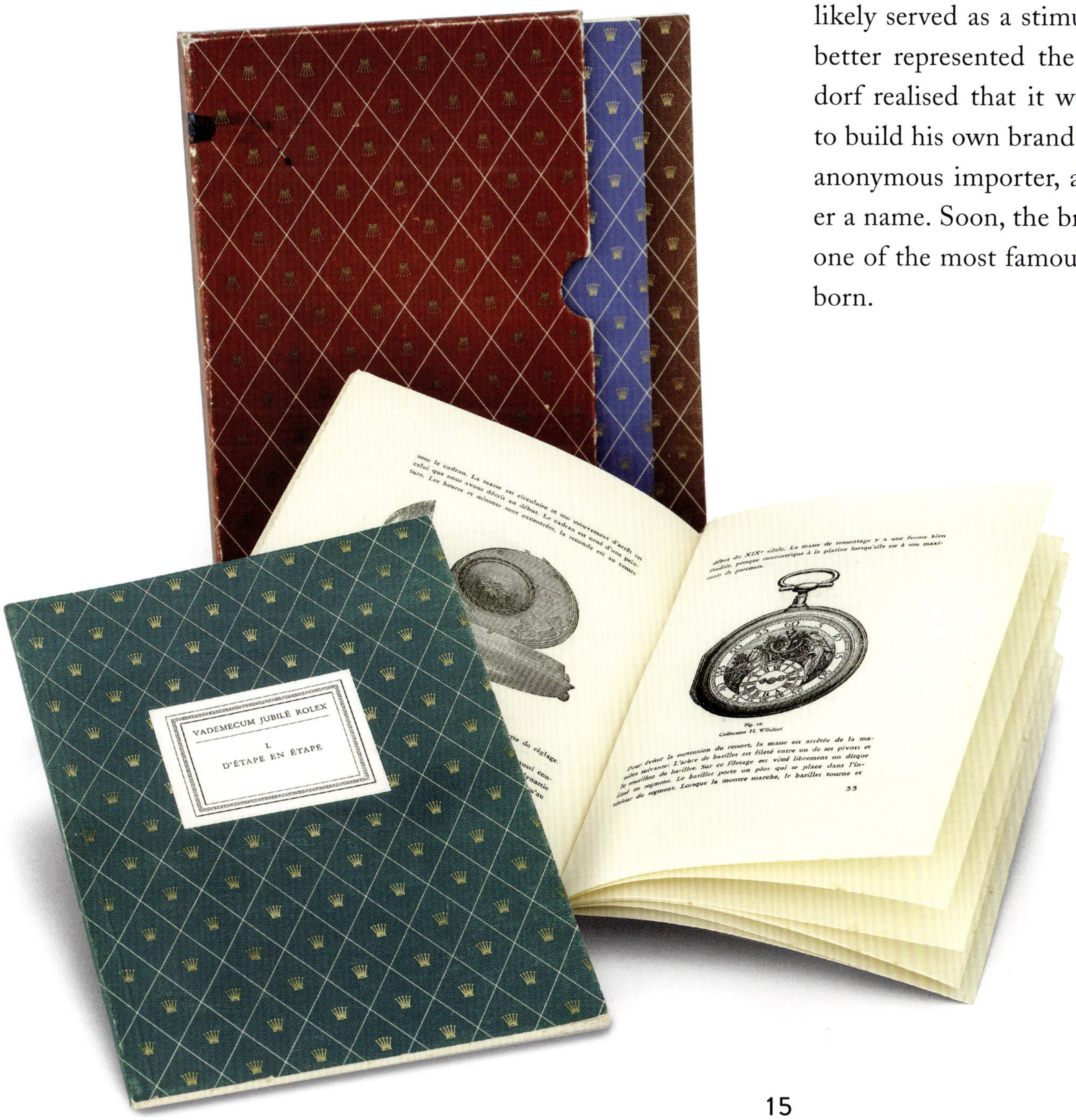

The Rolex *Vade Mecum*, a collection of booklets published in 1945 telling the story of Rolex's founding.

The Mysterious Birth of the Rolex Brand

Wilsdorf's modern mind also expressed itself in choosing a brand. Historic high-end watchmaking *maisons* typically bear the names of the founder or a pair of entrepreneurs who created the company. Wilsdorf, whose surname was of German origin, and not easy to pronounce, never considered this an option.

In November 1907, Wilsdorf registered the names of Lusitania and Mauretania, both the names of large luxury liners that sailed international waters, as well as The Eastern Watch. None of these three was ever used, however: in the *Vade Mecum*, the entrepreneur stated that he had wanted a name that was easy to remember and to pronounce in as many languages as possible but had no specific meaning. It had to be short enough while also composed of letters of the same size, in order to harmonise aesthetically with the watches' dials and movements.

'I tried combining the letters of the alphabet in every possible way,' Wilsdorf wrote in his memoirs. 'This gave some hundred names, but none of them felt quite right.'

This is where the logical narrative ends. According to Wilsdorf's story, however, the brand first came about as a sudden thought that occurred to him one morning while in the city of London, traveling on a horse-drawn omnibus: 'Rolex'. The account seems nearly improbable. Yet anyone who has taken on creative work understands perfectly well that this is how the mind functions: following a lengthy period of concentration and a series of logical elaborations, the sudden springing forth of an idea is not at all unusual; it is actually the result of a collaboration between the conscious and unconscious mind, and often provides excellent results. It was certainly extraordinary in the case of Rolex, which in 1908 Wilsdorf & Davis registered as a trademark in La Chaux-de-Fonds. Some scholars have speculated that the name was derived from the words '*hoROLogie d'EXcellence*' – which is possible, but not so significant. Much more important was the strategy involved, one that would bring the brand closer to the spotlight.

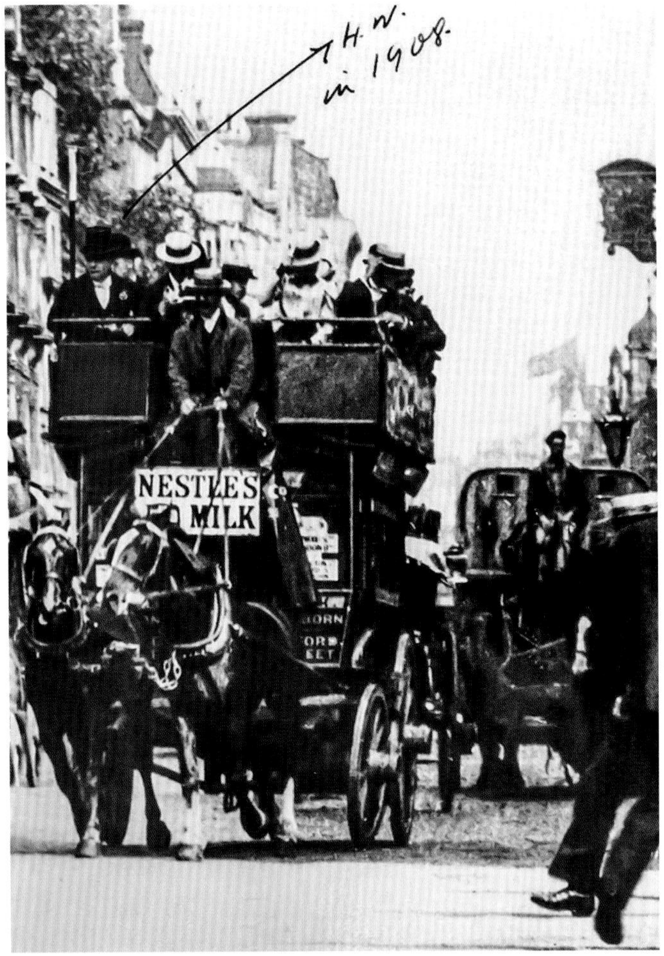

1908: Hans Wilsdorf (left) in a horse-drawn omnibus.

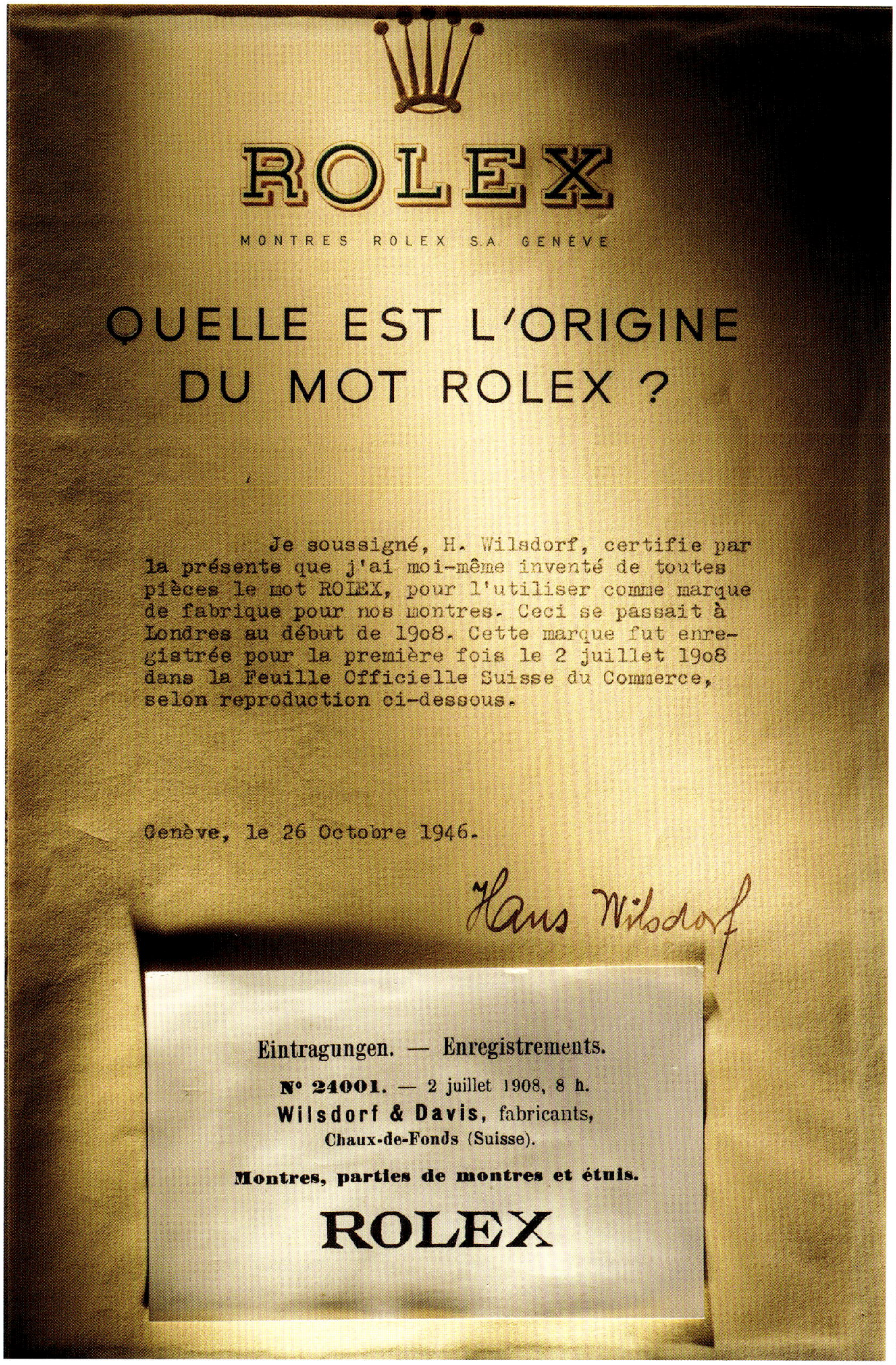
ROLEX

MONTRES ROLEX S.A. GENÈVE

QUELLE EST L'ORIGINE
DU MOT ROLEX ?

Je soussigné, H. Wilsdorf, certifie par la présente que j'ai moi-même inventé de toutes pièces le mot ROLEX, pour l'utiliser comme marque de fabrique pour nos montres. Ceci se passait à Londres au début de 1908. Cette marque fut enregistrée pour la première fois le 2 juillet 1908 dans la Feuille Officielle Suisse du Commerce, selon reproduction ci-dessous.

Genève, le 26 Octobre 1946.

Hans Wilsdorf

Eintragungen. — Enregistrements.

Nº 24001. — 2 juillet 1908, 8 h.

Wilsdorf & Davis, fabricants,

Chaux-de-Fonds (Suisse).

Montres, parties de montres et étuis.

ROLEX

1908: The Rolex trademark is registered.

Rolex became the trademark of Wilsdorf & Davis, and Wilsdorf immediately asked Aegler to use the new logo on all his watches. Aegler was reluctant at first, wishing to continue using his own name, but he accepted ultimately, in part due to the fact that he was not an exclusive supplier of Wilsdorf & Davis. Wilsdorf sought to create a brand that distinguished his products from all others, though at the time they might as well have housed the same movements.

Return to Switzerland

London was the export centre for Wilsdorf & Davis watches, both for overseas markets and for distribution in Great Britain. By 1914, the company had grown to such an extent that it filled large offices and had over sixty employees. The Rolex Watch Company, Limited was incorporated as a private limited company in London in November 1915.

In September 1915, due to restrictions imposed by the global conflict devastating Europe, the British government imposed a duty of over 33% on imported luxury goods, which included watches: any watch that entered Great Britain, if only for inspection before subsequent export abroad, would be subject to the high taxation rate. In December 1916, the importation of all gold and silver items, including gold watches and watch cases, was prohibited; the standard did not, however, cover silver watches or watch cases.

In the face of duties and the ban on imported gold watch cases, Wilsdorf's entrepreneurial spirit once again made itself manifest. He responded in various ways: he expanded the Swiss office by transferring it from La Chaux-de-Fonds to Bienne, ensuring that all watches destined for countries other than Britain did not need to pass through London and were sold directly from Switzerland instead. He also began to work with English producers of gold cases, which were not to be imported – only the watch movements came from Switzerland and would be placed in cases on English soil. In 1919, he decided to move the Rolex headquarters from London to Geneva, while still considering Britain to be the company's target market.

The Rise of the Rolex Brand

The Rolex name, however, only gradually began to appear on watch dials. Brand awareness was especially difficult in Great Britain, where it was traditional for the retailer's brand to appear on the dial. Though the movement and case were marked with the Rolex brand, the customer rarely – if ever – looked inside the watch, and thus would never identify it. On the other hand, at the start of its business, the Rolex name was new and unknown, yet most of the shops that sold watches had a long history. So, of course, people would have more confidence purchasing a watch marked 'Asprey' or 'Harrods' than one with the unknown 'Rolex'.

Wilsdorf knew, however, that he would receive more orders and gain greater control over the pricing of his watches if he could increase demand, so customers would then enter a store to request – or, better yet, insist on – a Rolex-branded watch.

To familiarise customers further with the brand, Wilsdorf opted to apply labelling on the backs of his watches, but this was also not enough. So, in 1921, he began to ship certain watches with the Rolex name printed on their dials. At that time, watches were shipped in small boxes, each containing six watches. At first, only one watch in each box had the Rolex name on the dial. Later, there were two watches, and slowly the number of branded watches increased. In 1925, Rolex registered the crown symbol as a trademark.

As he explains in the *Vade Mecum*, Wilsdorf had significant difficulties persuading British dealers to accept 'Rolex' being printed on the dials.

'Despite the qualities of this trade name, it took 20 years of hard work to make the idea acceptable in England. At first, I ventured to inscribe it on one watch in every six; then it appeared on two, and later three, in every six. This half victory was still unsatisfactory and we knew that it would take many more years to obtain the desired results.

'Tired of waiting, in 1925 I decided to launch the "Rolex" trademark by means of an intensive advertising campaign. The policy entailed an annual expenditure of more than 12,000 British pounds – not for one year alone, but for several in succession. One of the results thus obtained was that dealers agreed first that four, and later five, out of every six watches should bear the name of "Rolex".'

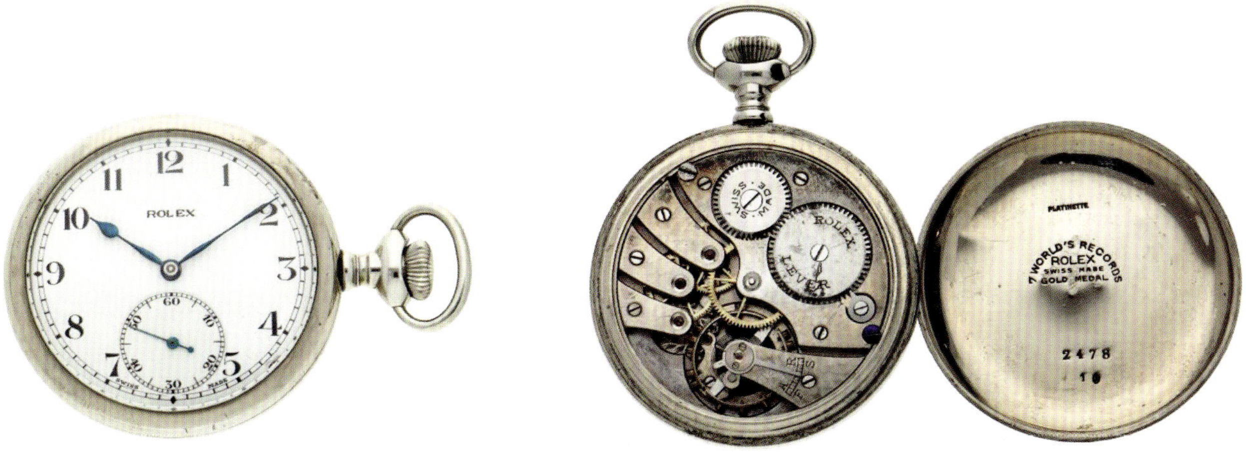

Early watches bearing the Rolex name, both on the dial and inside the case.

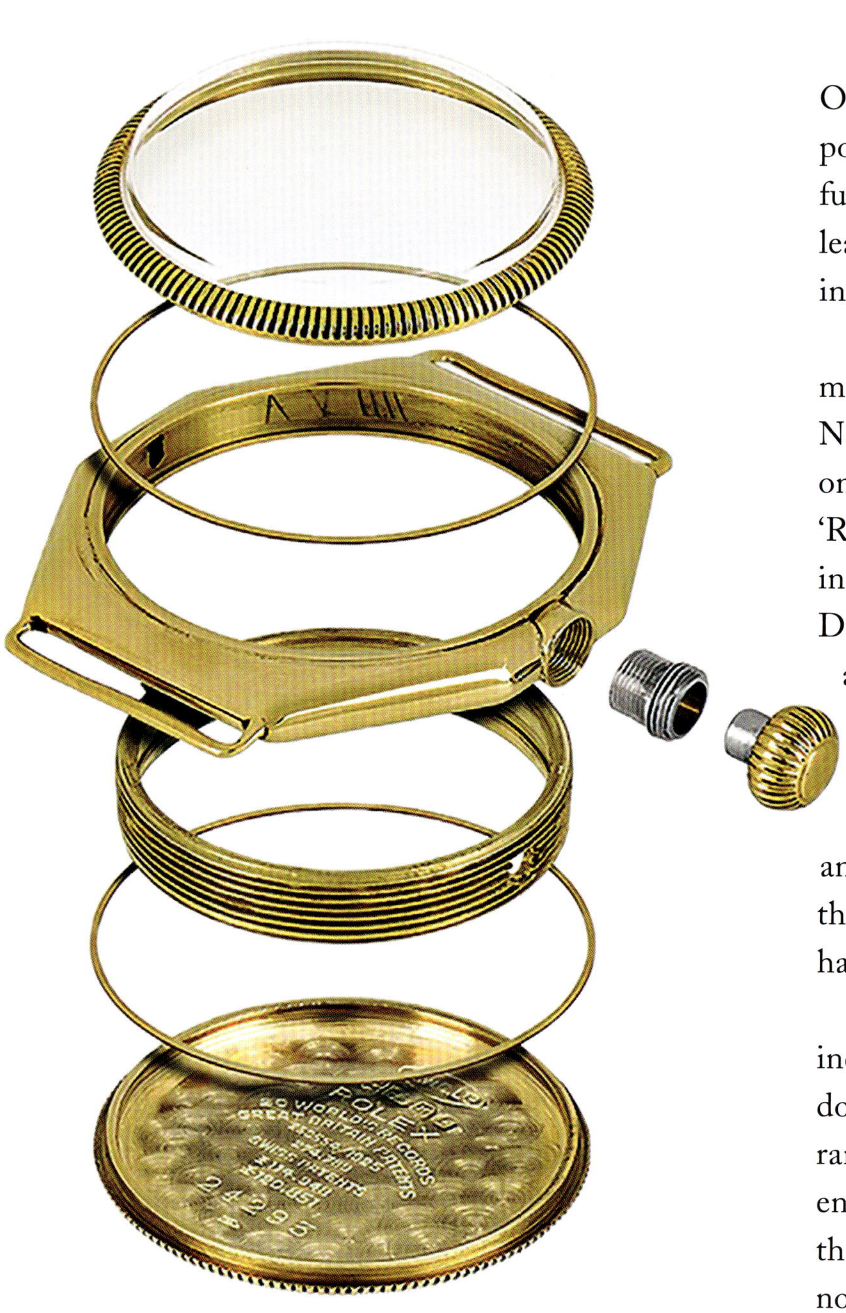

Top: Architecture of the Oyster case, 1926.

Right: An Oyster model. The Oyster, the world's first waterproof wristwatch, played a pioneering role in the development of the modern wristwatch.

'At last, in 1927, the waterproof 'Rolex-Oyster' was launched, and we were then in a position to announce definitively that in the future not one of our watches would ever again leave our work without our name on the dial, inside the case and on the movement.'

Wilsdorf initiated a brand-advocacy communication strategy that reached its peak on 24 November 1927, when a full-page ad appeared on the front page of the *Daily Mail* advertising 'Rolex Introduces [...] the greatest Triumph in Watchmaking – The Wonder Watch that Defies the Elements.' With his branding and advertising campaigns, Wilsdorf effectively turned the tables: he would no longer have to ask dealers to sell his watches; customers would request Rolex-branded watches, and dealers would have to come to him to get them. It was the beginning of a strategy that has been successfully carried into the present.

Wilsdorf had also created other brands, including Unicorn, Marconi, Rolco and Tudor to be applied to products in a lower price range, reserving the Rolex name only for high-end products. Over time, however, he realised that creating awareness of these brands was not a good idea, as it took sales away from the Rolex brand. As people became more aware of the brand, it diluted the effects of the expensive marketing, which aimed to build prestige around ownership of a Rolex. The company's name was thus gradually dissociated from the other brands.

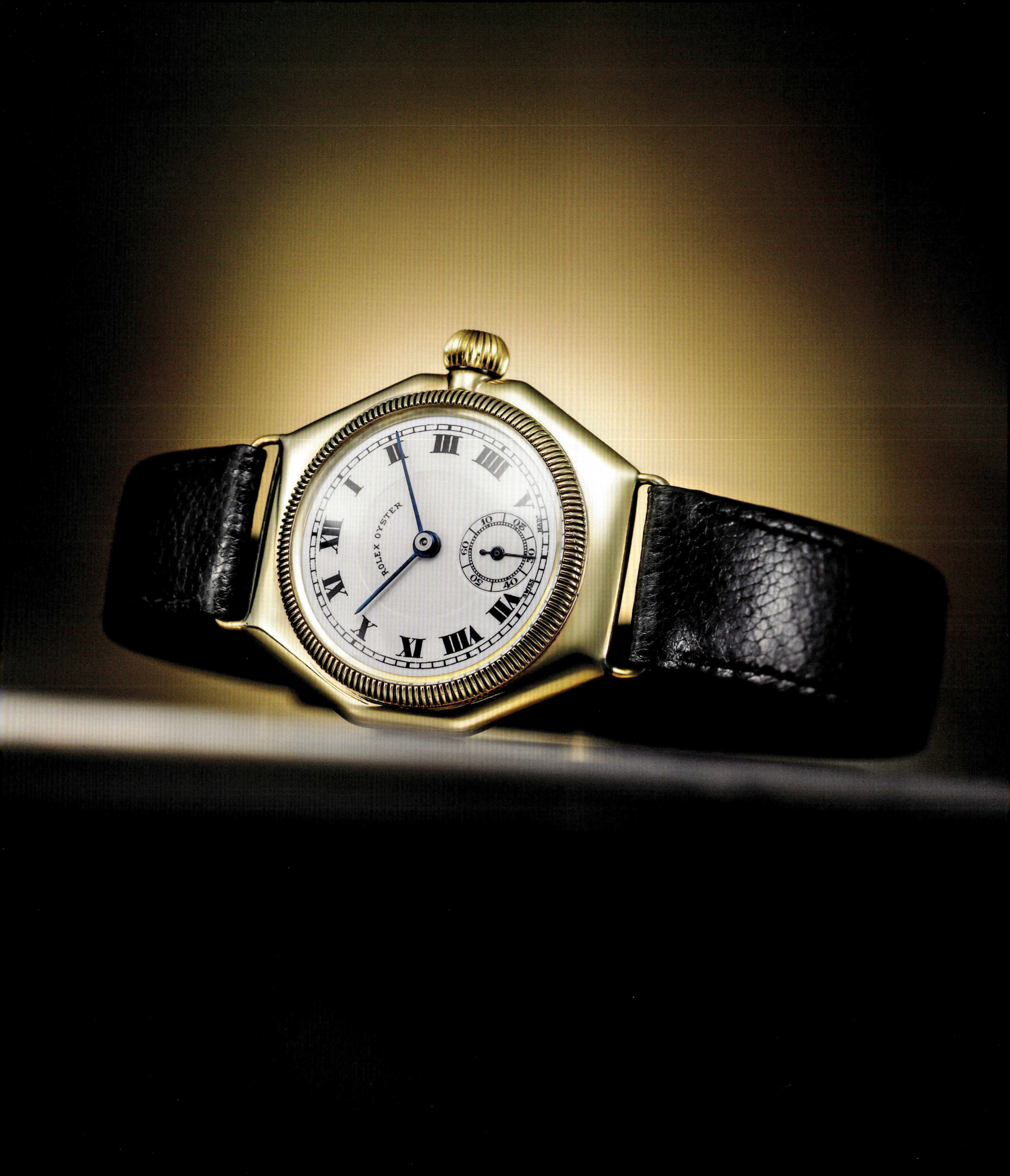
ROLEX OYSTER

The Success of a Foolhardy Idea

In the *Vade Mecum*, Wilsdorf writes: 'Next came the idea of expanding bracelets, which an important jewelry firm invented and launched in about 1906. This too won the approval of our British clientele. From that time on, our little gold watch became increasingly popular throughout the Empire.'

The major jewelry company the Rolex head mentions was Edwin Harrop, who had patented and named the extendable Britannic bracelet.

Wilsdorf understood perfectly that this type of bracelet was useful for distributing wristwatches, observing that '[...] a new fashion and a great commercial success sprang from an apparently foolhardy idea. Soon we were placing orders for tens of thousands of pieces and, to the uninformed, must have appeared extremely audacious young people.' The initial success of Rolex wristwatches was due, at least in part, to this bracelet, which was produced for many years, until at least 1964.

The bracelets were primarily designed for women's wristwatches. An attempt by Harrop to introduce a similar 'Army'-branded extendable bracelet during the Great War, emphasising the advantages of a metal bracelet over leather straps, especially when used in wet conditions and muddy environments, was not well received by the public, who likened it to feminine products.

Very few wristwatches were sold to men before the Great War, as many men essentially considered them bracelets, and therefore women's watches. The military, who found it convenient to tell the time on wristwatches, were an exception, but there are no documents testifying to Rolex producing watches for the military, nor that Wilsdorf was interested in this market niche.

Still, convinced that wristwatches would be successful, Wilsdorf concentrated on developing and marketing ever more technically complex models. From the start of the business, he asked his technicians to focus on the key elements that still characterise a Rolex today: robustness, precision and quality.

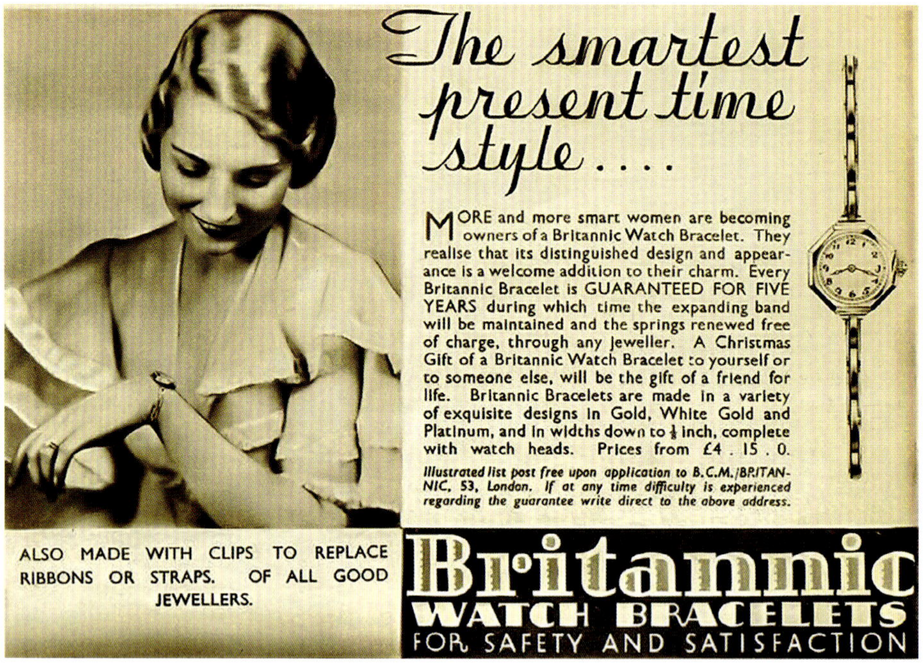

Above: A newspaper ad for new women's wristwatches marrying a Rolex with Edwin Harrop's Britannic bracelet model.

Right: 1914. A Rolex wristwatch analogous to the specimen certified a class 'A' chronometer by Kew Observatory in 1914.

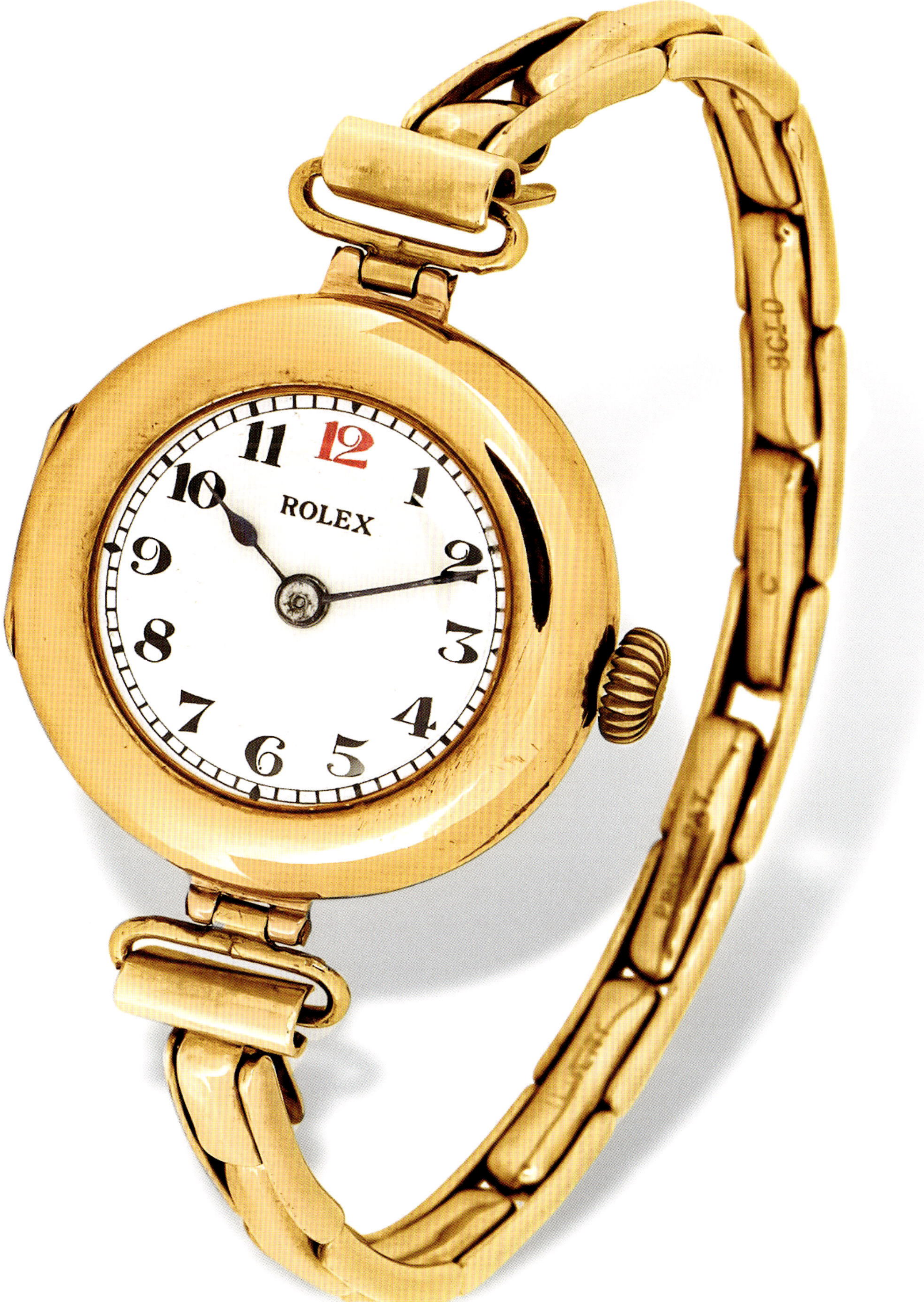
ROLEX

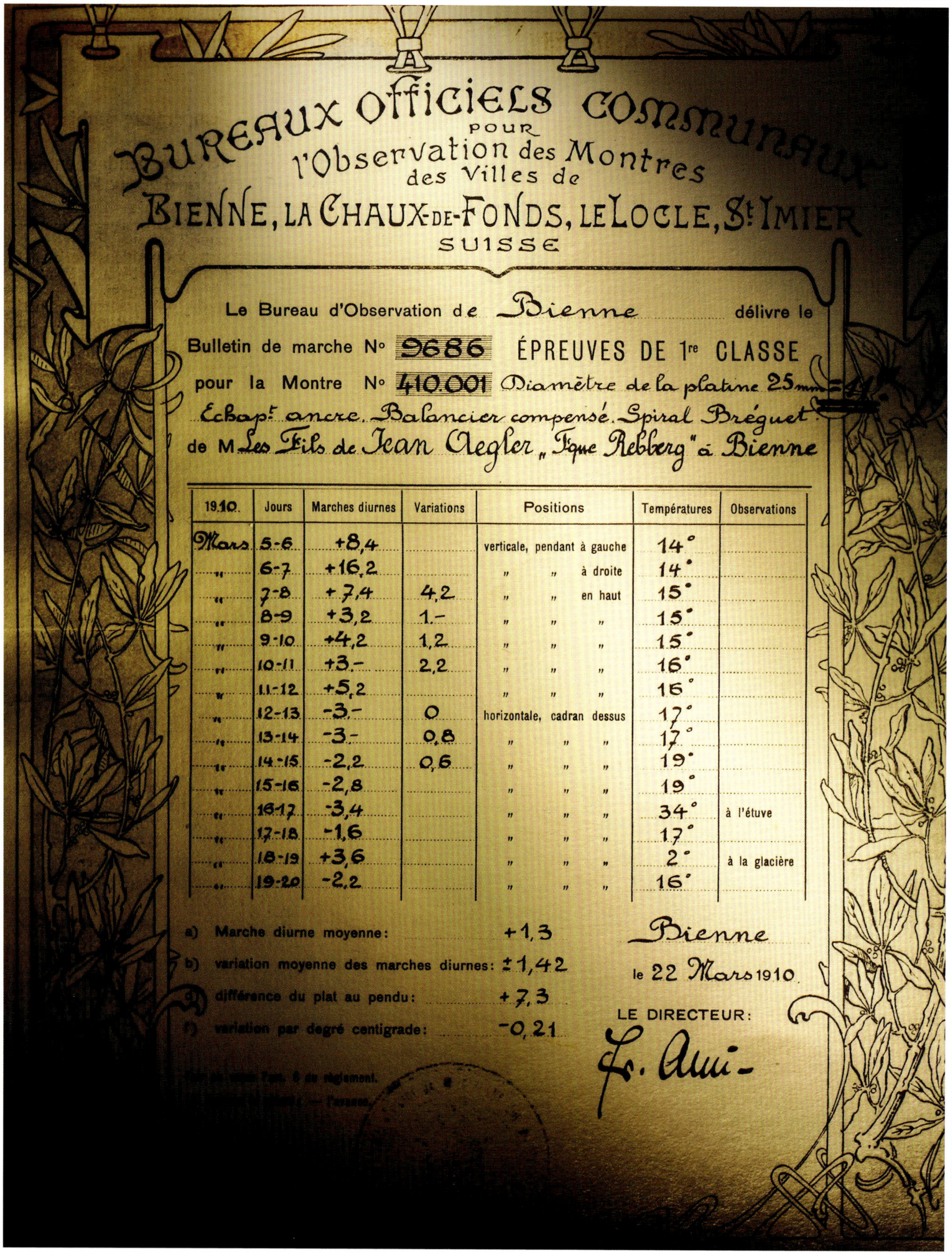

Bureaux Officiels Communaux pour l'Observation des Montres des Villes de Bienne, La Chaux-de-Fonds, Le Locle, St Imier

SUISSE

Le Bureau d'Observation de *Bienne* délivre le Bulletin de marche No 9686 ÉPREUVES DE 1re CLASSE pour la Montre No 410.001 *Diamètre de la platine 25mm*

Echapt ancre. Balancier compensé. Spiral Bréguet

de M. *Les Fils de Jean Aegler „Fque Rebberg" à Bienne*

1910	Jours	Marches diurnes	Variations	Positions	Températures	Observations
Mars	5-6	+8,4		verticale, pendant à gauche	14°	
„	6-7	+16,2		„ „ à droite	14°	
„	7-8	+7,4	4,2	„ „ en haut	15°	
„	8-9	+3,2	1.-	„ „ „	15°	
„	9-10	+4,2	1,2	„ „ „	15°	
„	10-11	+3.-	2,2	„ „ „	16°	
„	11-12	+5,2		„ „ „	16°	
„	12-13	-3.-	0	horizontale, cadran dessus	17°	
„	13-14	-3.-	0,8	„ „ „	17°	
„	14-15	-2,2	0,6	„ „ „	19°	
„	15-16	-2,8		„ „ „	19°	
„	16-17	-3,4		„ „ „	34°	à l'étuve
„	17-18	-1,6		„ „ „	17°	
„	18-19	+3,6		„ „ „	2°	à la glacière
„	19-20	-2,2		„ „ „	16°	

a) Marche diurne moyenne: +1,3

b) variation moyenne des marches diurnes: ±1,42

c) différence du plat au pendu: +7,3

f) variation par degré centigrade: -0,21

Bienne le 22 *Mars* 1910

LE DIRECTEUR:

[signature]

The Quest for Precision

Wilsdorf was very insistent on obtaining certificates from independent testing institutes, which had been established in the 19th century to test and certify the performance of Navy chronometers. They later also began to verify the functionality of other types of watches.

Wilsdorf was a perfectionist and never ceased to push Aegler specialists to improve the chronometric precision of the watches they produced for Rolex. In 1910, the Bureau Officiel de Contrôle de la Marche des Montres in Bienne, Switzerland, issued a Rolex wristwatch with the first chronometry certificate ever obtained in the world for a timepiece of this type.

On 15 July 1914, a small 11-ligne (25-millimetre diameter) Rolex wristwatch received a Class 'A' certificate for precision from the prestigious Kew Observatory in England – the first time a Class 'A' Kew certificate had been awarded to a wristwatch. Wilsdorf said this had been an important day in the development of the company, one that he would never forget. From then on, Wilsdorf reportedly asked Aegler to submit all Rolex calibres to chronometric tests. The ability of a wristwatch to keep accurate time could no longer be questioned.

In 1941, the Rolex Watch Company in Bienne received a letter dated 3 October from the Swiss Bureau Officiel de Contrôle de la Marche des Montres confirming that on that day, a 9¾-inch chronometer bracelet no. 96717 had passed its test: the 20,000th Rolex Watch since June 1927 had finally been approved.

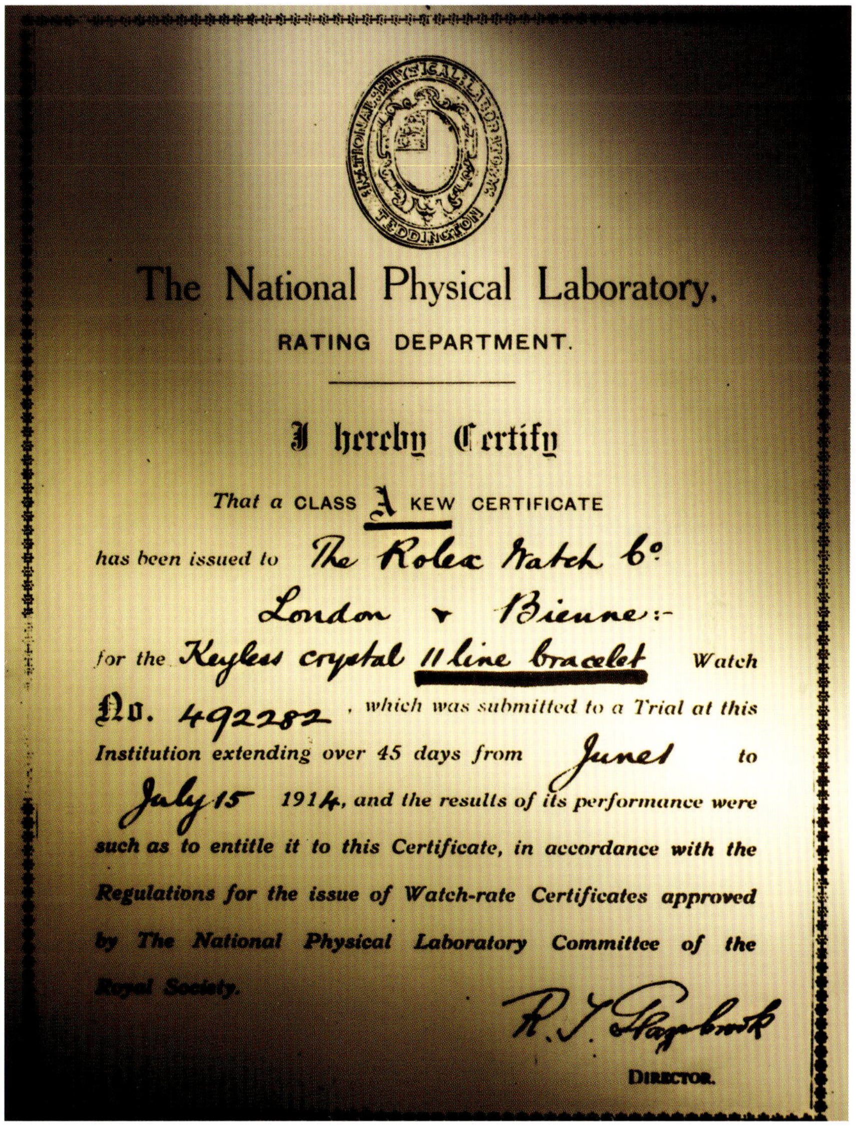

The National Physical Laboratory.

RATING DEPARTMENT.

I hereby Certify

That a CLASS A KEW CERTIFICATE has been issued to The Rolex Watch Co London & Bienne:- for the Keyless crystal 11 line bracelet Watch No. 492282, which was submitted to a Trial at this Institution extending over 45 days from June 1 to July 15 1914, and the results of its performance were such as to entitle it to this Certificate, in accordance with the Regulations for the issue of Watch-rate Certificates approved by The National Physical Laboratory Committee of the Royal Society.

R. J. Glazebrook

DIRECTOR.

Left: 1910. The first wristwatch to receive the Swiss certificate of chronometric precision issued by the Bureau Officiel de Contrôle de la Marche des Montres in Bienne was a Rolex.

Above: 1914. Four years later, Kew Observatory then awarded a Rolex wristwatch a Class 'A' certificate for precision, a privilege that until then was the exclusive preserve of marine chronometers.

Ambitious Aims

Obtaining the Class 'A' precision certificate from the Kew Observatory spurred Wilsdorf to tackle other complex characteristics pertaining to wristwatches. Unlike a pocket watch, a wristwatch is exposed to shock, humidity and dust.

In his writings, one reads: 'To my technical assistants, my constant refrain was, from the earliest days: "We must succeed in making a watch case so tight that our movements will be permanently guaranteed against damage caused by dust, perspiration, water, heat and cold. Only then will the perfect accuracy of the Rolex watch be secured.' "

Wilsdorf was not the first person to attempt to create a waterproof watch, but he was the first who was able to launch one on the market and achieve enormous commercial success.

In January 1921, Jean Finger, a watch-case maker from Lengnau, near Bern, obtained a patent for a *'montre à remontoire avec boitier protecteur'*, literally 'a manually-wound watch with a protective case'. This solution had the virtue of simplicity: the watch was placed inside a larger case with a screw-down bezel that protected it from the external environment. While this case achieved the waterproof effect desired, it had the serious drawback that the bezel of the outer case needed to be unscrewed each day to wind the watch.

The 'Submarine' trademark was registered by the Rolex Watch Company on 31 March, 1922, presumably soon after Wilsdorf bought the rights to Finger's patent. This project, however, marked just the first phase of research, which aimed to create a completely waterproof watch that was easy to use. To achieve this, Wilsdorf was inspired by other designers' projects, significantly improving them to render them more practical and reliable. In 1903, the Swiss watch-case maker François Borgel had designed a three-piece screw-down case; but to make the case waterproof, a crown was also necessary to guarantee watertightness. Wilsdorf referred to research conducted by Paul Perregaux and Georges Perret, two watchmaker-craftsmen from La Chaux-de-Fonds who had developed the screw-down crown in 1925. Having acquired the rights to Perregaux and Perret's inventions, Wilsdorf had a screw-down crown designed and patented with an outer seal guaranteeing its water resistance. In 1926, Rolex launched a new watch with a screw-down crown and waterproof case. It was called the 'Oyster' because, as Wilsdorf said, 'Like an oyster, it can remain an unlimited time under water without detriment to its parts.'

1927: The first face of Rolex, Mercedes Gleitze, and her swim across the English Channel on the front page of the *Daily Mail.*

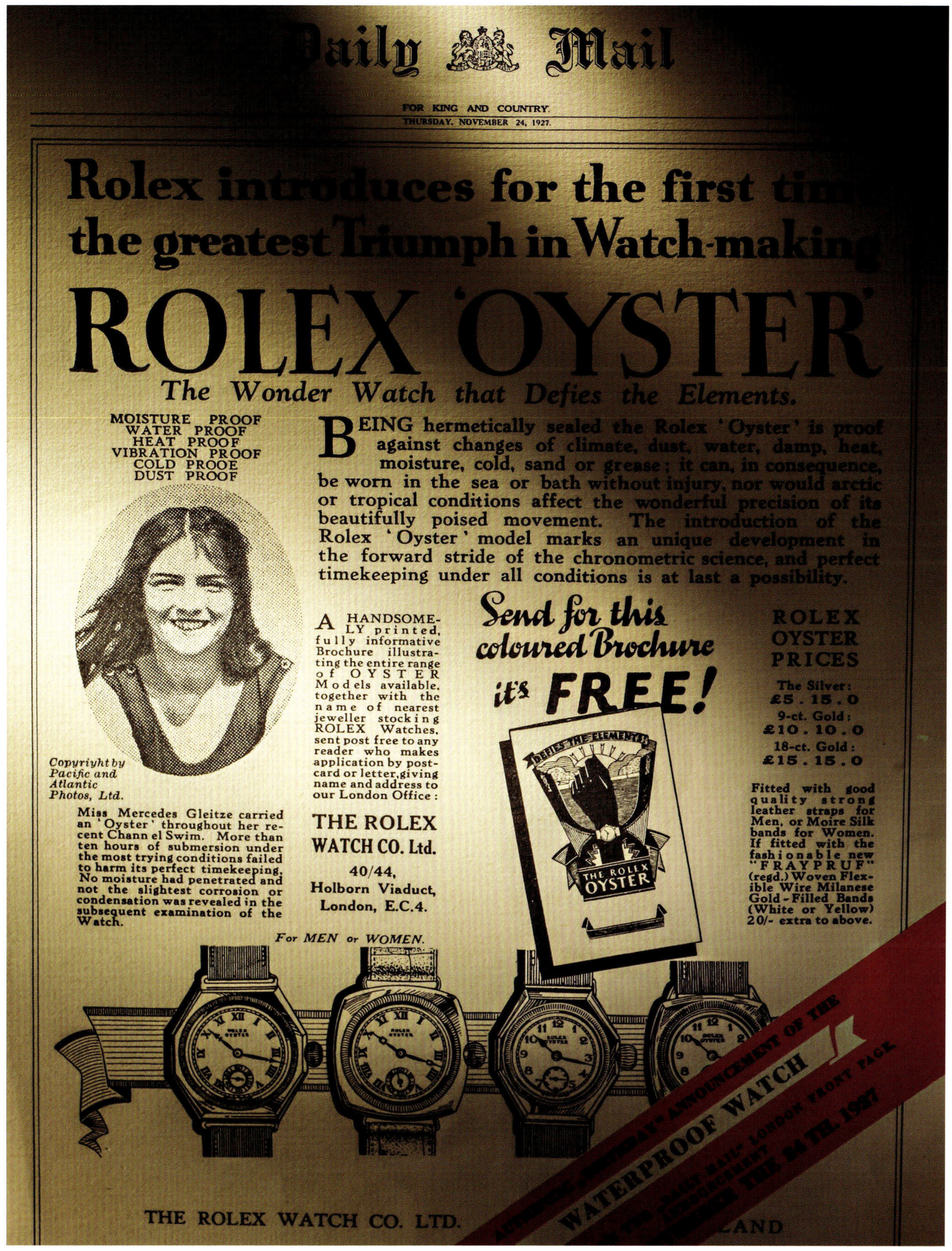
Daily Mail
FOR KING AND COUNTRY.
THURSDAY, NOVEMBER 24, 1927.
Rolex introduces for the first ti
the greatest Triumph in Watch-makin
ROLEX 'OYSTER'
The Wonder Watch that Defies the Elements.
MOISTURE PROOF
WATER PROOF
HEAT PROOF
VIBRATION PROOF
COLD PROOF
DUST PROOF
BEING hermetically sealed the Rolex 'Oyster' is proof against changes of climate, dust, water, damp, heat, moisture, cold, sand or grease; it can, in consequence, be worn in the sea or bath without injury, nor would arctic or tropical conditions affect the wonderful precision of its beautifully poised movement. The introduction of the Rolex 'Oyster' model marks an unique development in the forward stride of the chronometric science, and perfect timekeeping under all conditions is at last a possibility.
Copyright by Pacific and Atlantic Photos, Ltd.
Miss Mercedes Gleitze carried an 'Oyster' throughout her recent Channel Swim. More than ten hours of submersion under the most trying conditions failed to harm its perfect timekeeping. No moisture had penetrated and not the slightest corrosion or condensation was revealed in the subsequent examination of the Watch.
A HANDSOMELY printed, fully informative Brochure illustrating the entire range of OYSTER Models available, together with the name of nearest jeweller stocking ROLEX Watches, sent post free to any reader who makes application by post-card or letter, giving name and address to our London Office:
THE ROLEX WATCH CO. Ltd.
40/44, Holborn Viaduct, London, E.C.4.
Send for this coloured Brochure
it's FREE!
DEFIES THE ELEMENTS
THE ROLEX OYSTER
ROLEX OYSTER PRICES
The Silver: £5.15.0
9-ct. Gold: £10.10.0
18-ct. Gold: £15.15.0
Fitted with good quality strong leather straps for Men, or Moire Silk bands for Women. If fitted with the fashionable new "FRAYPRUF" (regd.) Woven Flexible Wire Milanese Gold-Filled Bands (White or Yellow) 20/- extra to above.
For MEN or WOMEN.
ROLEX OYSTER
THE ROLEX WATCH CO. LTD.
LAND
ANNOUNCEMENT OF THE
WATERPROOF WATCH
LONDON FRONT PAGE
THE 24 TH. 1927

To promote the extraordinary qualities of the Oyster the following year, Wilsdorf chose a new and original approach for the time. Having learned that a young secretary named Mercedes Gleitze from Brighton was preparing to swim across the English Channel, and she would become the first British woman to accomplish the feat if successful, he asked her to wear an Oyster to demonstrate its flawless waterproof properties.

Following Gleitze's gruelling swim through the icy waters, a reporter for British newspaper *The Times* wrote that 'Miss Gleitze carried a small gold watch, which was found [...] to have kept good time throughout.' A month later, on 24 November 1927, Wilsdorf launched the Rolex Oyster watch in Great Britain with an advertisement on the front page of the *Daily Mail*, and the Oyster began its rise to fame.

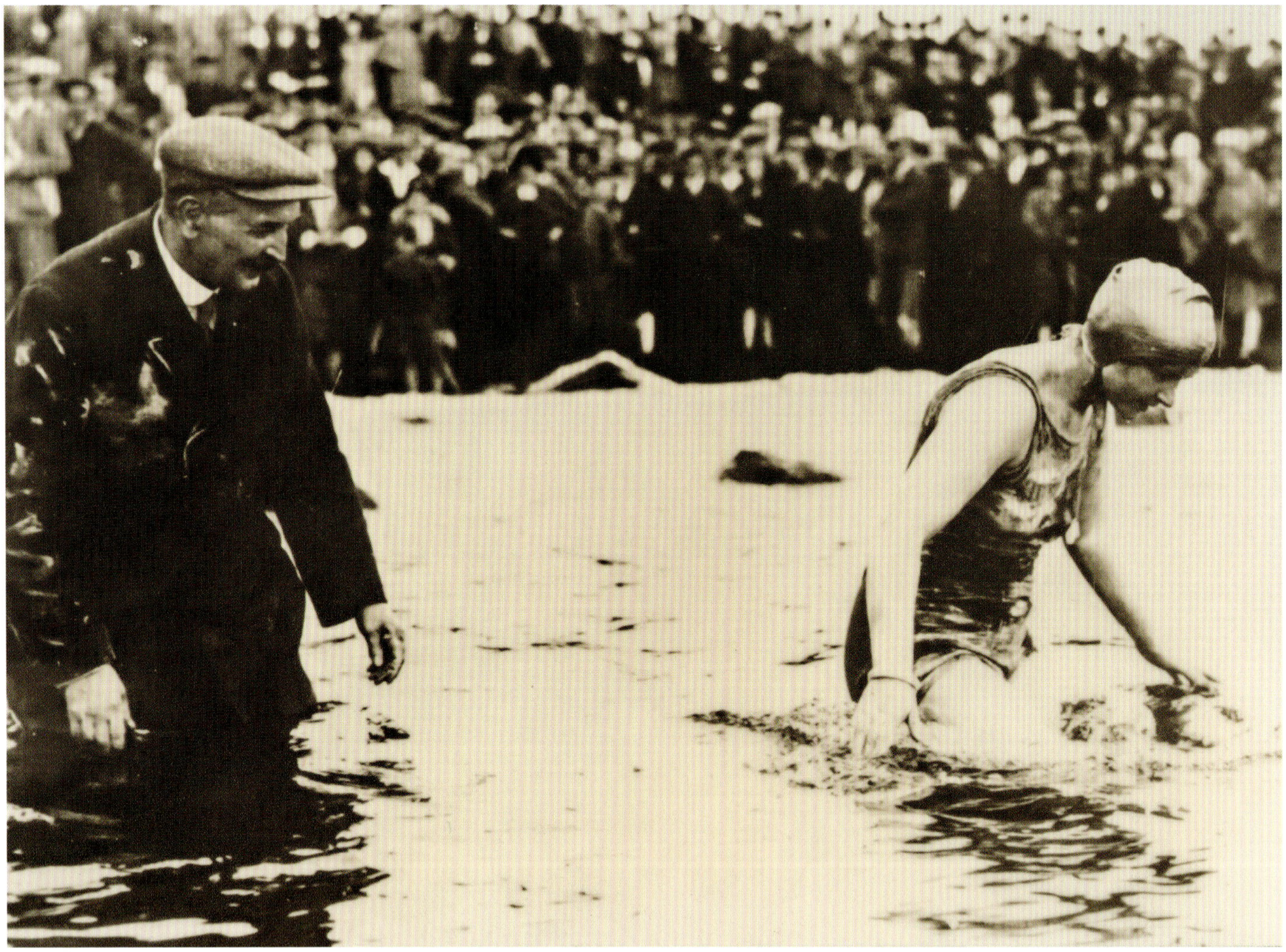

1927: A Rolex Oyster accompanied young British swimmer Mercedes Gleitze as she crossed the English Channel. The swim lasted more than 10 hours, with the watch staying in excellent condition throughout.

Wilsdorf had also made an unconventional choice in this case, turning to a female athlete to endorse his products.

This event marked the birth of the very concept of a celebrity product endorsement. It also began a long and fruitful partnership between Rolex and extraordinary individuals whose exploits, time and time again, demonstrated the excellence of Rolex watches and lent credence to a 1927 advertisement, which stated '[The Oyster] can be worn in the sea or bath without injury, nor would Arctic or tropical conditions affect the wonderful precision of its beautifully poised movement.'

Wilsdorf was not a watchmaker, but he was certainly a marketing genius, willing to invest so much in advertising the Rolex name that by the 1920s he had created a brand that had become famous despite its not having existed prior to 1908.

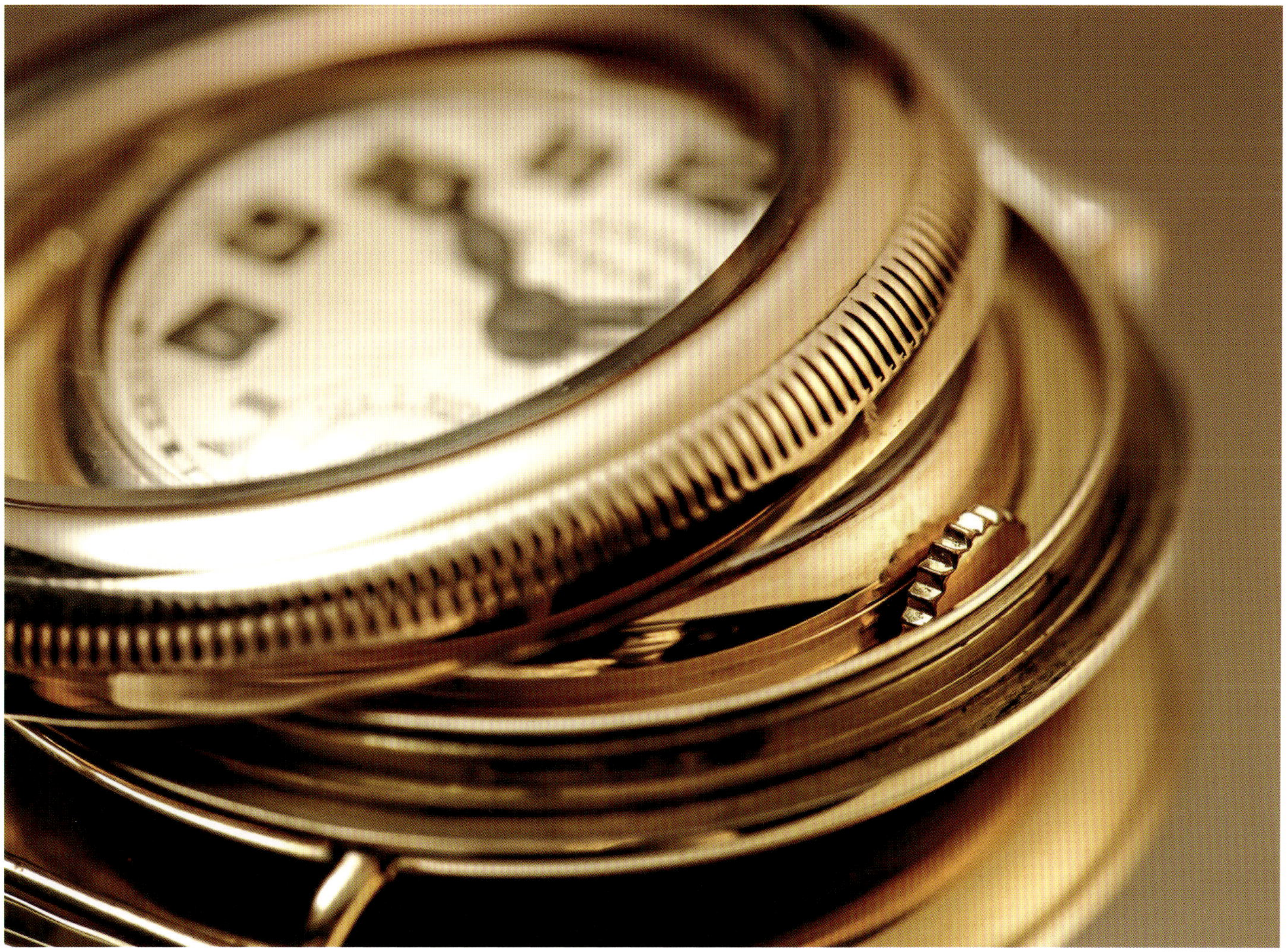

The Oyster case was patented in 1926. Rendered completely hermetic thanks to its system comprising a bezel, case back and winding crown that all screw to the middle case, it protects the inside of the watch from being damaged by exterior elements.

A Perpetual Watch

Once the matter of waterproofing was resolved, Wilsdorf faced the challenge of winding. With traditional watches, the owner needed to unscrew the crown each day to wind the watch, which presented two key challenges: firstly, a risk of not screwing the crown back in tightly, so the watch would therefore no longer be hermetically sealed; and secondly, the seals would wear over time, making the watch lose its impermeability. Rolex resolved these issues in 1931 with the introduction of its 'Perpetual' self-winding movement.

The concept of the self-winding watch was first introduced in the second half of the 18th century. Swiss watchmaker Abraham-Louis Perrelet had designed a self-winding watch in late 1775 or early 1776. The winding mechanism worked on the same principle as a pedometer, using a reciprocating weight inside the watch which moved up and down as the owner walked and which, through a set of gears, wound the mainspring. Parisian watchmaker Abraham-Louis Breguet improved upon the mechanism, calling his watches *perpetuelles*.

The first self-winding wristwatch was invented in 1923 by a watchmaker on the Isle of Man named John Harwood. The Harwood system used a semicircular weight that pivoted at the centre of the movement and swung in a 300-degree arc when the wearer moved a wrist or arm and through a train of gears wound the mainspring. Fully wound, the watch could run for twelve hours. It did not have a traditional crown, so the hands were moved by rotating the bezel around the dial.

Harwood's watches went on sale in 1928. They were not very successful on the market, however, and only about 30,000 were produced. The Harwood company closed in 1931 during the Great Depression, and while the patent still existed, there was no one who owned it, so other companies were free to develop their own versions of the self-winding movement.

Emile Borer, nephew and successor of Hermann Aegler, who was head of research and development at the Bienne factory, took up the Harwood design and used it as the basis for a self-winding mechanism. Presented and patented by Rolex in 1931, the Perpetual rotor consisted of a crescent-shaped mass which swung in both directions at the slightest movement of the wrist due to the effect of Earth's gravity. Thanks to automatic winding, it was no longer necessary to wind the watch every day by hand, and the crown was now chiefly used to set the time. Since this did not have to be done often because of the precision of the watch, the level of waterproofing was not jeopardised. As a result, Rolex Oyster Perpetual watches were not only accurate but also durable.

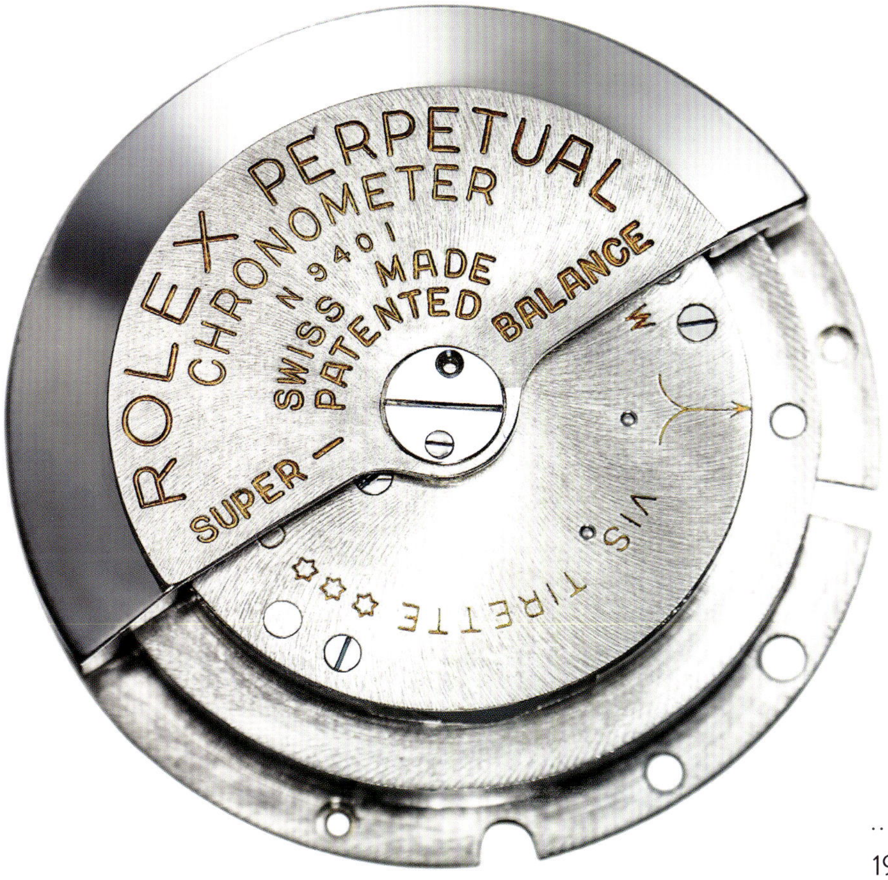

1931: The Perpetual movement.

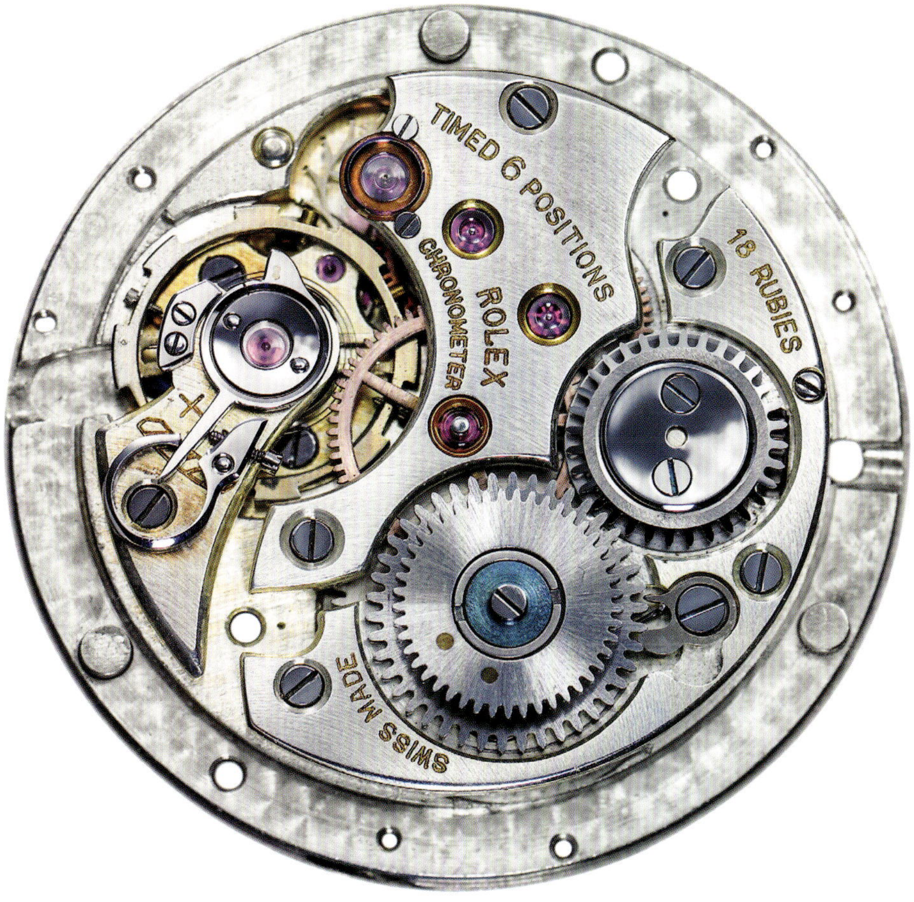

Record-Breaking Watches

After the success of the advertisements linked to Mercedes Gleitze's feat, Wilsdorf engaged various sports personalities in support of the Oyster. An ad that appeared in *Punch* on 18 June 1930 replicated a letter from famous racing driver Captain Malcolm Campbell, who had used a Rolex Oyster in a race: 'Dear Sirs, I have now been using my Rolex Watch for some little while, and it is keeping perfect time under somewhat strenuous conditions. I was wearing it on the occasion of the J.C.C. Double 12 Hours Race on Friday and Saturday last, and the vibration which this watch had to withstand during this long period has not upset its time-keeping properties in the least. I would like to congratulate you on having produced a very first-class watch, suitable for really rough treatment. Yours faithfully, M. Campbell'.

A Rolex Oyster was on Campbell's wrist in 1935 when the driver reached a speed over 300 miles per hour, setting the world land speed record in his race car at the Bonneville Salt Flats.

When Mount Everest was first flown over in 1933, the crew were equipped with Rolex watches, which also proved their worth in flight. Over the years, Rolex has linked its name to great feats, records and sporting events.

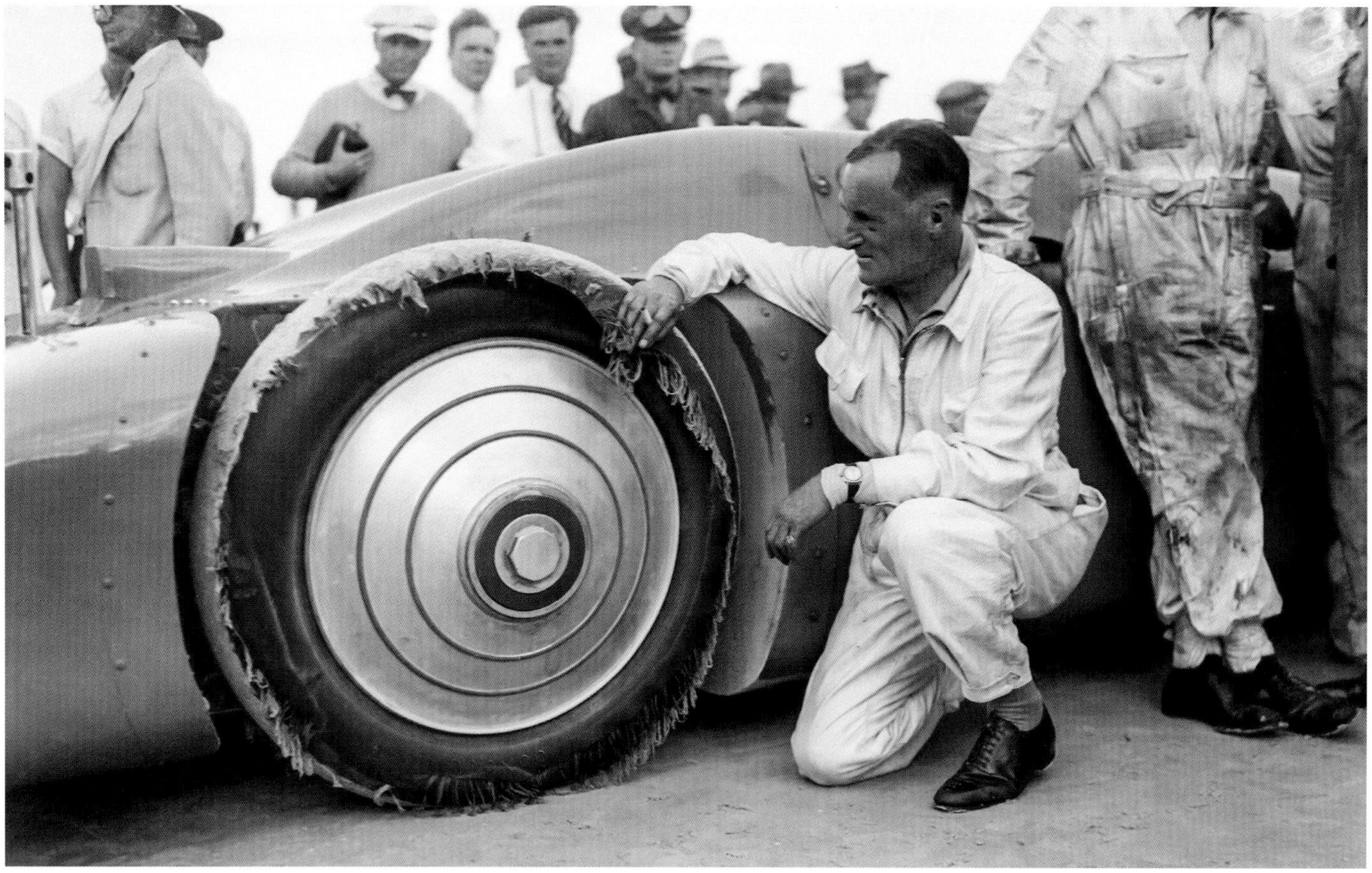

The Oyster is adopted by British aviators and motor speed-racing champions.

Two advertisements for the Oyster model. One celebrates the feat of mountaineer Hugh Ruttledge, who led Everest expeditions in 1933 and 1936. The second features a letter from British Army Lieutenant-Colonel Stewart Blacker, who in 1933 was the first to fly over Everest. The latter's message to Rolex following his exploit is meaningful: *'I can hardly imagine that any watches have ever been subjected before to such extremes under practical conditions.'*

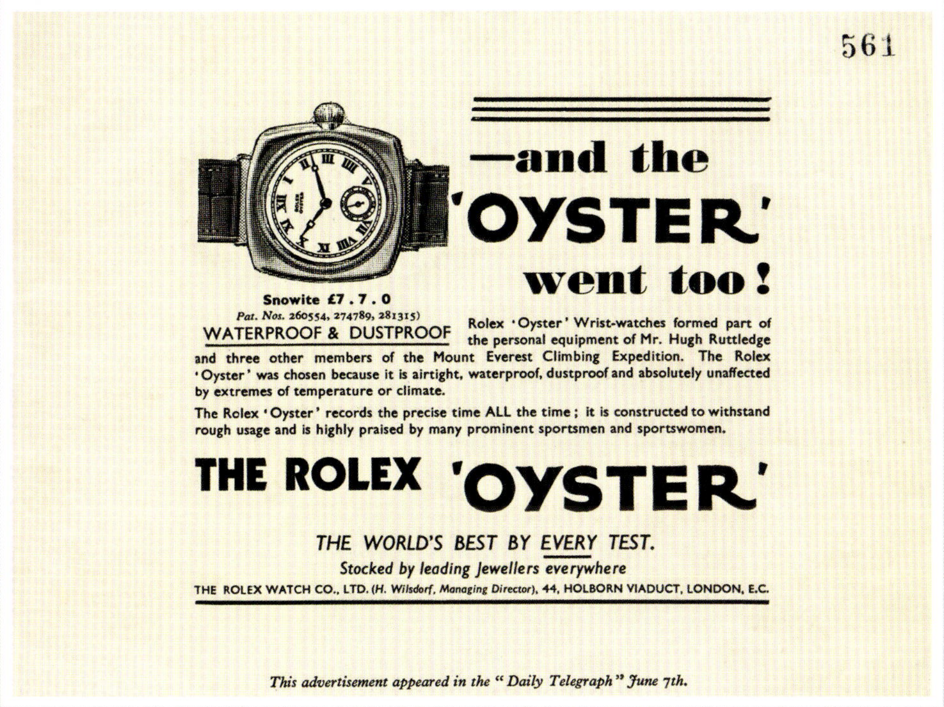

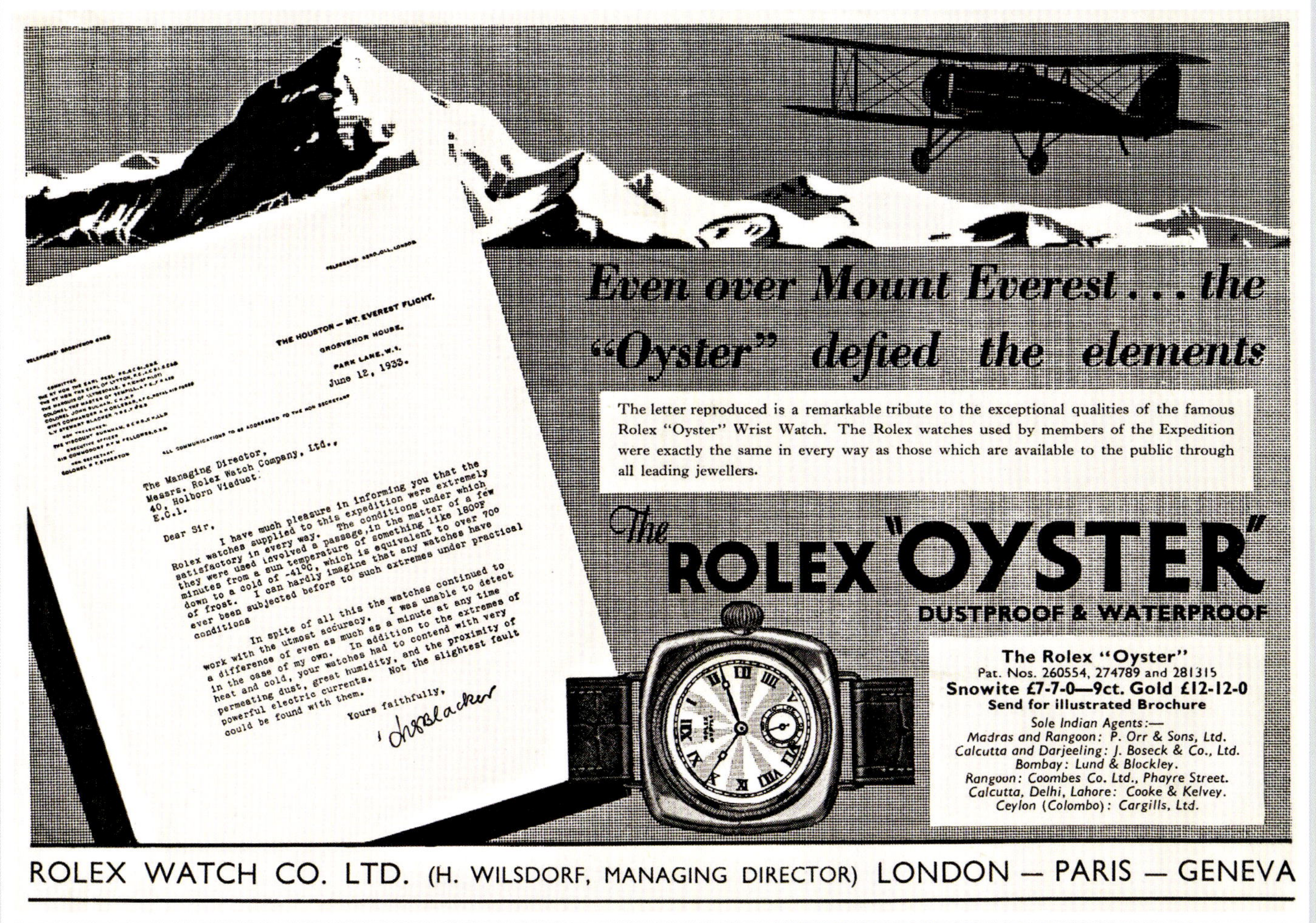

From Professional to Iconic

Over the decades that followed, Rolex developed a series of timepieces that became horological milestones. In the 1950s, thanks to a growing economy, industrial development and progress in the sciences, the Rolex workshops developed a series of watches with original designs to meet the needs of a new world, one of different lifestyles and professional fields that would later become watchmaking classics. Collaboration with institutions and industrial sectors further underscored one's ability to respond to the requirements of an evolving society and to keep in unison with modern life.

To celebrate the 40th anniversary of the founding of Wilsdorf & Davis (the company that later became Rolex), the Oyster Perpetual Datejust was introduced in 1945. This was the first waterproof automatic wrist chronometer that had a date display via a window on its dial.

In 1953, two models were introduced that were destined for enormous success. On 29 May 1953, when New Zealand mountaineer and explorer Sir Edmund Hillary and Nepalese Sherpa Tenzing Norgay climbed Everest, reaching the highest peak in the world, on their wrists they each wore a Rolex Oyster Perpetual – a prototype made expressly for the circumstance. These were equipped with a leather strap instead of the steel bracelet used in the public version to follow. For Rolex, the Everest climb was an excellent opportunity to test their watches in extreme conditions (temperature and pressure changes at high altitudes) and was an important promotional launch. The same year, the company registered the 'Explorer' trademark and began to market the model to the public.

The Rolex Submariner was also introduced

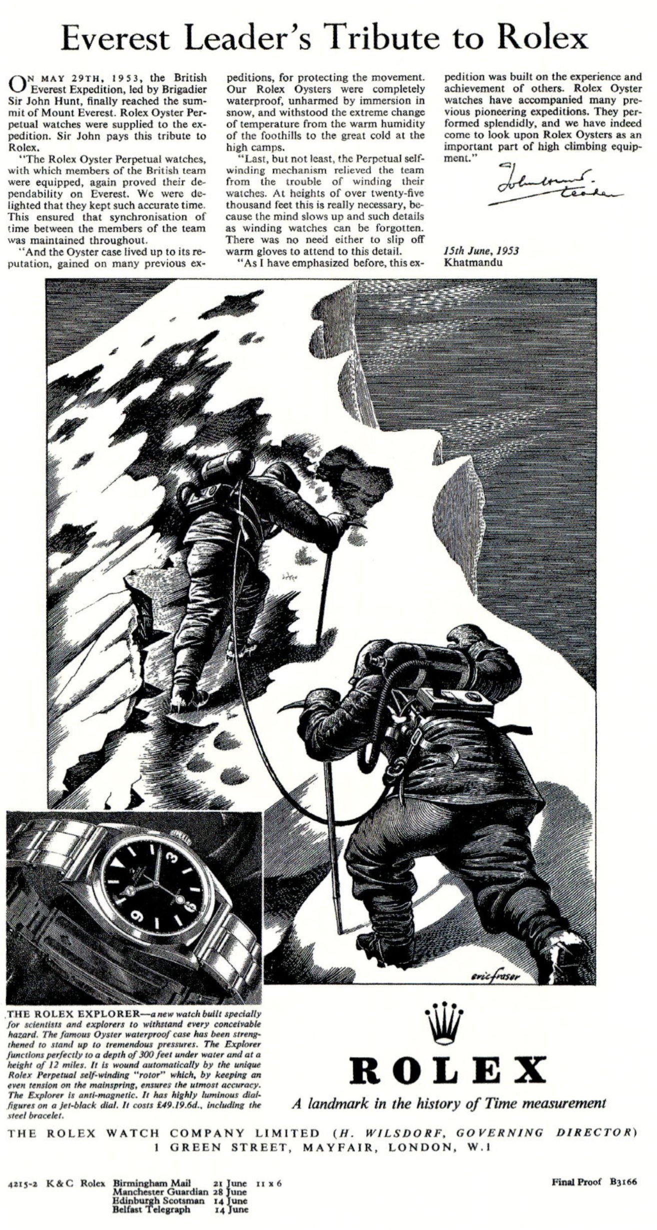

Sir Edmund Hillary and Tenzing Norgay's ascent to Everest's summit in 1953 had worldwide impact, and Rolex took part in it by supplying the expedition with Oyster Perpetual watches.

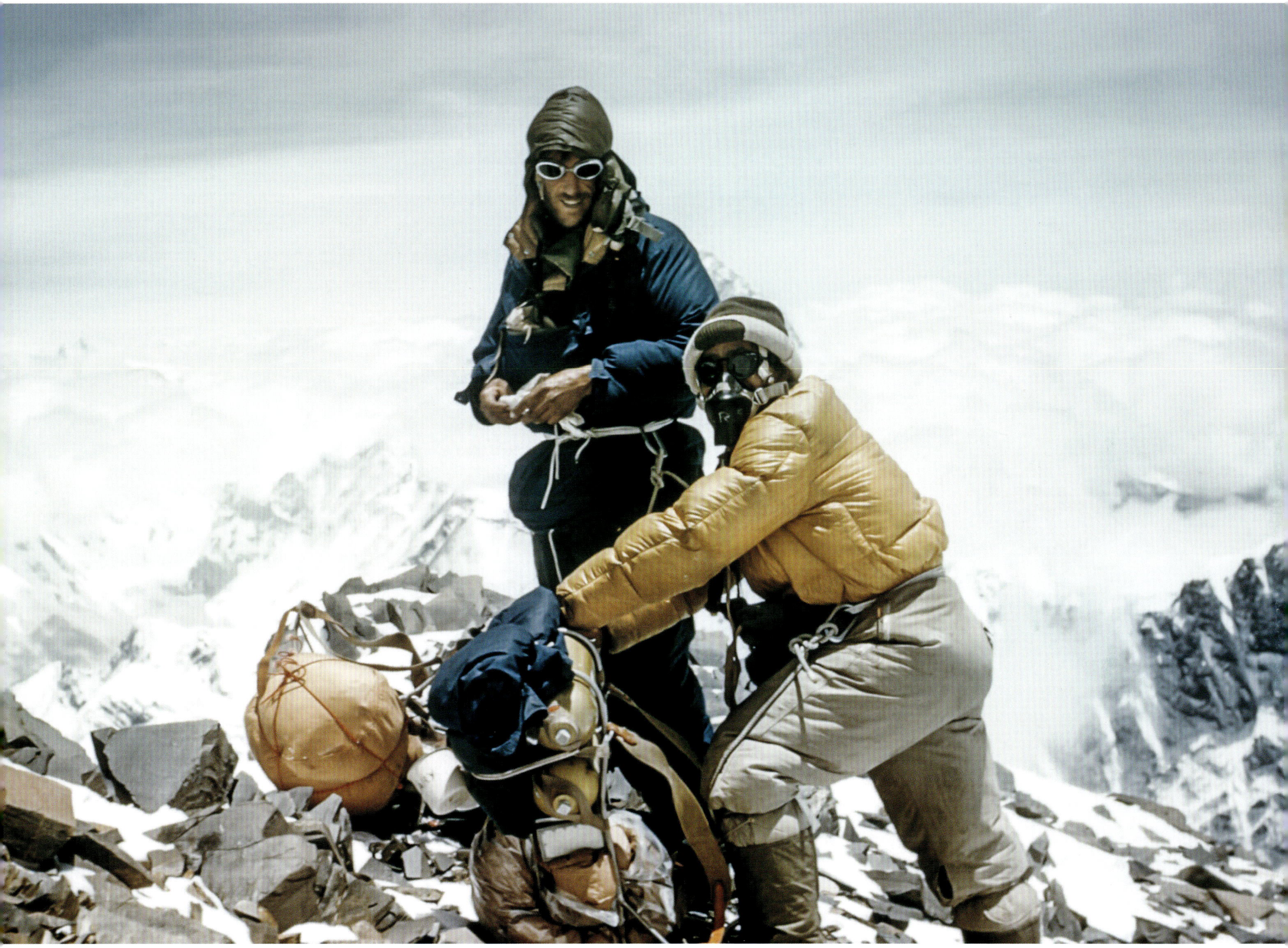

A vintage photo of New Zealander Sir Edmund Hillary and Nepalese-Indian Sherpa Tenzing Norgay, the first two men to conquer Everest in 1953, equipped with Rolexes.

in 1953. This was the company's first diving watch, waterproof up to a hundred metres deep, targeting the large number of passionate sport diving enthusiasts.

1955 was the year of the GMT-Master, which allowed for a second time zone to be displayed, thanks to a fourth hand added to the standard hour, minute and seconds hands. This 24-hour hand was born of a collaboration between Rolex and Pan Am World Airways, the first airline to operate nonstop intercontinental flights.

One important design and technical innovation was the Rolex Day-Date, launched at the Basel Fair in 1956. It was the first watch to indicate the entire day of the week in two small windows at 3 and 12, and the first at Rolex to boast a highly prestigious acknowledgment: the official 'Superlative Chronometer' certification of precision. The Rolex Milgauss was also presented in 1956, its name deriving from a combination of the words *mille* ('one thousand' in French), and 'gauss', a unit for measuring magnetic induction. It was guaranteed to function perfectly, even in the presence of strong magnetic fields. Rolex also developed the Air-King model in this period, which was sold from 1957. Its design referenced models produced during the Second World War, which were mounted in a 34-millimetre case (larger than traditional cases, with a diameter of no more than 33 millimetres), and its dial's graphic design was extremely legible. The model enjoyed great success both in and out of the professional sector.

Toward the end of the 1950s, Wilsdorf began to consider the future of the company he had founded with such passion and had led to success. After the death of his wife Florence Crotty (Geneva, 1882–1944), he established the Hans Wilsdorf Foundation in 1945, to which Rolex still belongs. Wilsdorf would die in Geneva in 1960 at the age of 79.

We can certainly recognise a 'strong will' in Wilsdorf, combined with an extraordinary aptitude for anticipating the future. He devoted himself with immense enthusiasm to producing an object commonly thought of as a 'charming feminine item': the wristwatch.

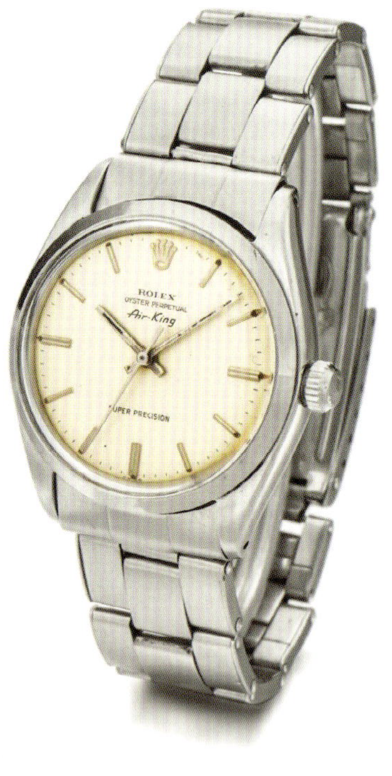

The Air-King watch model.

Right: Rolex advertising booklet from the 1950s inspired by the cover of *Time* magazine.

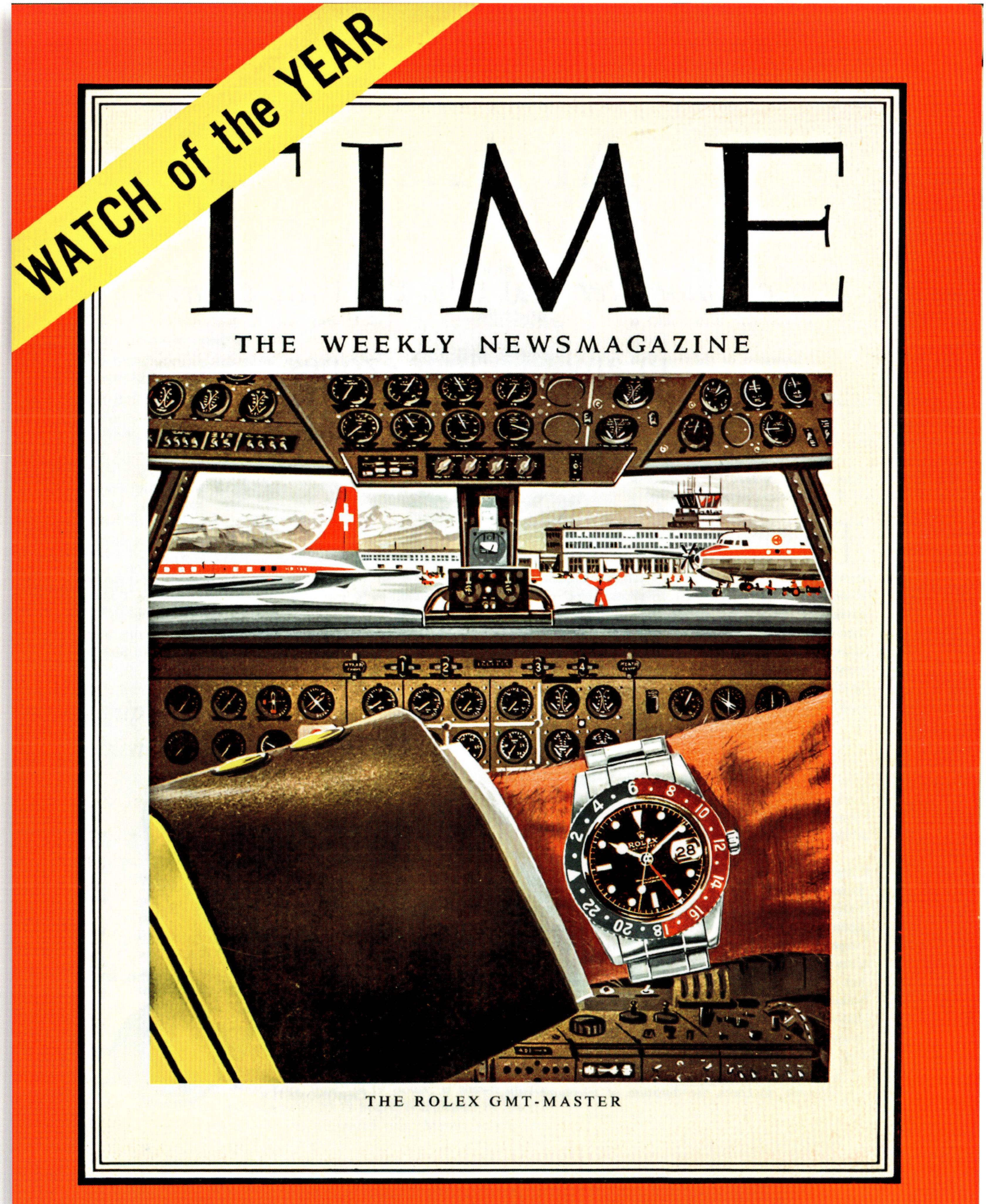
WATCH of the YEAR
TIME
THE WEEKLY NEWSMAGAZINE
ROLEX
THE ROLEX GMT-MASTER

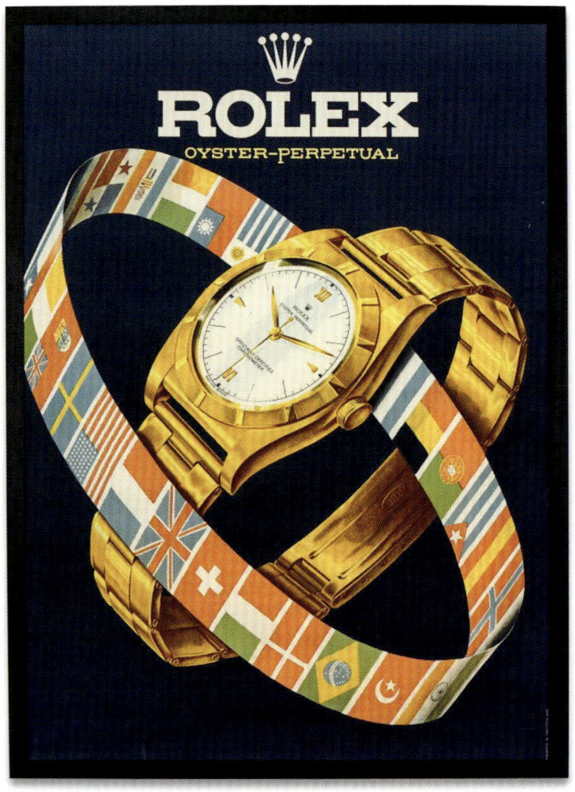

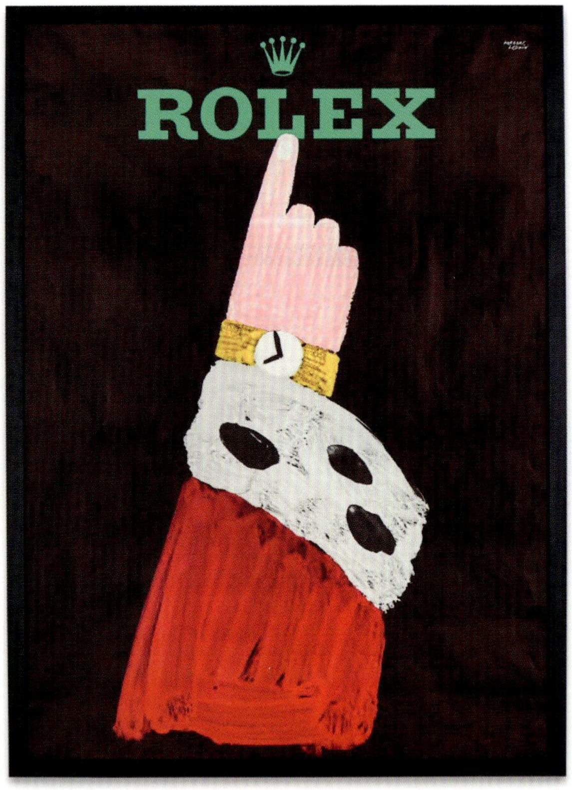

Vintage Rolex posters.

Enlightened Direction

With no heirs, Wilsdorf searched among his collaborators for the right person to keep the company going. His choice fell on André Heiniger, who became chairman of Rolex in 1963 and left his position in 1998. It was in 1948, however, that Heiniger had begun his career at Rolex. After graduating in business administration in Switzerland and specialising in England, Heiniger, who was fluent in many languages including French, German, Spanish, Portuguese, Italian and English, began working for a watch company in La Chaux-de-Fonds. He remained there for six years, specialising in manufacturing and business administration. After a few years, he met Wilsdorf, who offered him the opportunity to develop the market for Rolex in South America: Heiniger stayed in Buenos Aires from 1948 until 1955, when he returned to the Geneva office with an executive position.

Wilsdorf was reportedly so impressed with the young manager that he selected him as his successor to run Rolex shortly before his death in 1960. Oddly, Heiniger was not appointed chief executive of Rolex until 1964 – he likely needed time to adequately prepare for replacing a brilliant leader. Rumors also circulate about his appointment being opposed by other managers. Whatever the reasons may be for the delay, once Heiniger was appointed, he consistently effected an enlightened and skillful management of the company for over thirty years.

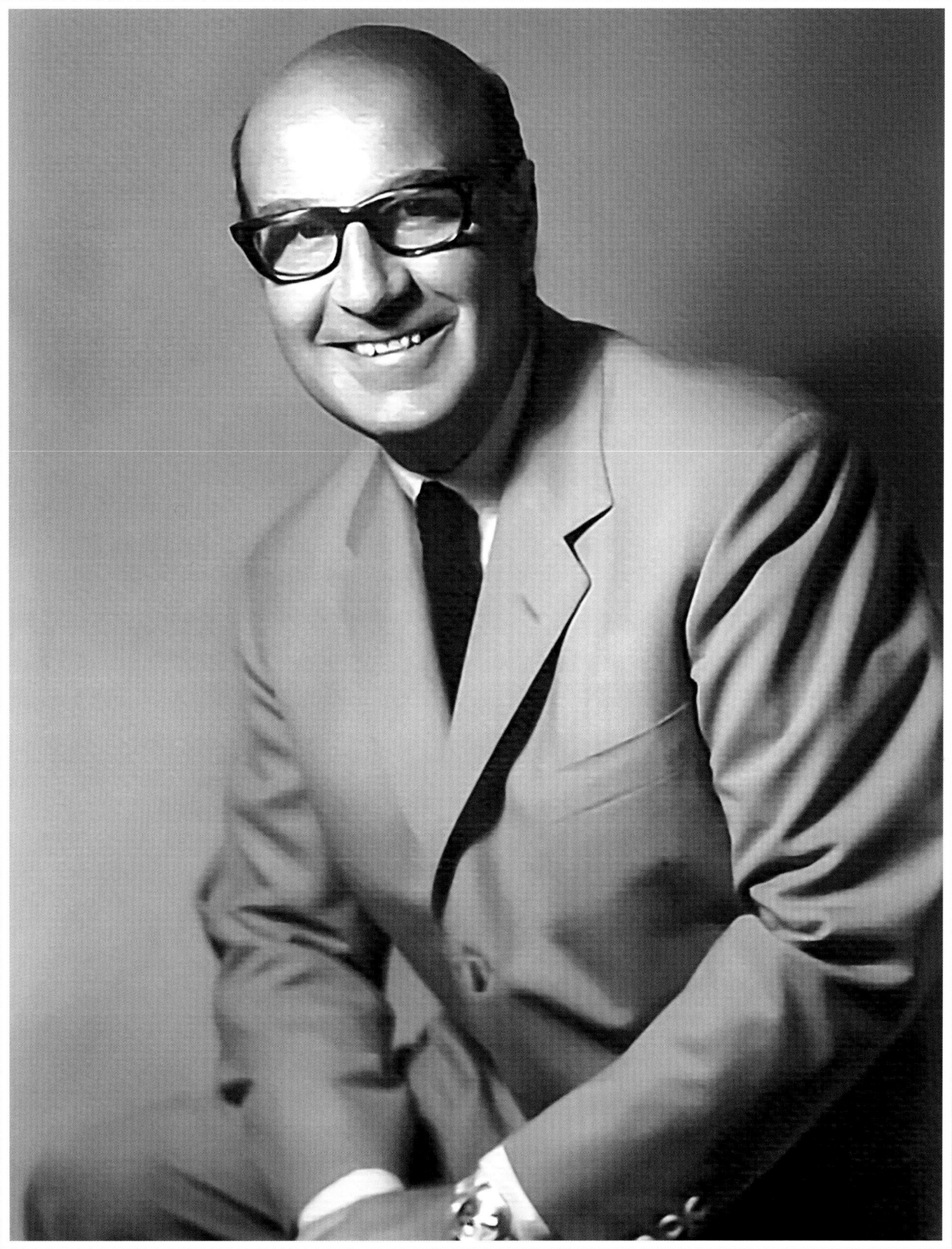

André Heiniger in 1963.

Daring Projects

A man of great commercial experience, Heiniger reinforced Rolex's image by not only continuing the manufacture of an excellent product, but also by boosting the administrative department running the business. He was responsible for the construction of an avant-garde headquarters near Geneva. The building's plans adhered to a design lexicon inspired by art deco and influenced by the Bauhaus canon, reflecting the style of the *maison*'s watches.

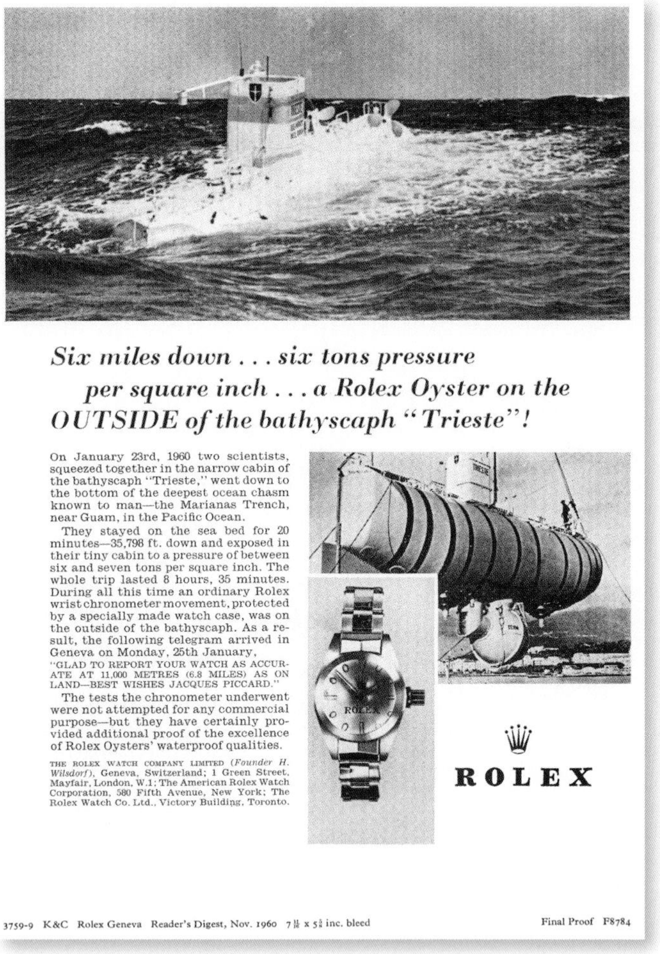

Above: The Deepsea Special celebrates its endurance during its dive into the Mariana Trench, fixed to the bathyscaphe *Trieste*.

Under Heiniger's stewardship, Rolex continued to experiment successfully, working with explorers and scientists as they accomplished daring feats.

In 1960, the Deepsea Special watch, specially designed to withstand strong water pressure, was fixed to the bathyscaphe *Trieste*'s exterior and descended into the Mariana Trench off the coast of the island of Guam to a depth of 10,916 metres. To withstand the pressure of more than one ton per square centimetre, a special, highly robust steel case was developed, with a convex crystal centimetres thick. When he returned to the surface, head of exploration Jacques Piccard confirmed that the watch 'worked perfectly'.

In 1967, the Sea-Dweller was born of a collaboration between Rolex and Comex (Compagnie Maritime d'Expertises), a French company specialising in submarine engineering. The French company's divers first used the Submariner, which was waterproof to a two-hundred-metre depth, but the helium used in the hyperbaric chambers, having penetrated the watches, risked creating such a pressure that the glass would shatter during decompression. In a quest for chronographic precision, another model conceived was destined to become a true icon: the Cosmograph Daytona. Launched in 1963 and designed to meet the chronometric needs of professional race-car drivers, this chronograph allows for times to be displayed and measured on the

Right: An advertisement celebrating the collaboration between Rolex and Comex, as well as the Sea-Dweller's reaching a depth of 520m.

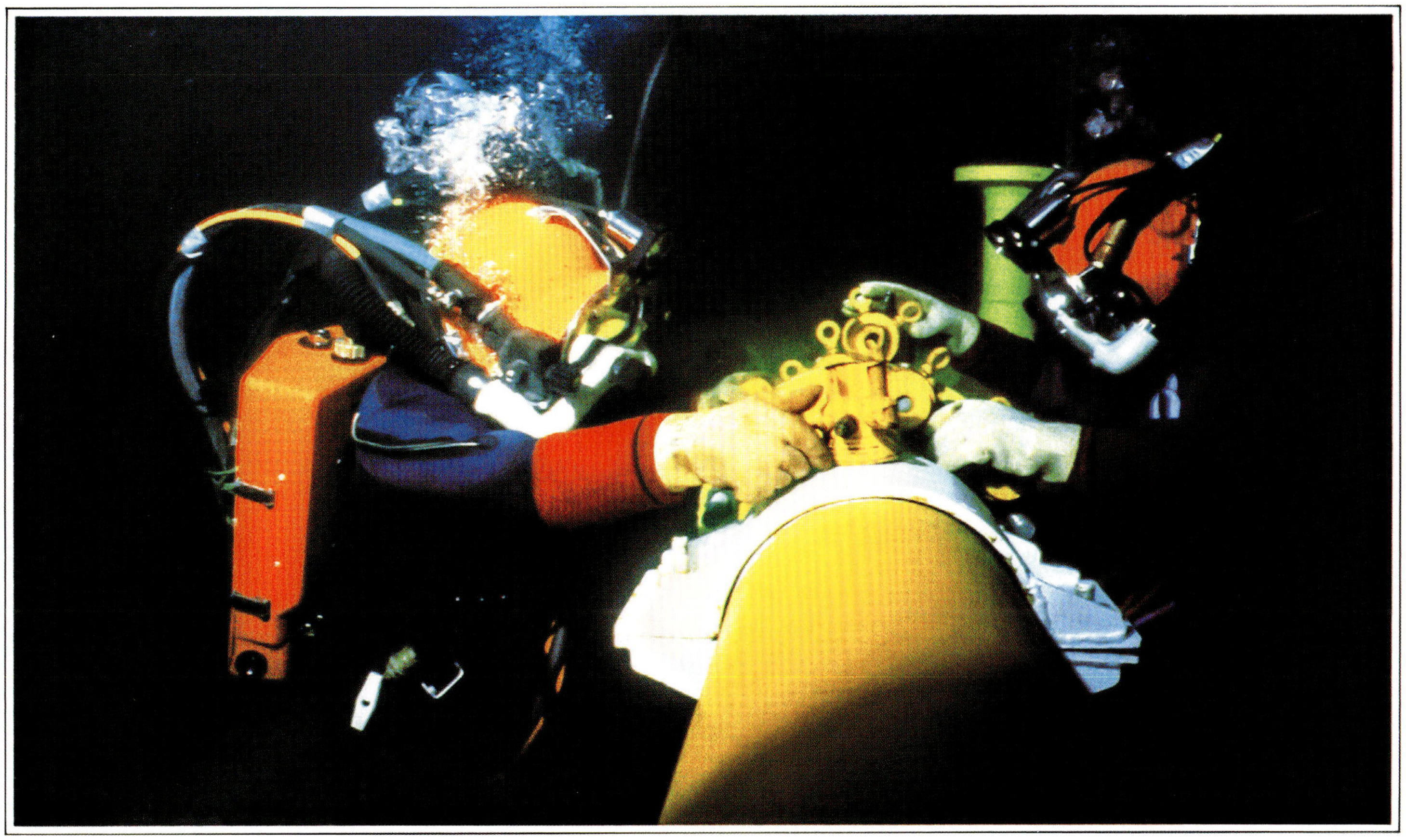

DIVERS CONNECTING A PIPELINE AT A DEPTH OF 520 METRES.

At a record 520 metres deep, the Sea-Dweller is still in its element.

The Comex divers have once more forced forward the frontiers of the impossible. Diving off Marseilles, a six-man team spent eight days performing complex underwater tasks at the unprecedented depth of 520 metres.

An astonishing breathing mixture, hydraliox, made it possible for them to live and operate at such inhuman depths. Hydraliox is a combination of hydrogen, helium and oxygen. It gives divers complete freedom of movement with minds unimpaired by pressure effects.

The Comex operation, known as Hydra VIII, is a world record, comparable in its own way to man's first moon landing. It has created huge possibilities for research.

Tomorrow, thanks to the divers' determination and skill, men will descend even deeper, to repair oil rigs and explore the ocean floor.

Beneath the pressure of over 500 metres of water, there is a ceaseless need for precision. At such astonishing depths divers must depend on something rather more than a mere watch. They need a veritable measuring instrument: the Rolex Sea-Dweller.

Like the Comex divers themselves, the Sea-Dweller is strong, precise and reliable. Guaranteed water-proof to 1,200 metres, it withstands the enormous pressures as well as decompression upon surfacing. This is ensured by a patented Rolex invention.

Men still have another 680 metres to descend before they rival the Rolex Sea-Dweller. But then, divers – even the most skilled – are always in an alien element. The Sea-Dweller is not.

ROLEX
of Geneva

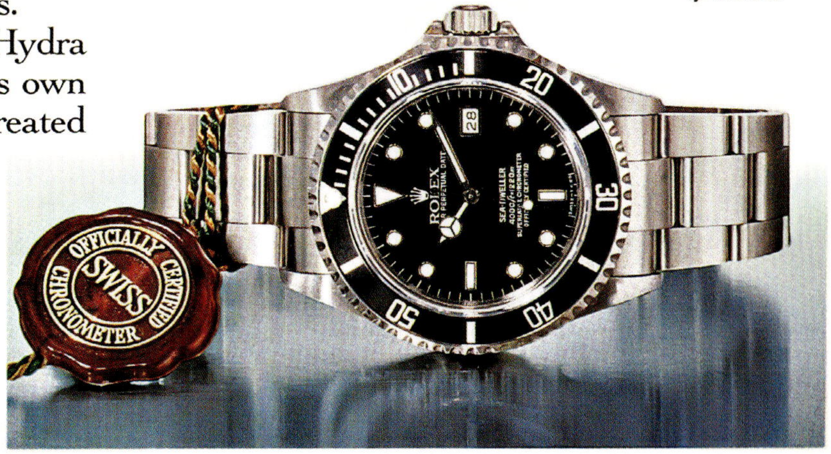

THE ROLEX SEA-DWELLER DATE CHRONOMETER IN STAINLESS STEEL.

dial in hours, minutes and seconds. Initially it was not very successful, and only in the early 1980s was it reevaluated, considered an 'object of desire' by thousands of people.

An advertisement from the 1960s reads that 'The Rolex watches of today have reached such heights of precision, accuracy and ruggedness that it is difficult to predict what the watches of tomorrow will be like. Whether they will be in cases of titanium or synthetic diamond. Whether they will be powered by atomic energy or by the sun. But one thing is certain. The finest watches of tomorrow will be built by the men at Rolex.'

The imagination of Rolex executives had not foreseen the imminent 'quartz crisis' that would overwhelm the market in the early 1970s.

A Crisis Skillfully Handled

In 1969, the Japanese company Seiko presented its Astron model to the market. The Astron was the first quartz watch and was advertised with the slogan 'Some day, all watches will be made this way.' Japanese computer manufacturer Casio began to make quartz watches in 1974 and soon became a top brand, revolutionising the watch market.

A number of famous manufacturers in Switzerland, including Rolex, formed a consortium dubbed CEH (Centre Electronique Horloger), to face this new technology. The centre worked on the development of quartz technology and in 1970 it presented the Beta 21 movement. It was used by manufacturers in the 1970s, but always in combination with analogue dials. Rolex included it in the reference 5100. Soon, however, the Geneva-based company decided to choose a different, autonomous path and in 1977 it presented the Rolex Datejust Oysterquartz reference 17000 to the public. The watch was powered by the quartz Calibre 5035, created by the Rolex laboratories and exclusive to its watches.

The method with which electromechanical Rolex Calibres 5035 and 5055 were conceived was a completely different way from movements by the CEH consortium, boasting various, highly significant technical innovations.

To achieve very high precision, Rolex increased the frequency of oscillations until it reached 32.768 Hz, about four times that of the Beta 21. A special sensor detected the ambient temperature and sent it to the electronic control module, which in turn proceeded to calibrate the rate of the quartz crystal; a manual adjustment system allowed for fine-tuning and, over the years, compensated for the quartz crystal's natural aging.

The quartz crystal was subject to further refinement in *c.*1978, meeting requirements for the prestigious chronometry certification. On 8 May 1978, an Oysterquartz would be on the wrist of climber Reinhold Messner when he conquered Everest without an oxygen mask.

Produced in limited numbers, this watch never became one of the company's flagship models.

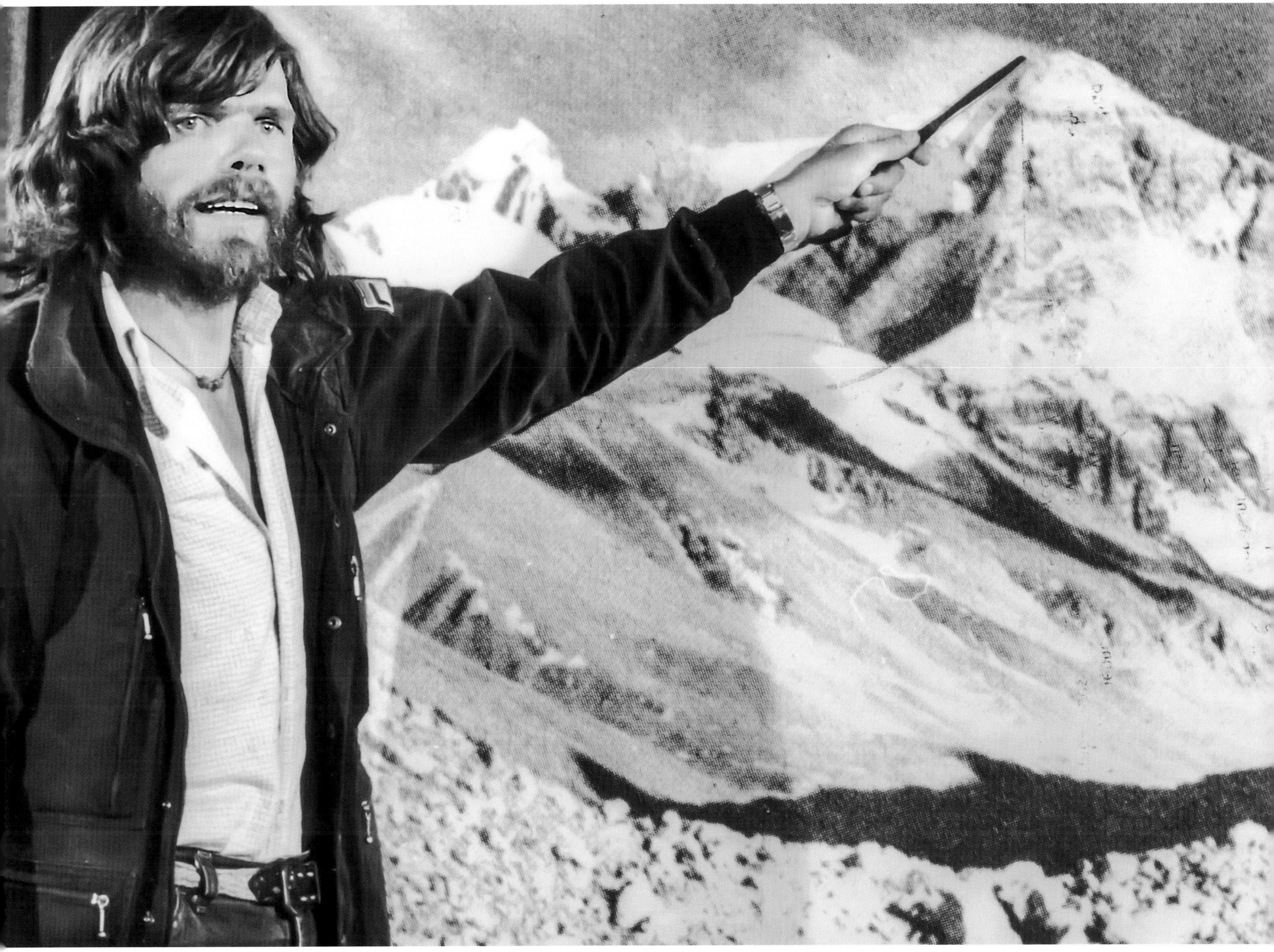

Reinhold Messner makes plans to climb Everest without an oxygen mask.
He would succeed in 1978, with an Oysterquartz on his wrist.

A New Positioning

Apart from this foray into the quartz sector, Heiniger's strategy was to garner a higher position for Rolex and its products, turning it from a company that produced 'reliable watches' into a luxury-goods company.

For Heiniger, communication was also fundamental. It was in the 1970s that advertisements appeared linking Rolex to political power ('Men who guide the destinies of the world wear Rolex watches') or to leadership roles ('If you were flying the Concorde tomorrow you'd wear a Rolex'). Heiniger's famous answer to the question 'How's the watch business?' dates to these years: 'I have no idea. Rolex is not in the watch business. We are in the luxury business.'

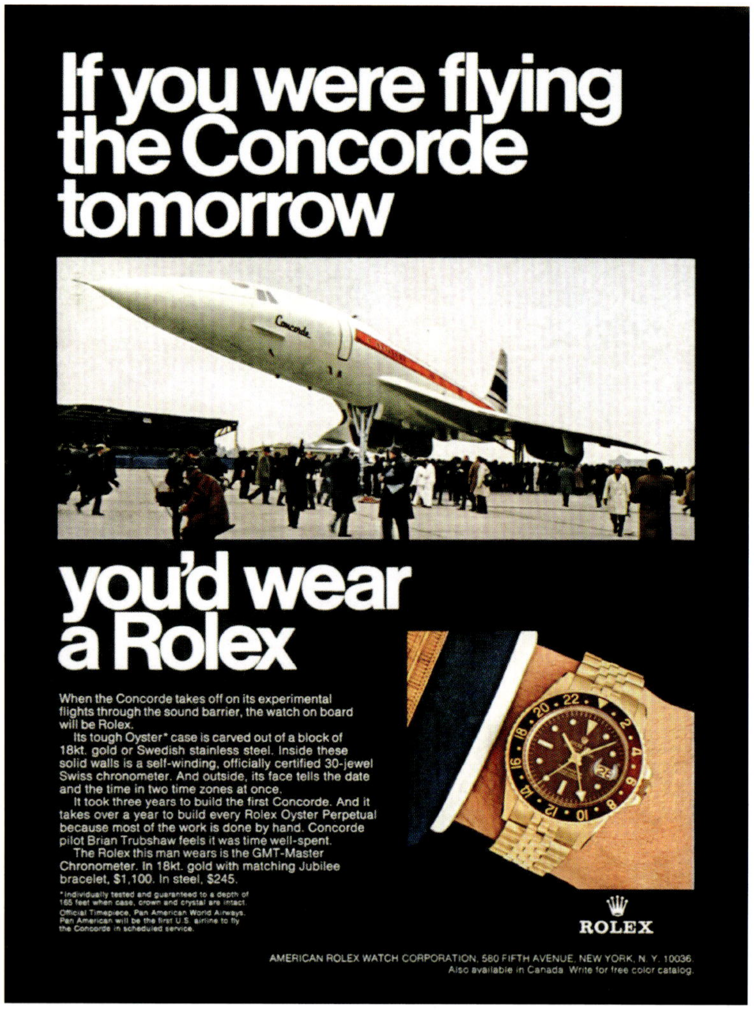

Above: When the first supersonic airliner, the Concorde, completed its final flight tests in the 1960s, Rolex proudly announced that both the British and French test pilots were wearing a GMT-Master, thus cementing this watch's status in the era of supersonic flight.

This decade produced such models as the Explorer II (1971) and the Sea-Dweller 4000 (1978), which would contribute to the success of the company. Congruent with an exuberant economy and an opulent aesthetic, jewelled editions of different references were born in the 1980s, produced in a limited number of specimens.

Although Heiniger was a very private man and had a very discreet public presence, it was not uncommon for him to be photographed with stars and celebrities, like tennis champions Björn Borg and John McEnroe, or attending the Maxi Yacht Rolex Cup in Porto Cervo, Sardinia.

In keeping with the philanthropic spirit of Rolex's founder, and to commemorate the Oyster's 50th anniversary, Heiniger established the Rolex Awards for Enterprise in 1976, created to support 'exceptional individuals around the world, pioneers who had no or little access to traditional funding and were responding to major challenges with original and innovative projects that advance human knowledge and well-being.'

Right: Björn Borg and John McEnroe, competitors on the playing field but both Rolex watch enthusiasts.

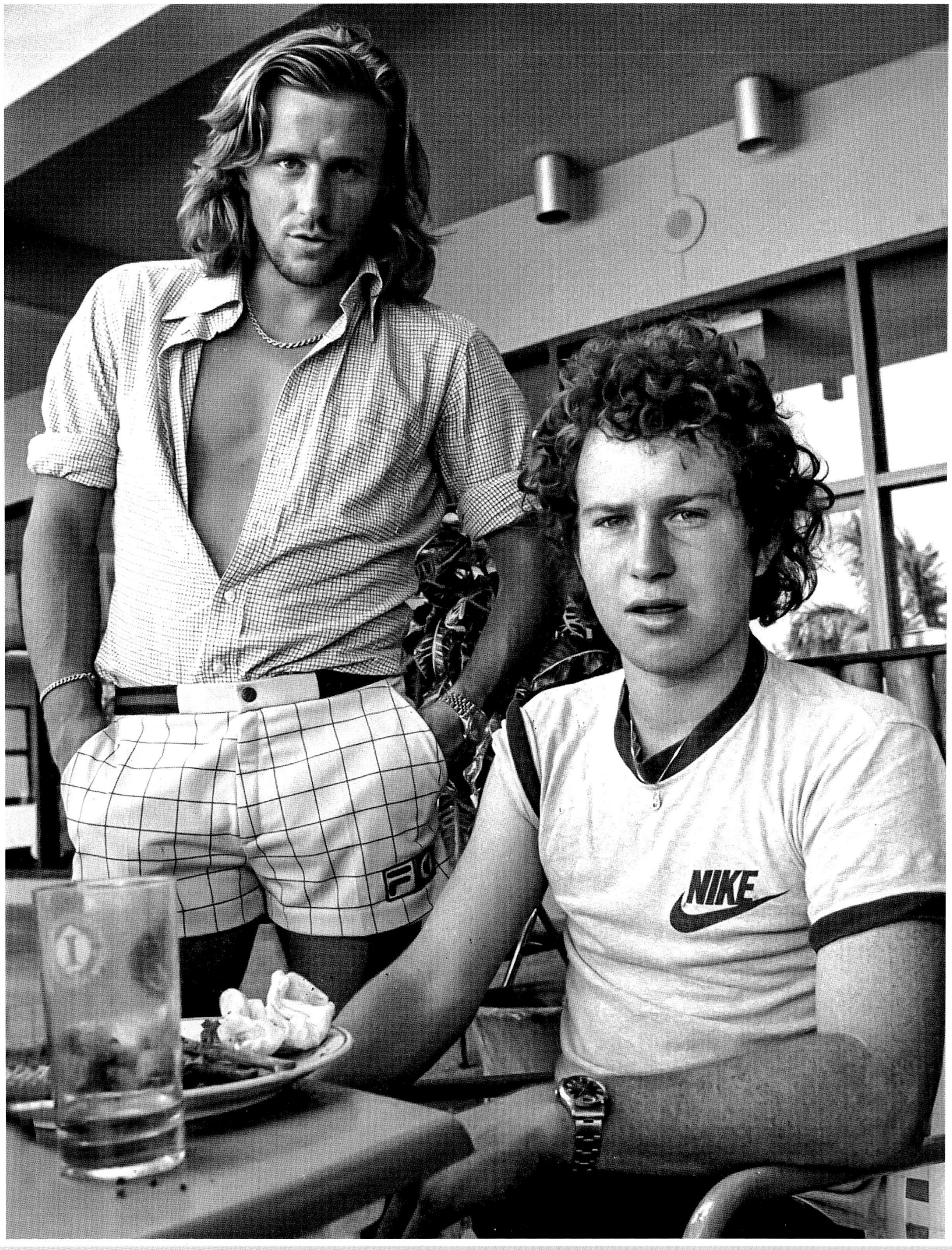
NIKE

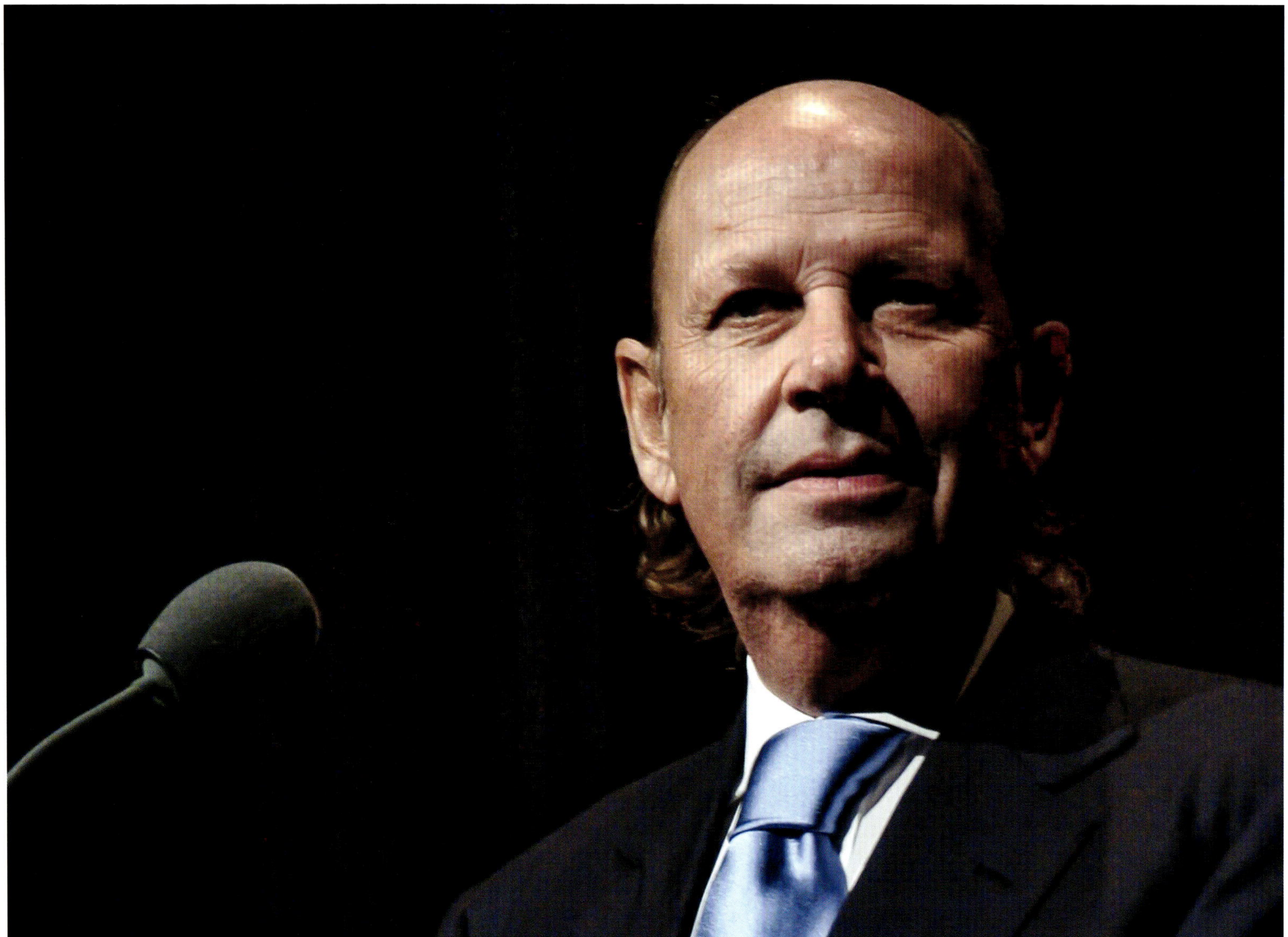

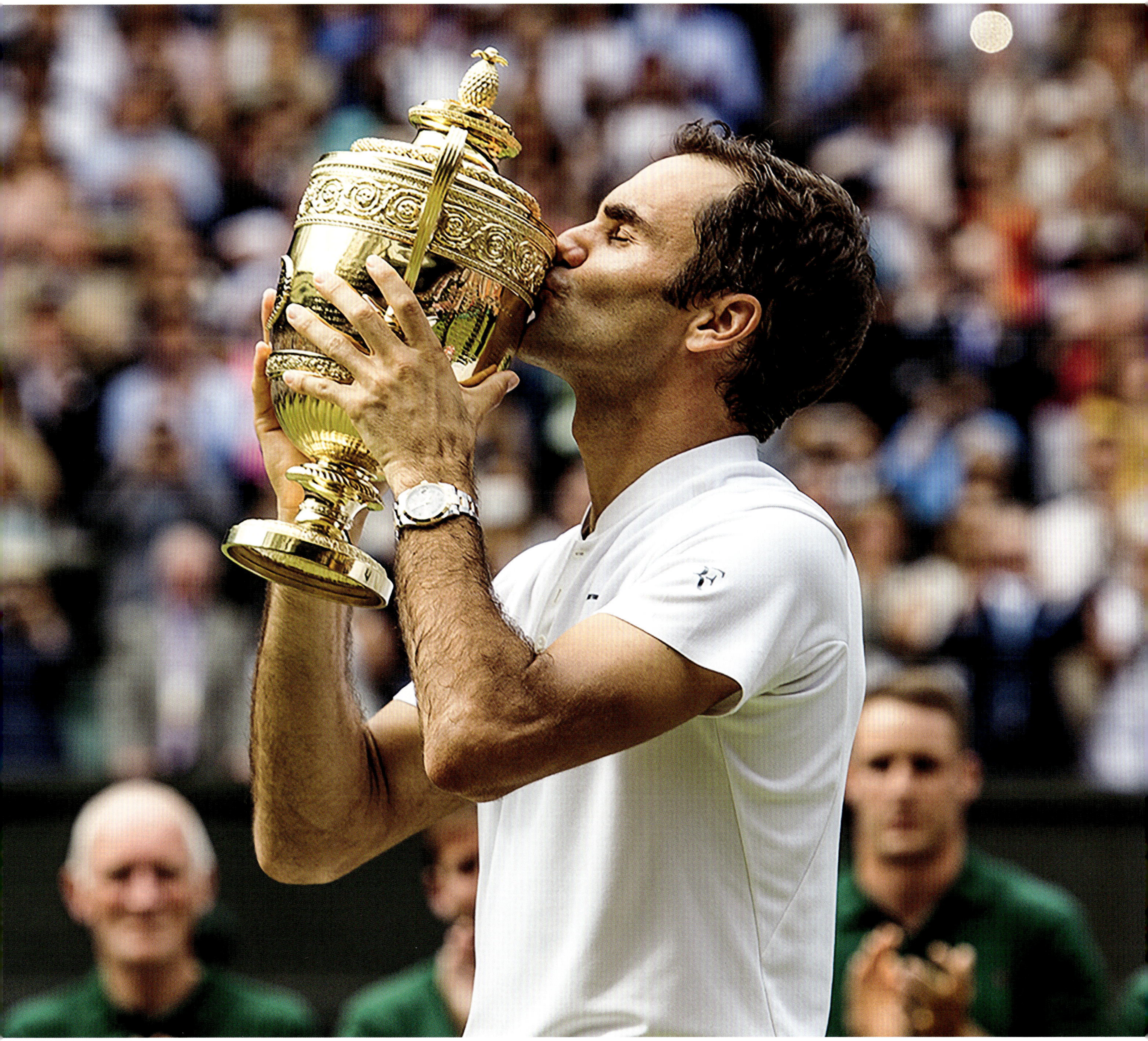

Tennis player Roger Federer is a Rolex brand ambassador.

It is the largest museum dedicated to cinema in the United States, housing over 50,000 square feet of gallery space dedicated to temporary and permanent exhibitions, two cinemas, including the David Geffen Theater with one thousand seats, an educational area and spaces for public events and ceremonies.

The Mentor and Protégé programme supports collaborations between emerging and established artists to discover and nurture new talent. Some of the greatest artists in the world have taken a turn in the role of Maestro over the years, including Sir Anish Kapoor, Sir David Adjaye, Sir David Chipperfield, David Hockney, Martin Scorsese, Alejandro G. Iñárritu, William Forsythe, Peter Zumthor, Margaret Atwood and Sir Colin Davis, as well as icons like Brian Eno, Gilberto Gil, Zakir Hussain and Youssou N'Dour. Of the 63 Master-Pupil pairs formed since 2002, nine concerned the world of cinema.

There is no shortage of partnerships with explorers and environmental advocates. There is the Mission Blue initiative, led by legendary underwater explorer Sylvia Earle, campaigning to establish a worldwide network of marine protected areas around the globe; it aims to protect species and critical ecosystems, inspire and promote local businesses, awaken public interest and encourage support of acts for ocean conservation around the world.

The Rolex Awards for Enterprise have received 35,000 nominations since the programme's inception in 1976, and Rolex has recognised 155 people presenting innovative projects. They have come from diverse fields, ranging from the creation of technologies to improve people's daily lives, the conservation of at-risk ecosystems, ocean protection and exploration of the planet's new frontiers to advanced discoveries in scientific and healthcare fields.

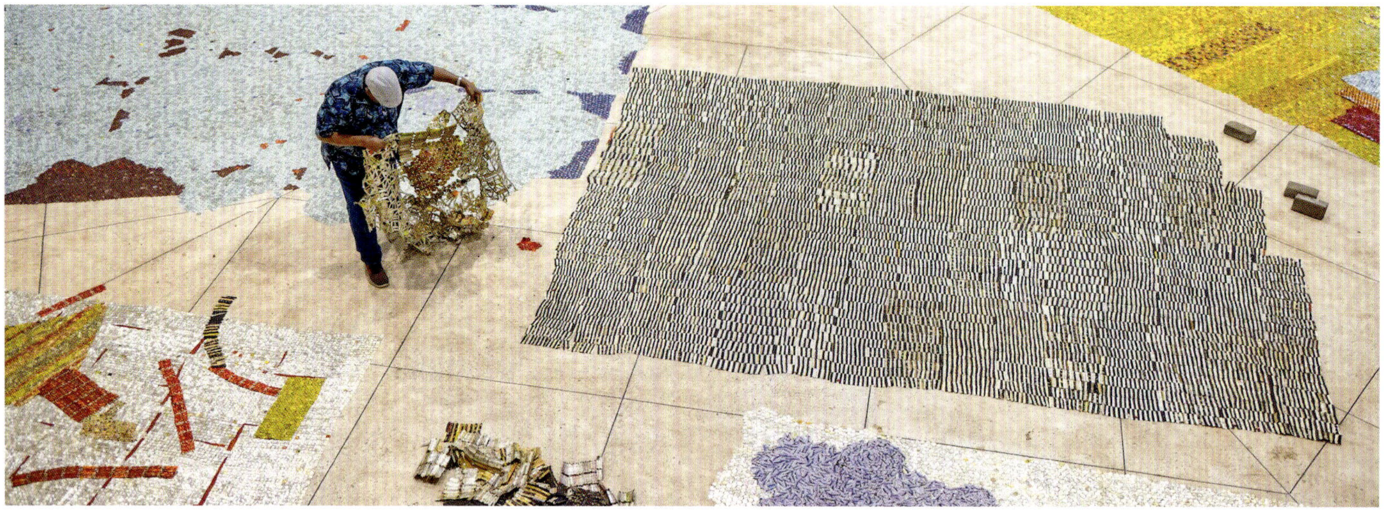

El Anatsui. The Ghanaian artist has helped redefine the global art world, bringing an African presence through his monumental sculptures and installations through the use of paint, textiles, sculpture and design, incorporating water, wind, wood, clay, stone, metal, printing plates, aluminium bottlecaps and other repurposed materials.

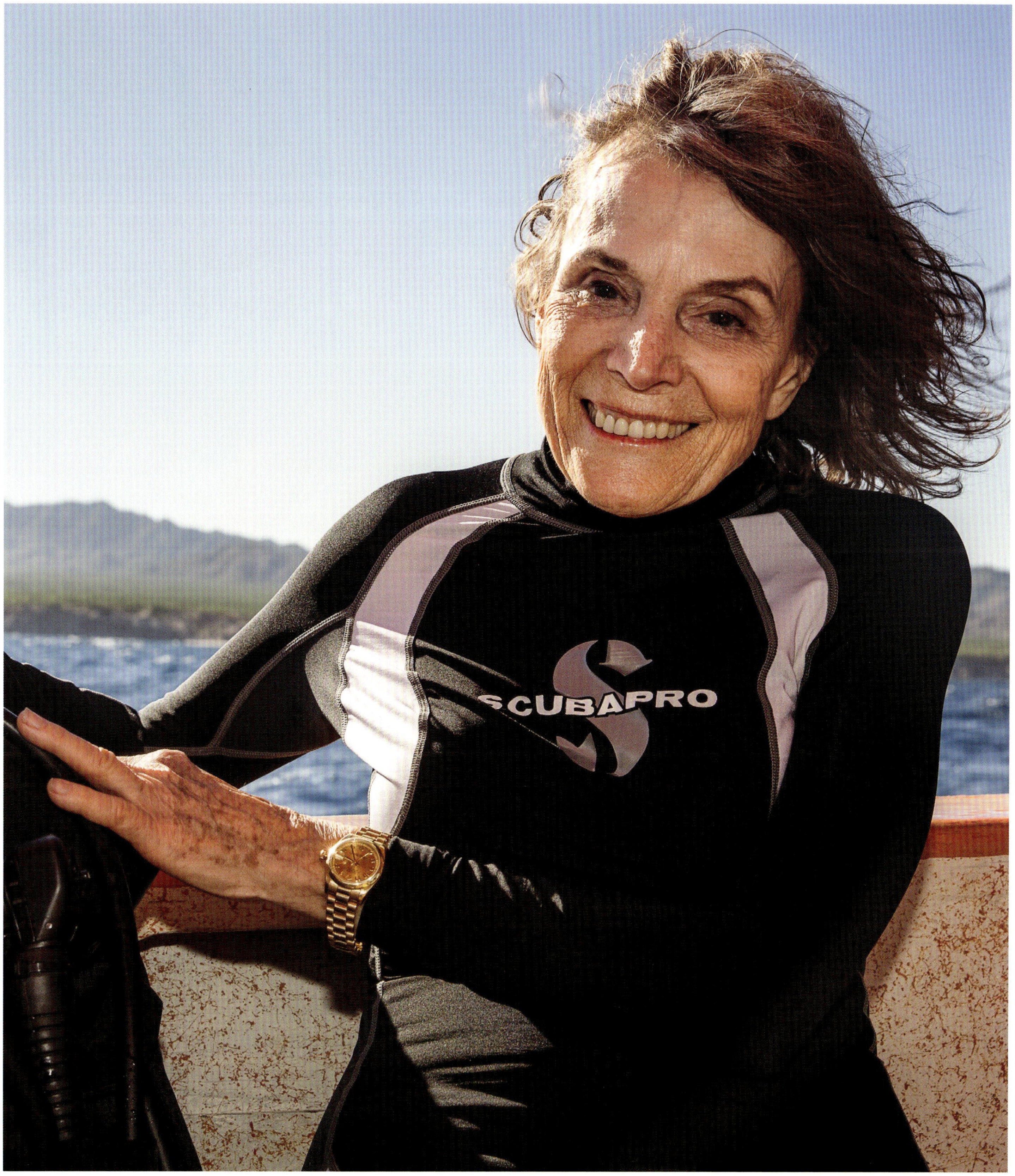

Sylvia Earle, marine biologist and oceanographer.

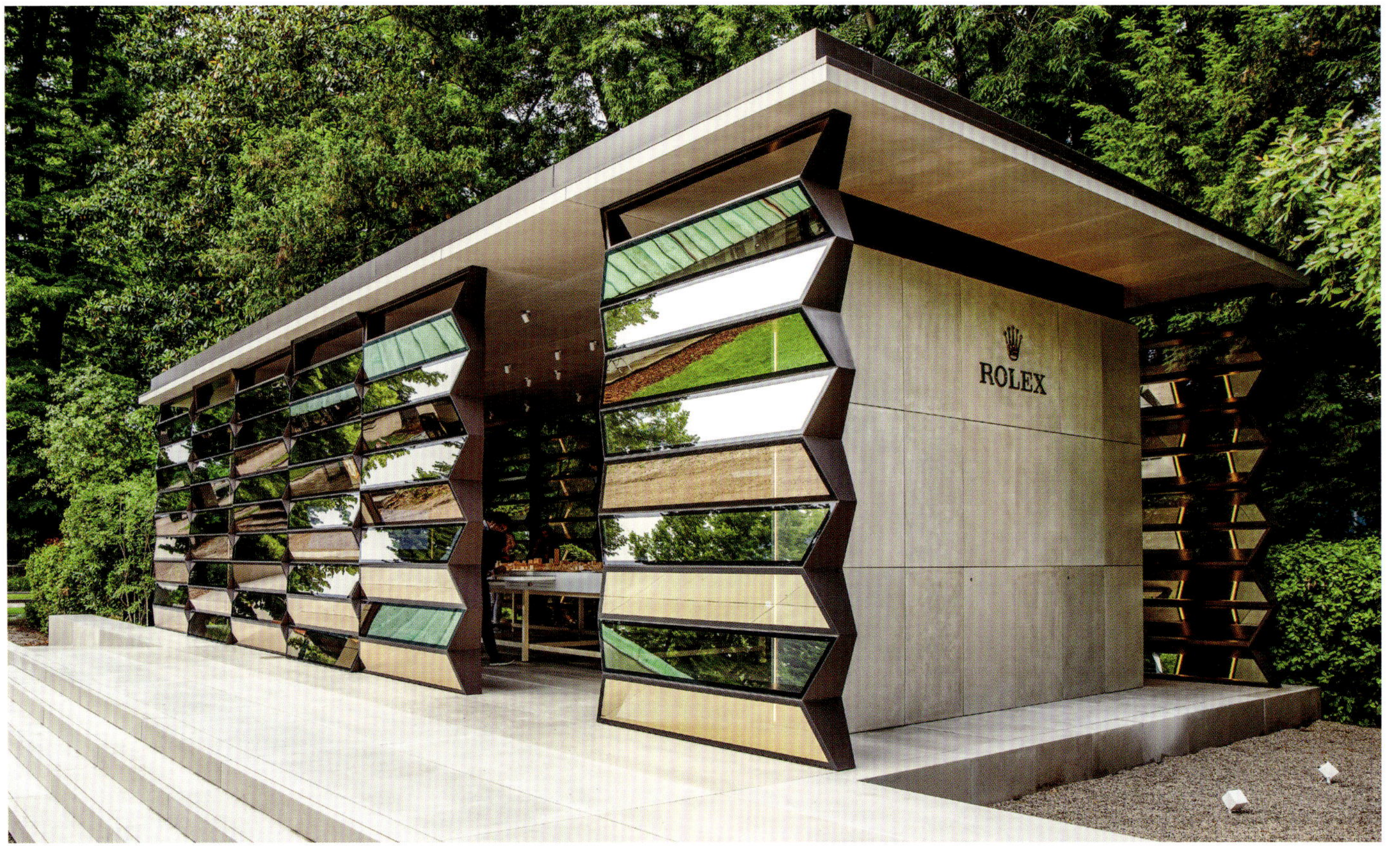

International Architecture Exhibition: Venice Biennale, Rolex pavilion.

The company's contributions embrace other forms of art such as architecture via the Venice Biennale International Architecture Exhibition. The Rolex Pavilion's design resembling a fluted bezel – an aesthetic signature visible on some of the brand's iconic watches – seeks to further emphasise an enduring link between architecture and watch-making.

In Dallas, Texas, Japanese architect Kengo Kuma designed a seven-story office tower that now houses Rolex's sales and service centre. The building features a series of planted terraces on each floor, and its form, in combination with its environmental qualities, has set new benchmarks in the city. The Rolex headquarters in New York City, meanwhile, was designed by Sir David Chipperfield, conceived as 'creating an exemplary building befitting the heritage and culture of the Rolex brand.' The 25-storey tower with overlapping, jagged volumes was designed to qualify LEED (Leadership in Energy and Environmental Design) Platinum certification, the highest level of accreditation for environmental friendliness and sustainable energy consumption. It is not only for the excellence of its products that Rolex is deemed valuable, but for its consistency and for its commitment to protecting the planet as well.

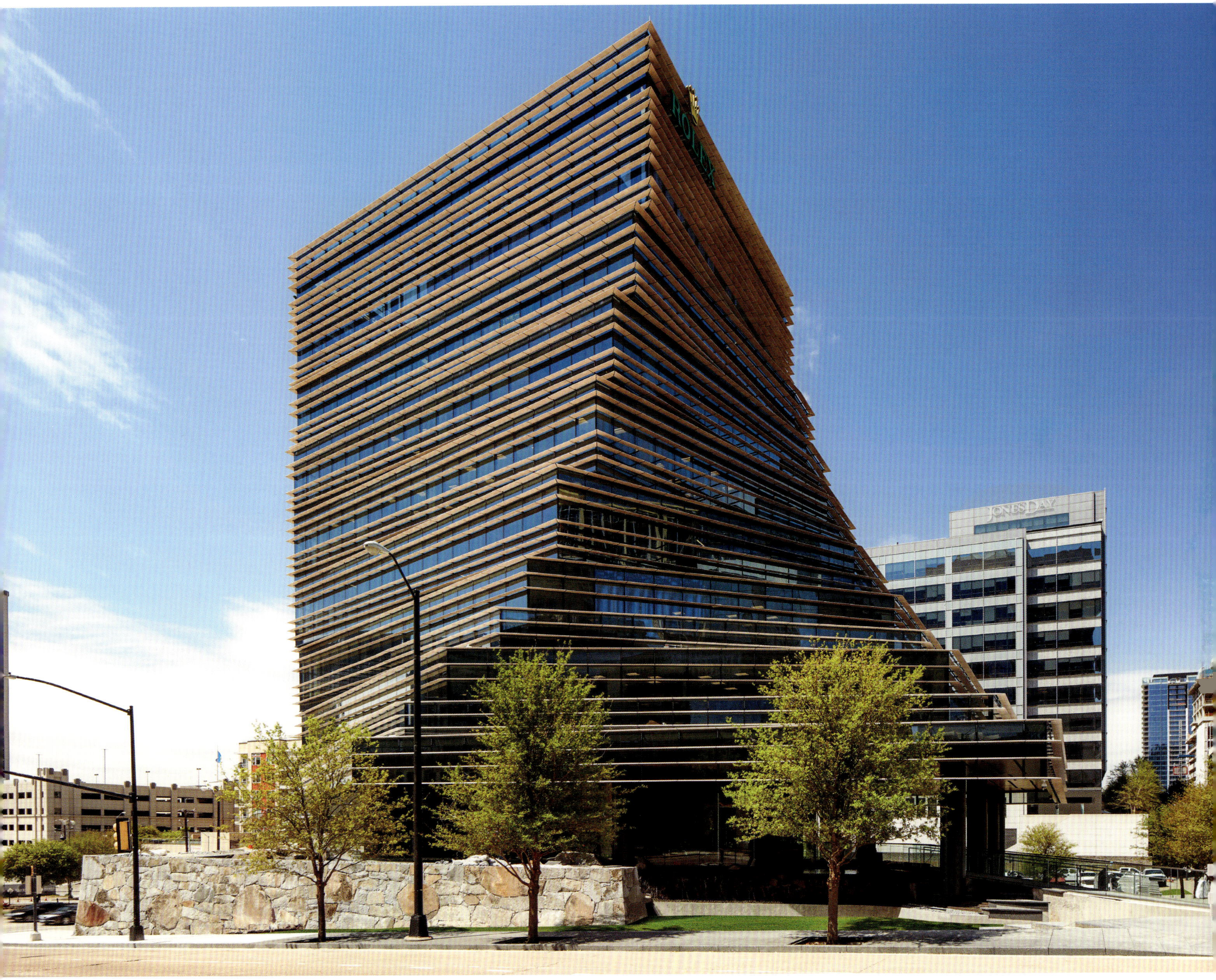

The twisted tower in Dallas, designed by architect Kengo Kuma, houses the Rolex sales and service centre.

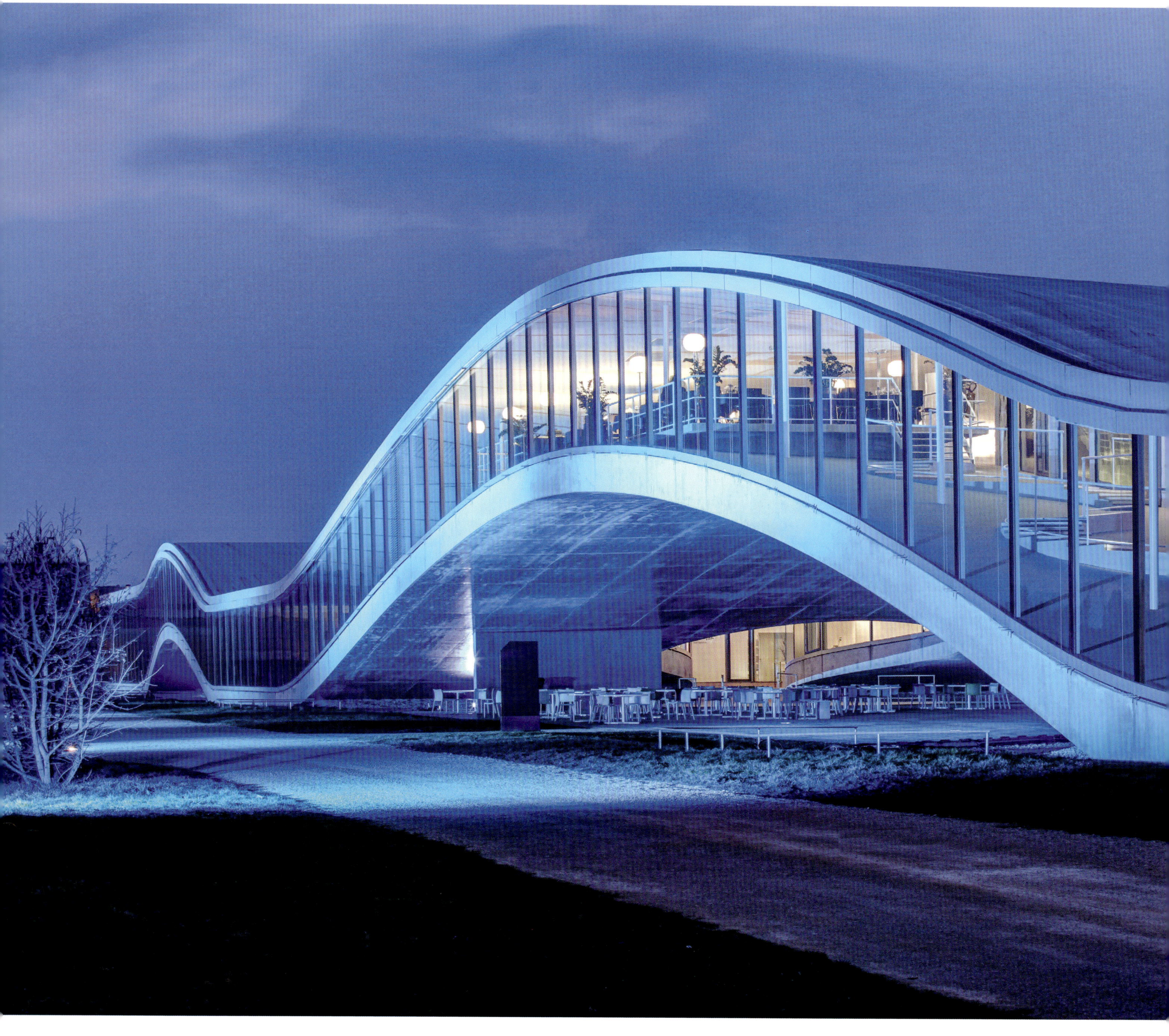

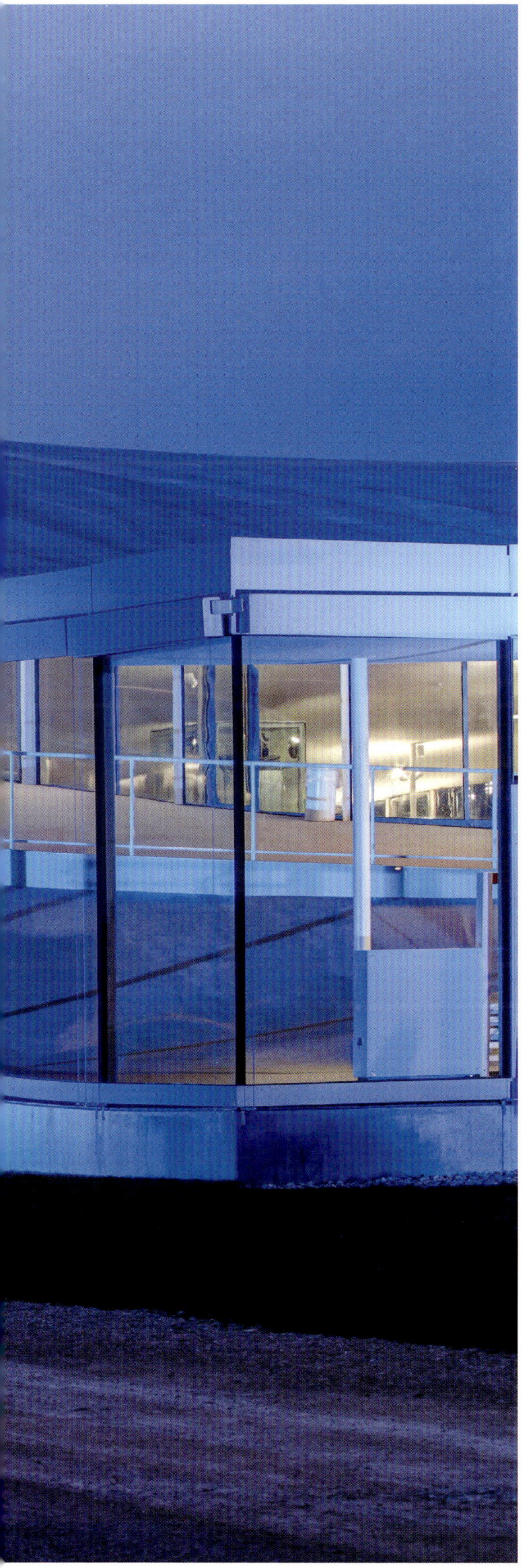

The Dream Lives On

Then again, a Rolex watch is not a mere watch. Nor is it a status symbol, in the old sense of the term: whoever wears it not only sends a signal of wealth but also seeks to establish that they belong to a world, a group of people, that knows how to appreciate quality and beauty, united under a successful international image linked to prominent personalities in fields of all types.

For Hans Wilsdorf, a visionary entrepreneur, making reliable watches that lasted over time was a matter of honour. It is in this spirit that, starting from the company's birth, Rolex's founder began to create a network of efficient workshops dedicated to after-sales service in countries where the brand had a presence. In doing so, he guaranteed that each watch produced by the manufacturer could be given maintenance service at any time and in the manner most appropriate to maintain its extraordinary performance over time.

The company's direction, administration and communication activities; those related to final assembly, final inspection and marketing of watches; and the Rolex World Service are all managed from the Rolex World Headquarters in Geneva. Creative as well as research and development activities are also carried out there, where the workshops are located and where all finished watches are tested before being shipped around the world. The Rolex Training Centre is also here, offering courses to company employees and managers, and trains apprentices in order to pass on its unique knowledge.

The Plan-les-Ouates site concentrates the activities involved in developing and producing Rolex watch cases and bracelets, from shaping and processing raw materials, up to the creation of finishes for the final pieces. The Chêne-Bourg site brings together activities associated with development and production of dials and bezels, as well as jewel setting. The Bienne site is entirely dedicated to the production of Rolex movements.

The Rolex Learning Centre in Lausanne, a symbol of the deep bond between Rolex and the Swiss Federal Institute of Technology Lausanne (EPFL).

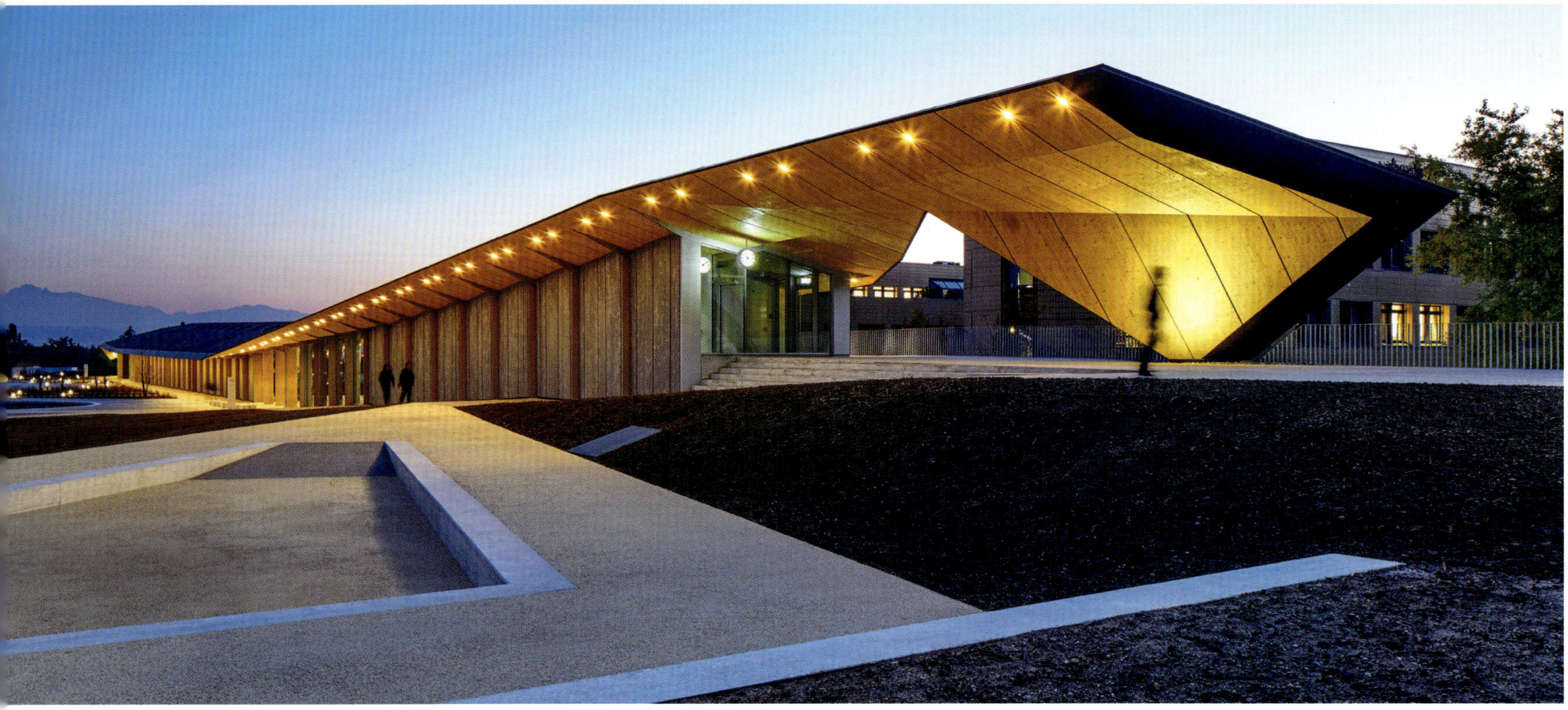

EPFL pavilions.

The components of the brand's movements – numbering from two hundred to around four hundred in more complex calibres – are produced with extreme precision and attention to detail. This is a robust effort on multiple levels to convey the message that a Rolex watch emerges from a careful study of technology and is a precious object – for consumers with good spending power, but also an object of desire to thousands of people, approaching one's *ideal* of a watch if not the *ideal watch*.

Today, the Swiss watch brand is one of the best known in the world and is first in terms of market share in the high-end range. According to a report by LuxeConsult and Morgan Stanley, Rolex had a market share of 29% and an estimated turnover of eight billion Swiss francs in 2021; the study also estimates that Rolex produced 1.05 million watches that same year.

The *maison*'s development in the spirit of Hans Wilsdorf, his entrepreneurial ability, and his out-of-the-box, ahead-of-his-time vision, as well as the great figures who succeeded him, ensured that Rolex remained an independent company, one of the few to not be publicly traded or belong to a finance group – over time, the strategy has proven successful.

After all, Coco Chanel was not trained as a dressmaker, and Hans Wilsdorf was not a watchmaker. Just as she changed the fashion world, he revolutionised the way people use watches for all time.

Right: The Rolex winding rotor (or oscillating weight).

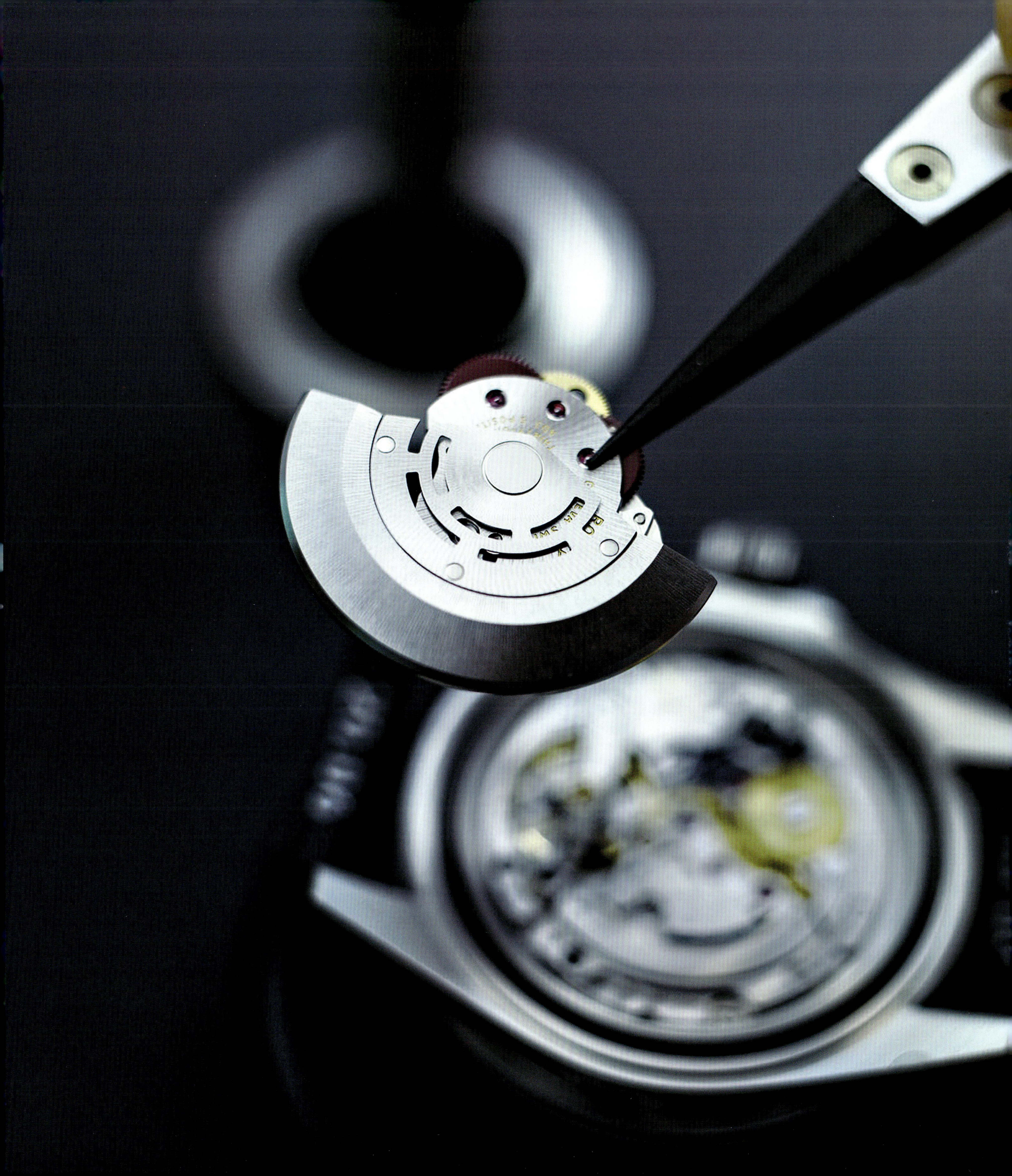

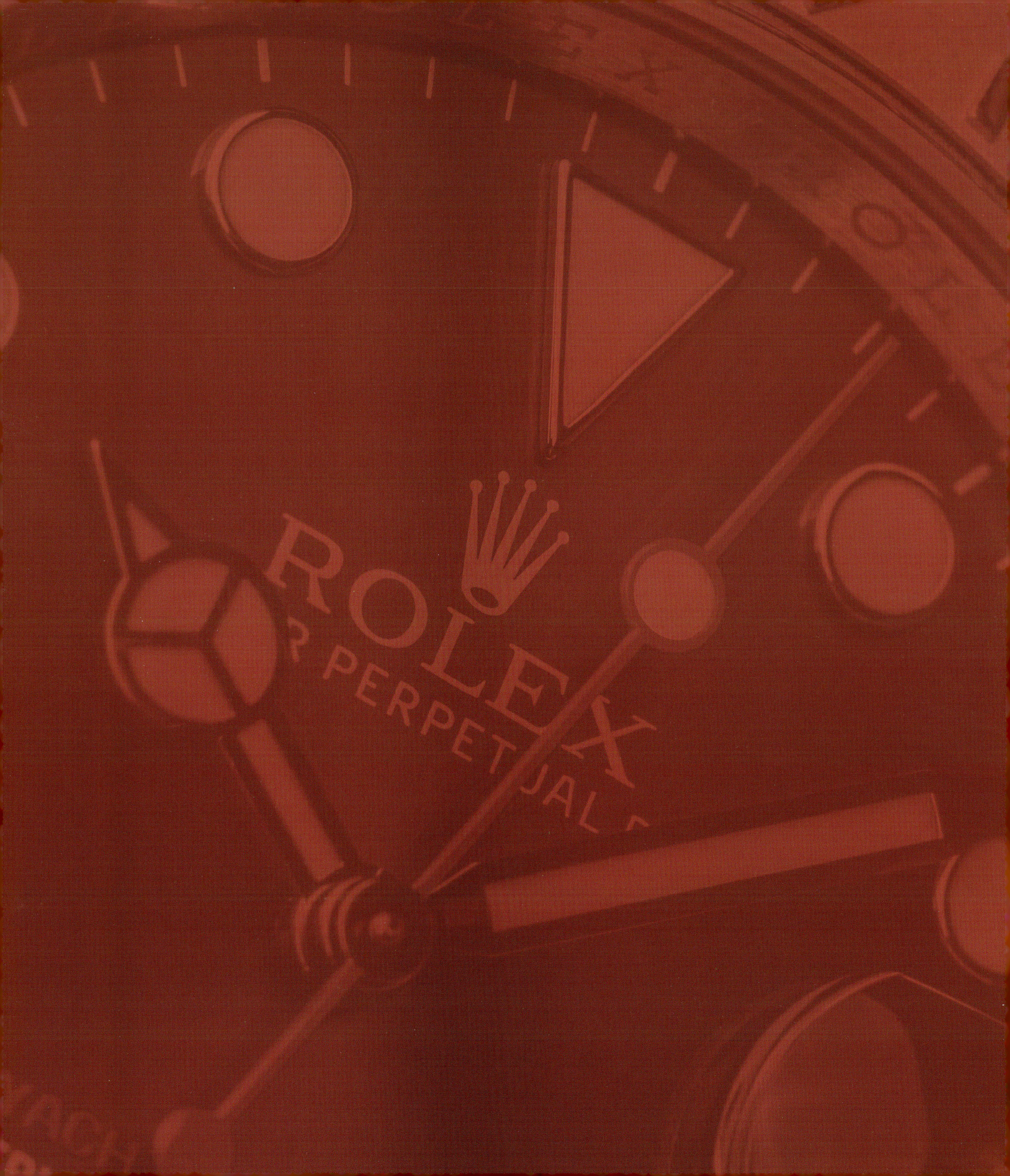
ROLEX
PERPETUAL

THE MODELS

HISTORY AND EVOLUTION

Oyster Perpetual

The Oyster Perpetual is the direct heir of the original Oyster, the first waterproof wristwatch in the world, on which Rolex developed its reputation for excellence in 1926.

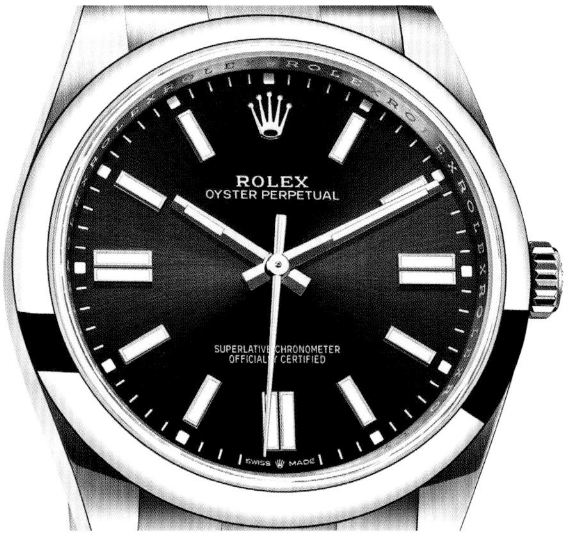

The birth of the Oyster Perpetual constituted a true revolution in the history of Rolex – there were two revolutions, in actuality, realised over the interval of a few years, but destined to change the history of horology, not just that of Rolex. The 'Oyster' name refers to the patent for the first waterproof case, made possible thanks to the invention of a screw-down winding crown in 1926. Rolex had developed the Oyster case at Hans Wilsdorf's behest. He was convinced of the invention's necessity in order to protect the timepiece from dust and humidity; they were considered the worst enemies of a watch, capable of damaging its interior permanently and compromising its ability to tell time. To be certain of its water resistance in any situation, Rolex regularly entrusted their watches to athletes and explorers who could test them in the field.

When the watch was presented, Wilsdorf revealed the reason that it was called the Oyster: It was 'a model housekeeper' that 'simply tolerates no dust or other impurities. [...] Gentlemen, we have borrowed these qualities and also her name. Here is a specimen of the Rolex Oyster – so called because it lives in water and shuts out all impurities.' With the Oyster, it was no longer necessary to remove the watch to wash one's hands or bathe, while one was at work in a dusty workshop or when one had perspired heavily. 'You just keep your Oyster on your wrist whatever happens, and it will never fail you.'

In 1927, Wilsdorf explained how the sealed case also guaranteed precision: 'We all know that the pivots must run in oil, and oil attracts all those fine particles of dust, which constantly, although in only very small quantities, penetrate into all watch movements, however well the cases are made. [...] Our Oyster excludes all dust and, consequently, it will always maintain perfect time.'

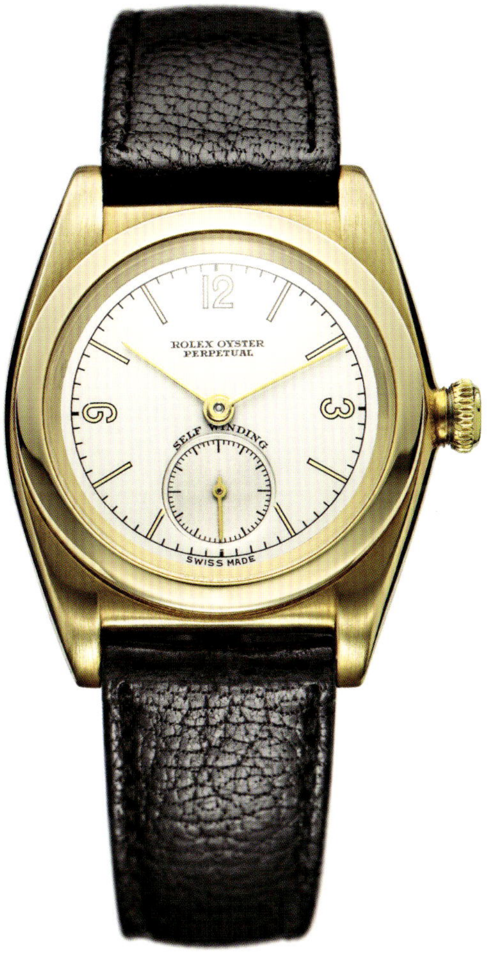

The Oyster Perpetual model.

The transition from the Oyster to the Oyster Perpetual occurred in 1931. This was the year the *maison* patented the automatic winding system known as the 'Perpetual'. With this new system, one could wind the watch through the mere natural movement of one's arm, without having to resort to daily manual winding. The Perpetual rotor consisted of a mass that rotated freely on an axis, in both directions, and ensured that the spring maintained constant tension, guaranteeing precision and reliability.

The direct descendants of the original Oyster, watches in the Oyster Perpetual range benefited from the collection's fundamental characteristics: namely, chronometric precision, the Oyster case's water resistance and the automatic winding of the movement via the Perpetual rotor. Characterised by a simple aesthetic, these stood out for their essential style, one that has been renewed over time, thanks to variations and evolutions in the dial's colour and design.

In 2020, the Oyster Perpetual family was

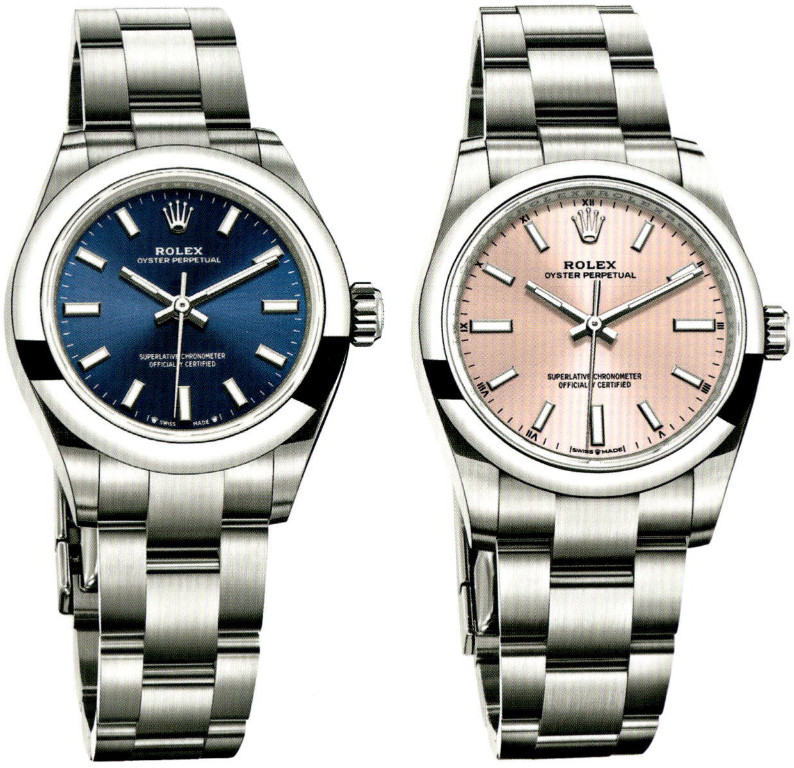

The Oyster Perpetual 36 model, shown with blue and candy pink dials.

enhanced with a new model, the Oyster Perpetual 41, which has been added to the Oyster Perpetual 28, 31, 34 and 36 models. New versions of the Oyster Perpetual 36 were also introduced, both featuring brightly coloured dials.

The first version of the Oyster Perpetual 41 has an atypical configuration: its silver dial with sunray finish is punctuated by yellow-gold hands and indices. The second version adopts a simple vivid black dial with sunray finish as well as white-gold hands and hour markers. The Oyster Perpetual 36, meanwhile, exhibits an energetic, colourful style, with a lacquered dial available in five new colours: candy pink, blue, yellow, coral red and green.

The generation of Oyster Perpetual watches features a Chromalight display: their hands and indices are coated or filled with a long-lasting luminescent substance that generates a blue light when the watch is in dark conditions. Like all Rolex watches, the Oyster Perpetual models have a Superlative Chronometer certification.

The Oyster Perpetual case, measuring 28, 31, 34, 36 or 41 millimetres in diameter, is guaranteed waterproof to a depth of one hundred metres. Its middle case is made from a solid block of Oystersteel, an especially corrosion-resistant alloy. The case back, edged with fine fluting, is hermetically screwed down using a special tool. Its Twinlock winding crown, fitted with a double waterproofing system, screws firmly to the case. The crystal is in scratch-resistant sapphire and benefits from an anti-reflective coating. Oyster Perpetual models are equipped with calibre 2232 (Oyster Perpetual 28, Oyster Perpetual 31 and Oyster Perpetual 34) or calibre 3230 (Oyster Perpetual 36 and Oyster Perpetual 41) self-winding mechanical movements entirely developed and manufactured by Rolex. The oscillator in both calibres has a balance wheel with variable inertia, regulated with extreme precision via gold Microstella nuts and mounted on high-performance Paraflex shock absorbers, also developed and patented by Rolex.

Calibre 2232 is fitted with a Syloxi hairspring in silicon, produced and patented by the

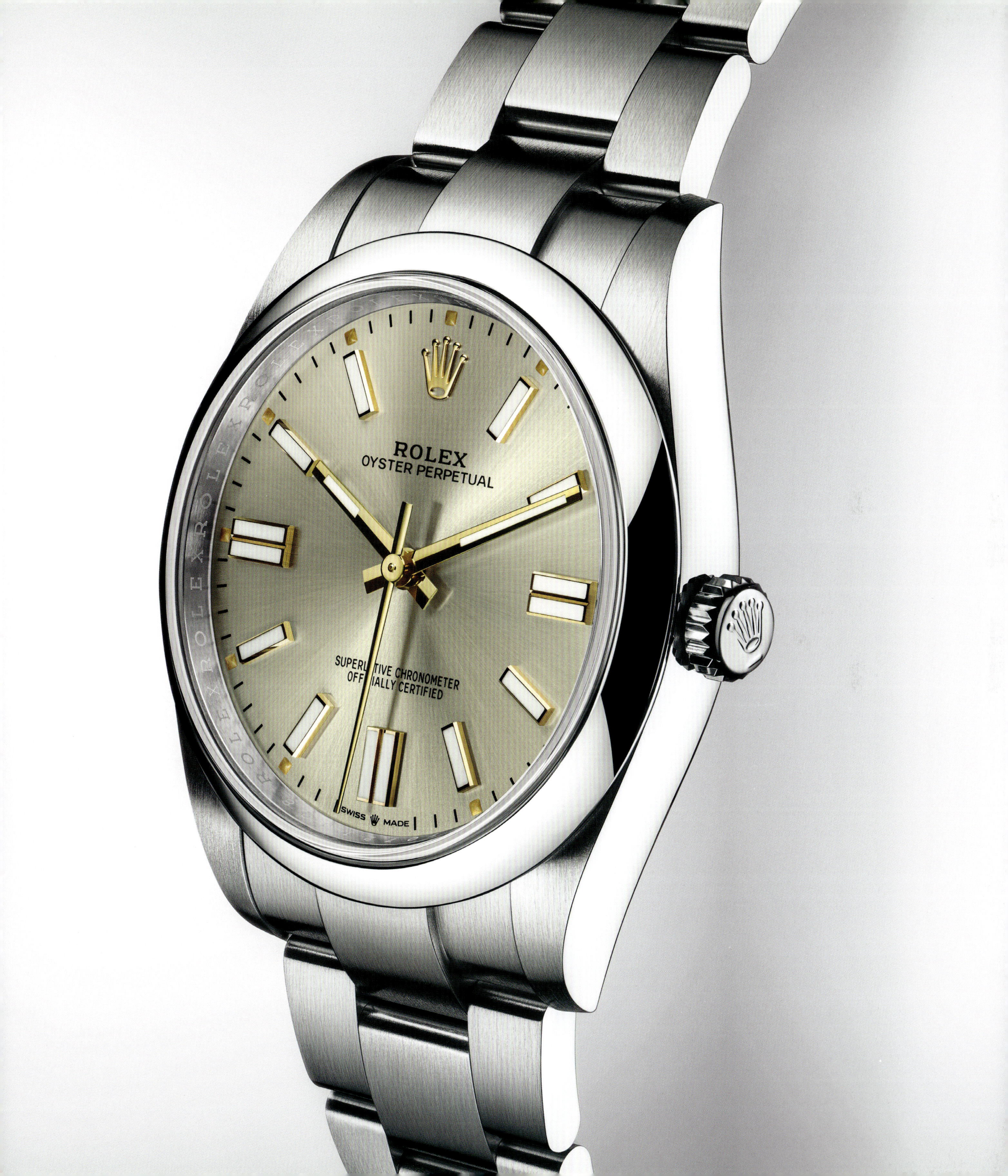
ROLEX
OYSTER PERPETUAL
SWISS MADE

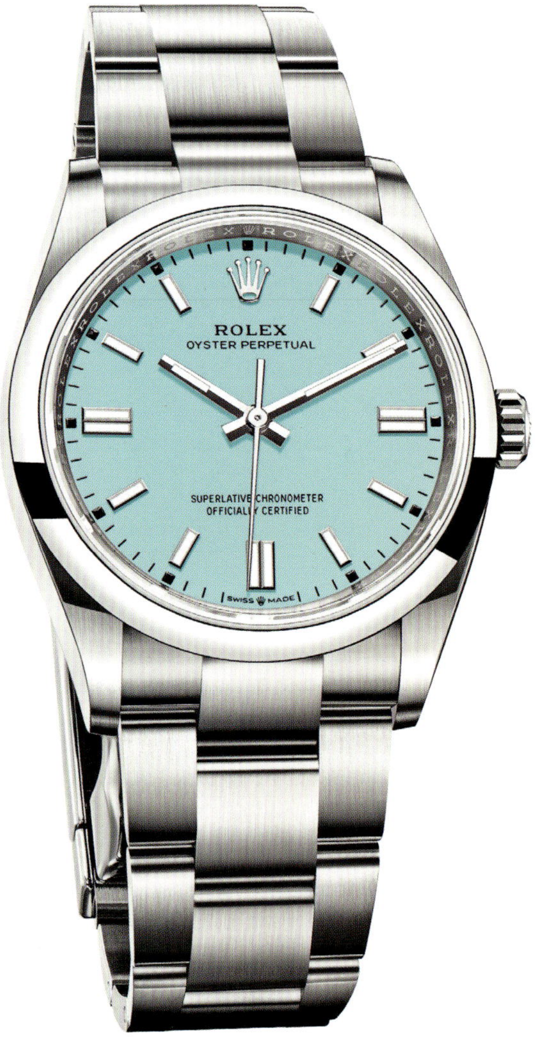

DIAL: These timepieces, made exclusively in Oystersteel, indicate the hour, minutes and seconds, constituting the most essential form of a wrist chronometer. The range of dials available is very wide.

CASE: The Oyster case is guaranteed waterproof to a depth of one hundred metres and provides optimal protection for the high-precision Rolex movement it houses against water, dust, pressure and shock. Its signature middle case is made from a solid block of Oystersteel, which is especially resistant to corrosion. In the most recent models the middle case reveals a satin finish with polished sides.

BRACELET: Models in the Oyster Perpetual range are fitted with a three-piece-link Oystersteel Oyster bracelet, featuring a Rolex-designed folding Oysterclasp and Easylink comfort-extension link for easy adjustment of its length (by about five millimetres).

brand, while calibre 3230 features an optimised version of the blue Parachrom hairspring, produced by Rolex in a paramagnetic alloy. In addition to being insensitive to magnetic fields, these two hairsprings are highly stable when experiencing changes in temperature. The geometry of the Syloxi hairspring in calibre 2232 guarantees regularity of the movement in every position, and the blue Parachrom hairspring in calibre 3230 has a Rolex overcoil that guarantees the same degree of regularity. Calibre 2232 is equipped with a paramagnetic nickel-phosphorus escape wheel, while calibre 3230 has the Chronergy escapement, patented and made in the same alloy. This escapement, which simultaneously ensures high energy efficiency and tremendous dependability, is also insensitive to magnetic fields. Both calibres have a self-winding module with a Perpetual rotor, which ensures continuous winding of the mainspring and thus supplies constant energy thanks to movements of the wrist. The first has a power reserve of approximately fifty-five hours, the second a power reserve of approximately seventy hours, thanks to its barrel architecture and the superior efficiency of its escapement.

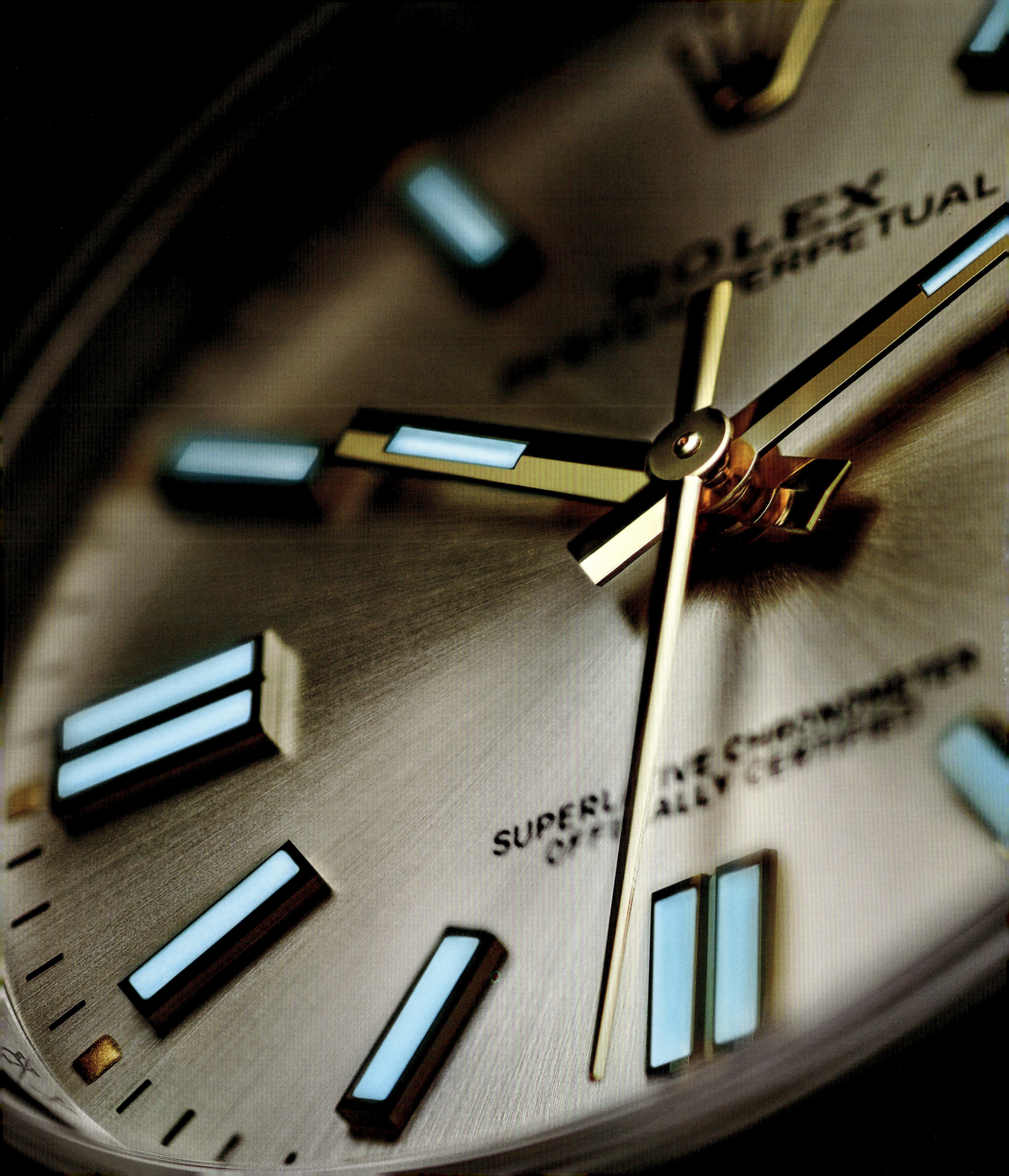

Datejust

Launched just after the Second World War, the Datejust was the first self-winding waterproof chronometer wristwatch in the world to display the date through an aperture on its dial.

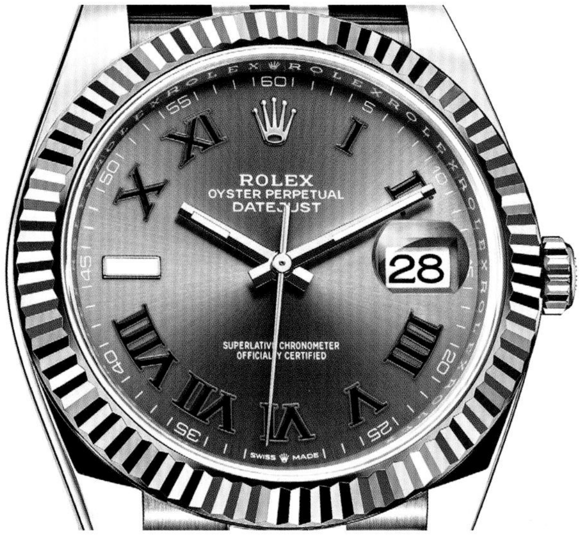

1945 marked the 40th anniversary of the founding of Rolex's progenitor company Wilsdorf & Davis, and the event was celebrated with the introduction of the Datejust model. The Oyster Perpetual Datejust revolutionised wristwatch date display. Traditionally, the day was indicated by a hand pointing to dates visible around the dial. With the Datejust, Rolex introduced a calendar indicating the date via a disc viewed through a window at 3 o'clock, a solution that would become the standard in watchmaking. Upon its launch, the Datejust concentrated all the chief innovations Rolex had brought to the modern wristwatch thus far: the precision of a chronometer on one's wrist (the brand having first received chronometer bulletins for wristwatches as early as 1910), water resistance (the invention of the Oyster case in 1926) and automatic winding (a mechanism with the Perpetual rotor, patented in 1931). The first model was distinguished by the reference number 4467 and produced only in 18-carat gold – most of them yellow gold, apart from some rare specimens in rose gold. The self-winding movement calibre A295 had a certain heft due to its date disc and its mechanism, necessitating a case back with a prominent dome.

The reference 4467 Datejust therefore had a domed appearance, with a thickness of about 14 millimetres. The date disc alternated between even days in red and odd days in black. Two more elements that distinguished the design of this watch were its unique coin-edge milled bezel and small winding crown. The Datejust ref. 4467 also marked the introduction of a new, dedicated bracelet, the Jubilee model. This was the first bracelet produced directly by Rolex without relying on external suppliers, available exclusively in 18-carat gold to emphasise its prestige. The successor of the Datejust ref. 4467 was the reference 6075, purported to have been introduced in 1950: it was from this specific model that the 'Datejust' name would be permanently displayed on the dial.

Aesthetically very similar to its predecessor, the 6075 features a push-in Super Oyster winding crown. It was available in both 18-carat gold and the steel-and-gold combination termed 'Rolesor', as well as a variant with a smooth bezel, identified by the reference number 6074.

In 1953, the Cyclops magnifying lens was added to the watch crystal, which made the date larger for easier reading. In 1956, the date change became instantaneous. As of 1977, the date could be set quickly and securely by

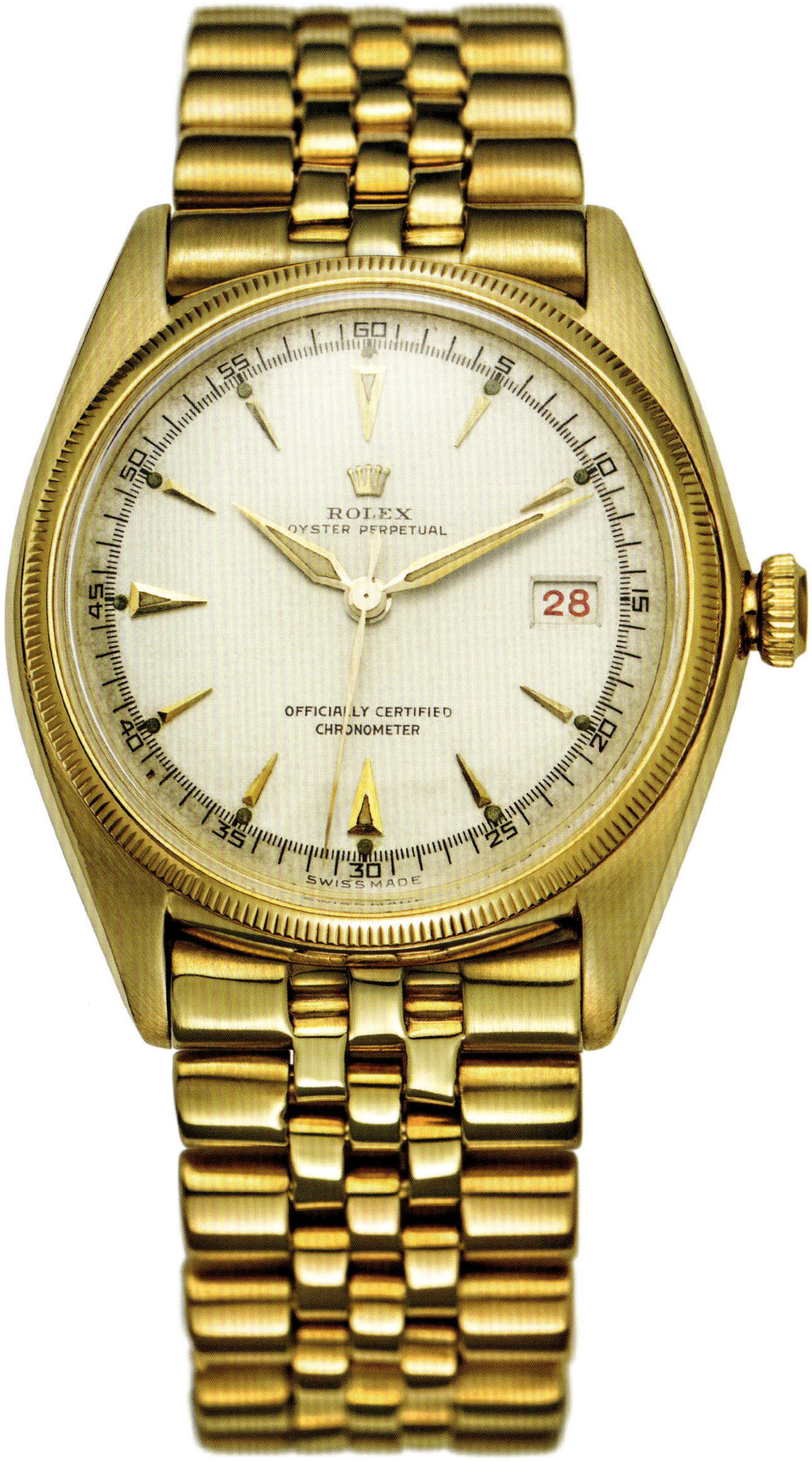

Datejust model 18ct gold version.

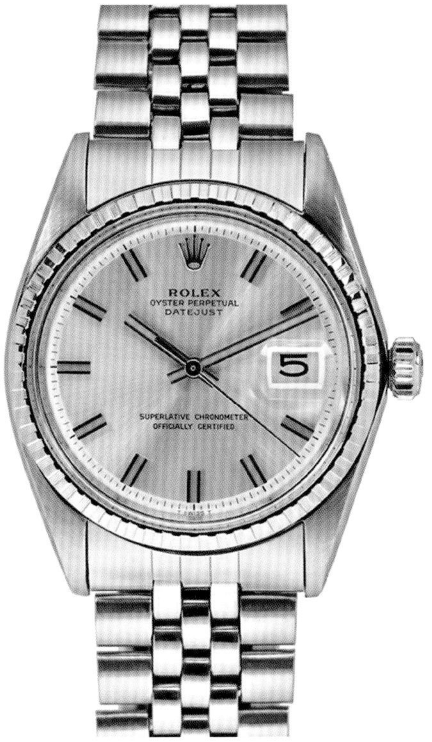

In 1953, the Cyclops magnifying lens was added to the watch crystal.

turning the winding crown in either direction; this was first introduced in Datejust models equipped with a medium- or large-sized calibre, and since 1983, models with a small movement have also had this feature.

Over time, the Datejust has kept to its continuous, instantly recognisable aesthetic codes. Today it is available in different diameters (from the Lady-Datejust's 28 millimetres, up to 41 millimetres) and in different materials – in Oystersteel; yellow, white or Everose gold; or in Rolesor (a combination of Oystersteel with yellow, white or Everose gold) – with a vast variety of dials, bezels and bracelets. The Datejust is also available in diamond-set editions.

The most recent Datejust models are equipped with the calibre 2236 (Datejust 31) or calibre 3235 (Datejust 36 and Datejust 41), both self-winding mechanical movements entirely developed and manufactured by Rolex. These movements boast several patents and guarantee high performance in terms of precision, autonomy, shock resistance, ease of use and reliability.

The oscillator of calibres 2236 and calibre 3235 features a balance wheel with variable inertia, with highly precise regulation via gold Microstella nuts. It is securely fixed by a height-adjustable traversing bridge for extremely stable positioning and increased shock resistance. The oscillator is moreover mounted on high-performance Rolex-developed and -patented Paraflex shock absorbers. Calibre 2236 is equipped with the silicon Syloxi hairspring, manufactured and patented by the brand, while 3235 includes an optimised version of the blue Parachrom hairspring, manufactured by Rolex in a paramagnetic alloy. In addition to being insensitive to magnetic fields, these two hairsprings boast high stability amid temperature fluctuations. The geometry of the Syloxi hairspring in calibre 2236 guarantees that the movement will run regularly in any position, while the blue Parachrom hairspring in calibre 3235 is fitted with a Rolex overcoil, ensuring the same level of regularity.

Calibre 2236 contains a paramagnetic nickel-phosphorus escape wheel. Calibre 3235 meanwhile features the patented Chronergy escapement made of the same alloy,

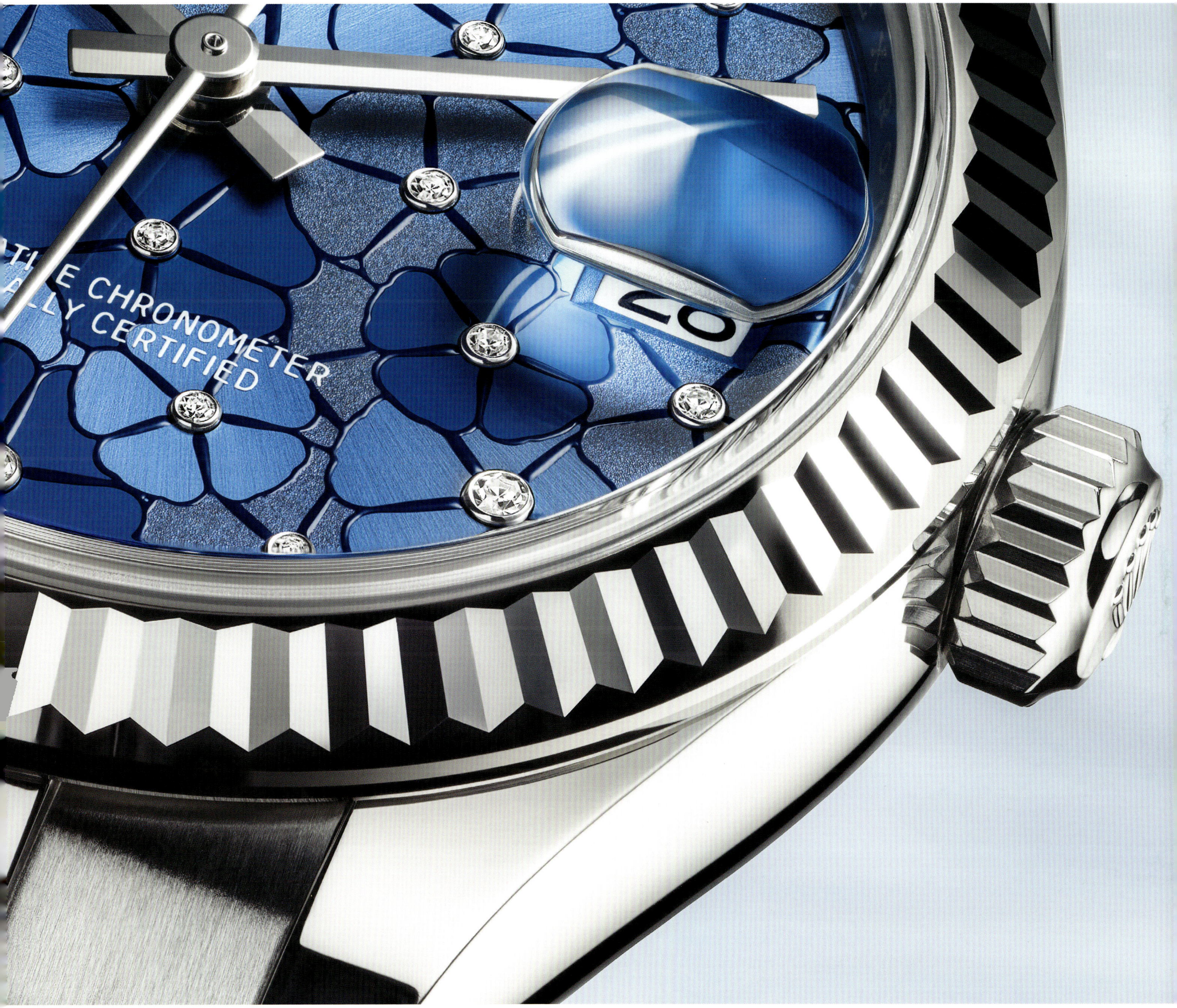
CHRONOMETER
CERTIFIED
20

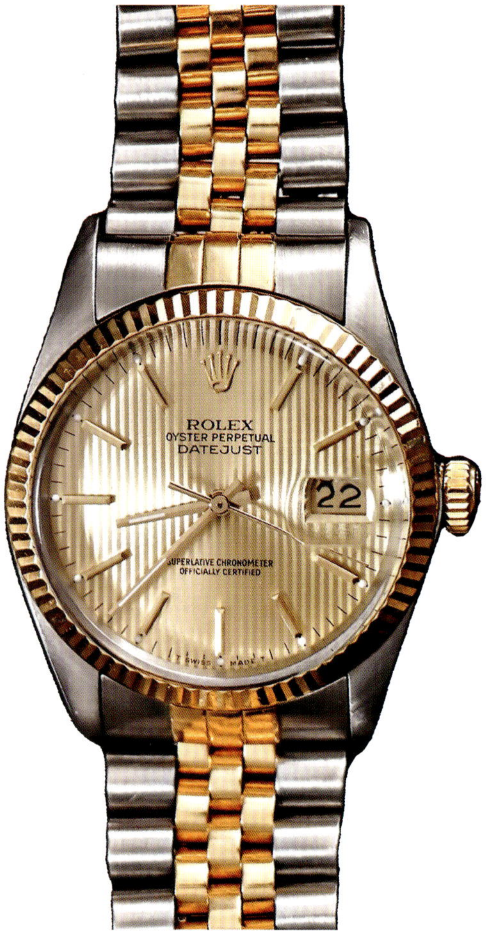

The Datejust 31 model.

DIAL: At the 1954 Basel fair, Rolex presented a significant innovation, its Cyclops lens. This is a small magnifying lens placed on the watch crystal over the date aperture, allowing for an easier view of the date beneath. The feature soon spread to all models displaying a date, becoming a hallmark of the maison.

CASE: The Datejust was introduced with a ribbed 'coin-edge' bezel, but in time references were offered with a smooth bezel or decoration via other processes, including scalloping, a linen-texture finish, or fluting.

BRACELET: The year 1956 saw the introduction of the President bracelet, offered in alternation with the Jubilee bracelet.

simultaneously guaranteeing high energy efficiency and extreme dependability, while also being unaffected by magnetic fields.

Both calibres have an automatic winding module with a Perpetual rotor, which ensures continuous winding of the mainspring and thus supplies continuous energy using movement of the wrist. Calibre 2236 has a power reserve of approximately 55 hours. The 3235, on the other hand, benefits from a reserve of approximately 70 hours due to its barrel architecture and the superior efficiency of its escapement.

The Datejust is certified as a Superlative Chronometer. Depending on the version, the Datejust 31 comes with a three-piece link Oyster bracelet with an Oysterclasp, a five-piece link Jubilee bracelet (created especially for the Datejust in 1945), or a three-piece link President bracelet. Jubilee and President bracelets feature a Crownclasp. The Datejusts 36 and 41 are combined, depending on the version, with a three-piece link Oyster bracelet or a five-piece link Jubilee bracelet, each of which comes with an Oysterclasp. A concealed attachment system creates perfect visual continuity between the bracelet and the case. Oysterclasp bracelets feature the Easylink comfort extension link, developed by Rolex, allowing for easy adjustment of the length by about 5 millimetres for greater comfort in any circumstance. The President bracelet, which comes with all gold versions of the Datejust 31, has ceramic inserts inside the links to improve its flexibility on the wrist and durability over time.

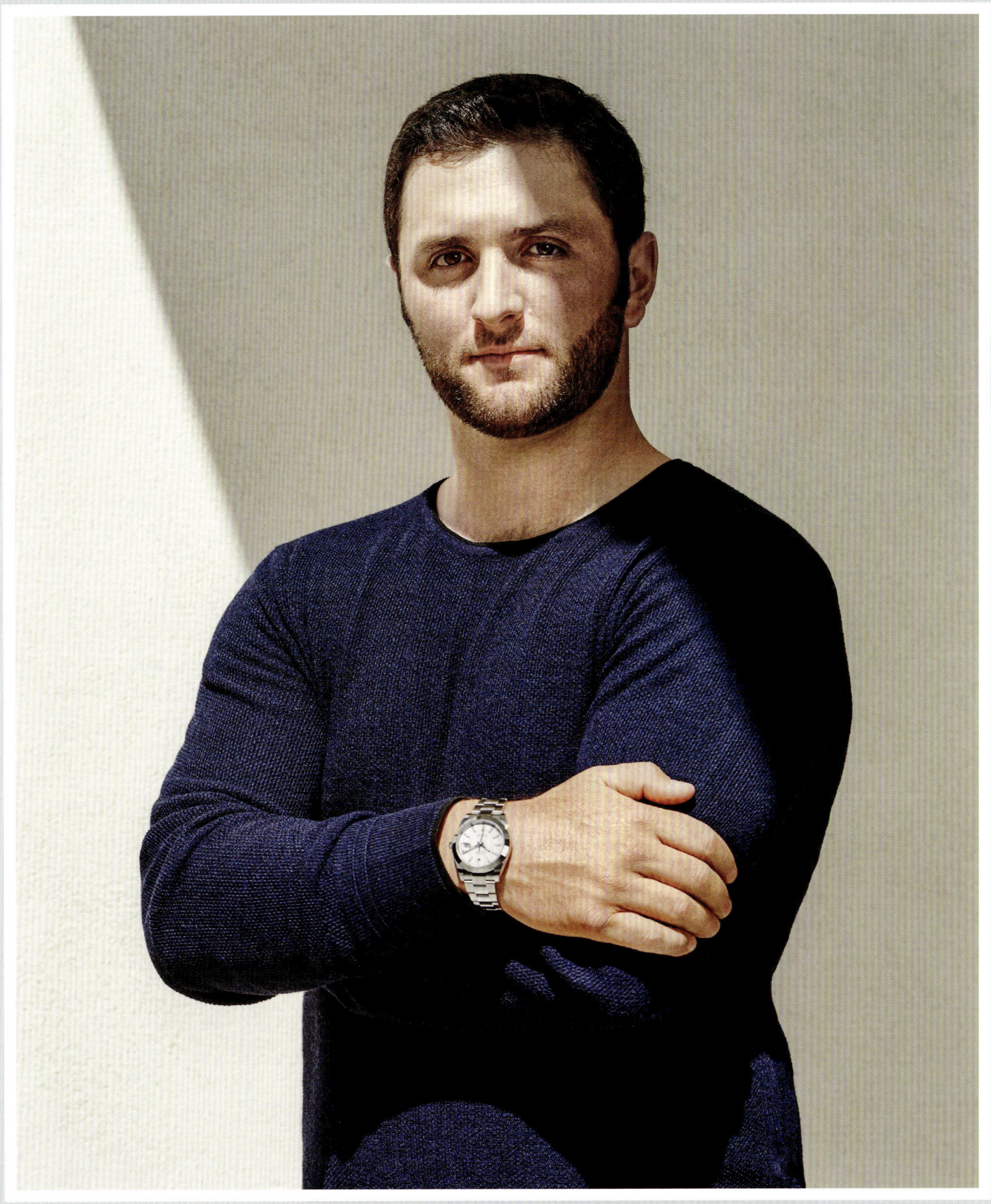

Golf champion Jon Rahm wears a Datejust 36 model.

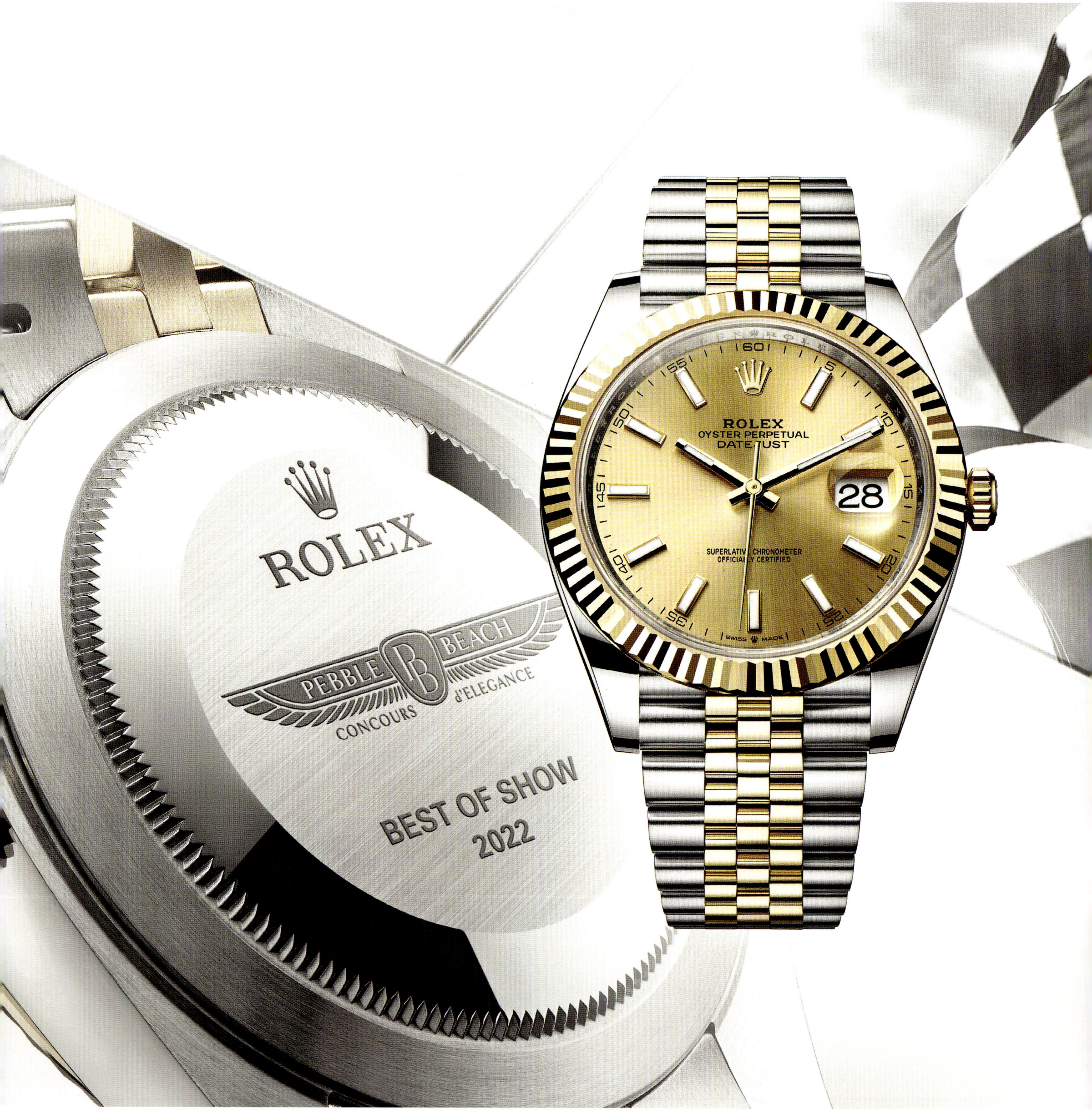
ROLEX
PEBBLE BEACH
CONCOURS d'ELEGANCE
BEST OF SHOW
2022
ROLEX
OYSTER PERPETUAL
DATEJUST
28
SUPERLATIVE CHRONOMETER
OFFICIALLY CERTIFIED

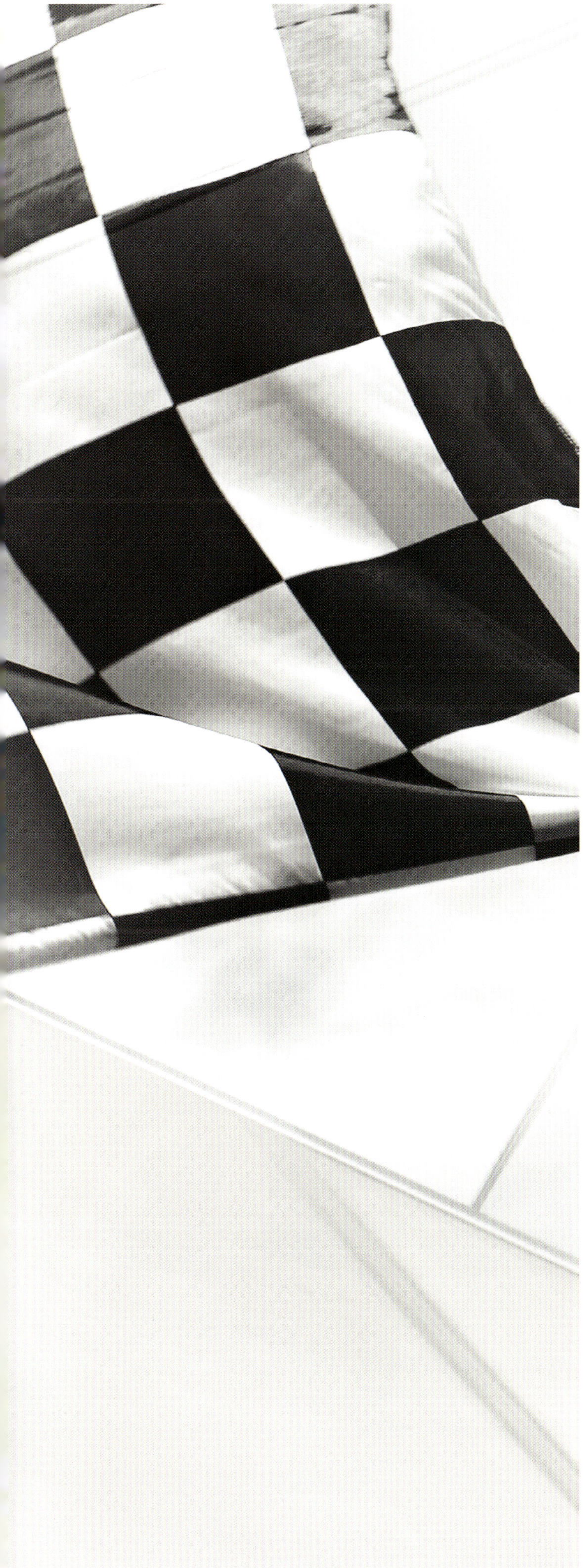

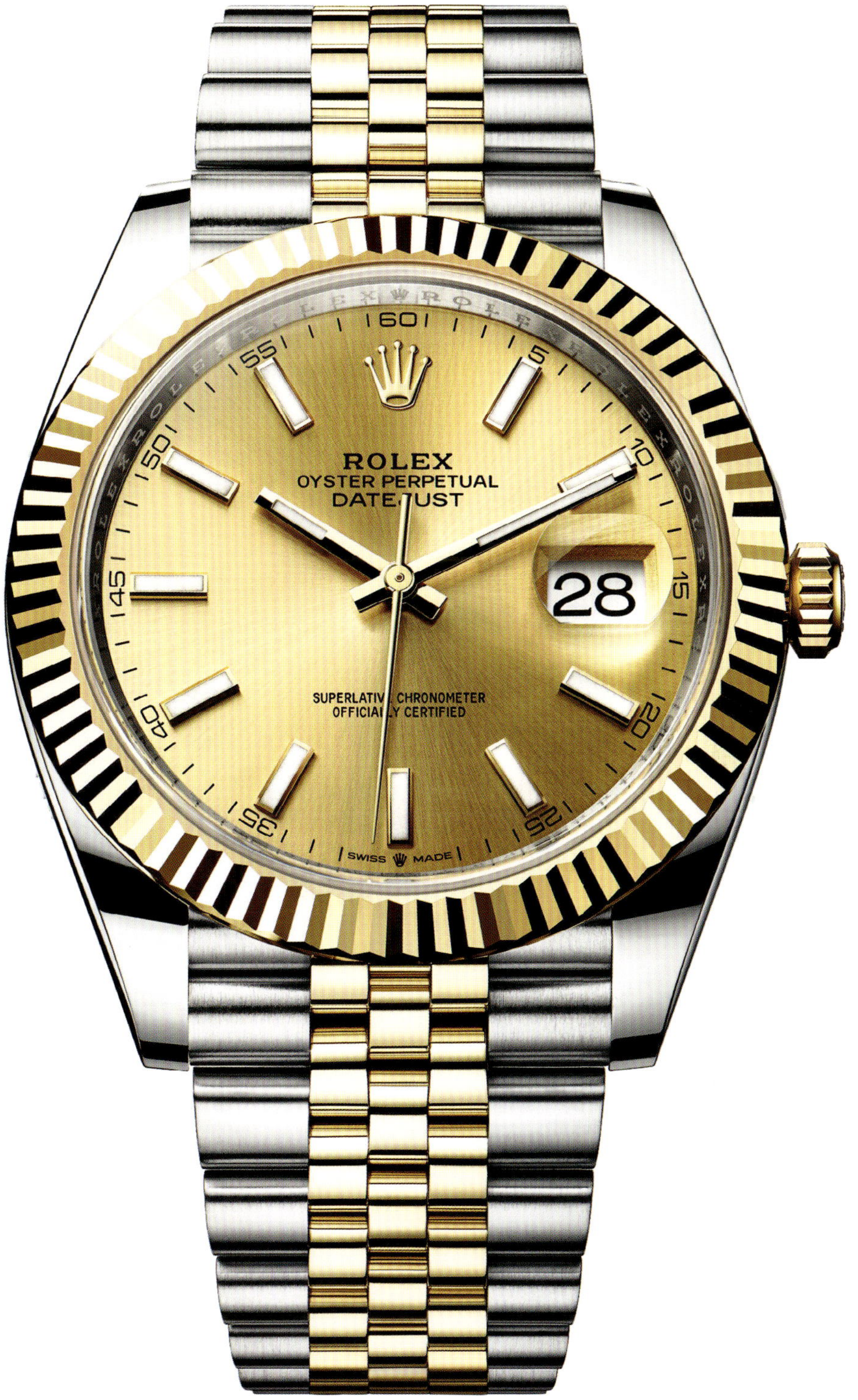

The Datejust 41 model.

Submariner

In 1953, Rolex unveiled the Submariner, the first diving wristwatch that was guaranteed waterproof to a 100-metre depth.

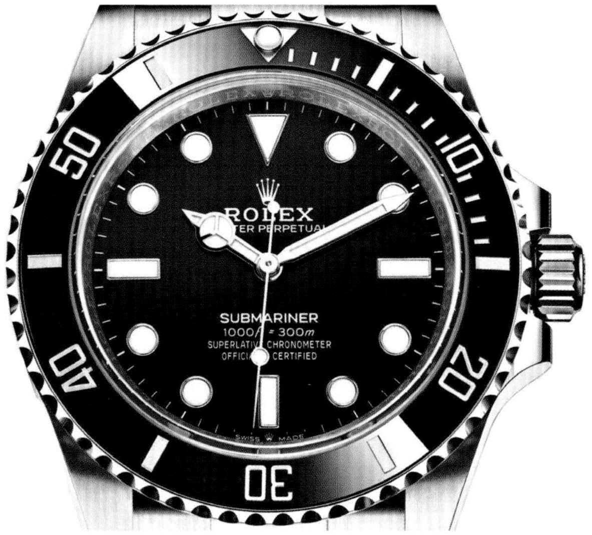

Due to their water resistance, Rolex watches have naturally been involved in diving ventures. The creation of the Submariner is profoundly linked to the development of diving. In the early 1940s, deep-sea exploration took a significant turn with the advent of scuba diving, and the development and manufacture of diving wristwatches responded to the needs of this discipline's professionals. Rolex's role in the manufacture and development of water-resistant wrist chronometers proved to be fundamental, especially due to the Oyster case, patented in 1926. Aiming to create the most suitable watch for divers' needs, the crown *maison* committed itself, beginning in the early 1950s, to a process of experimentation that would entail many often-successful collaborations with pioneers in scuba diving.

The experiments conducted, and the consequent technical developments, led to the launch in 1953 of the Submariner, the first diving wristwatch that was waterproof to a depth of up to 100 metres.

As soon as it was developed, the Submariner immediately enjoyed a certain success among pioneering deep-sea divers. The first to certify its performance was the French engineer and explorer Dimitri Rebikoff. Over the course of five months, Rebikoff carried out 132 diving missions with a Submariner watch, taking it to depths between 12 and 60 metres. The report he provided in 1953 was extremely positive: 'We are able to confirm that this watch has not only given entire satisfaction in diving conditions which were extremely tough and particularly dangerous for the material used, but that it has proved an indispensable accessory for all diving with independent equipment.' Rebikoff specifically emphasised the usefulness of its graduated rotatable bezel, which greatly contributed to divers' safety, allowing for time spent underwater to be kept under control. He also praised the robustness of the watch, which had been subjected to several shocks and spent a considerable number of hours in sea water.

Just one year following its launch, the Submariner became waterproof up to the depth of 200 metres. The timepiece quickly incorporated further technical innovations, including the addition of a luminescent disc on its hour hand – to clearly distinguish it from the minute hand – and a guard protecting the winding crown.

In 1969, Rolex introduced a model equipped with a date function: the Submariner Date. Its water resistance, at the time guaranteed up to a depth of 200 metres, increased to 300 metres in 1979. The Submariner's water resistance

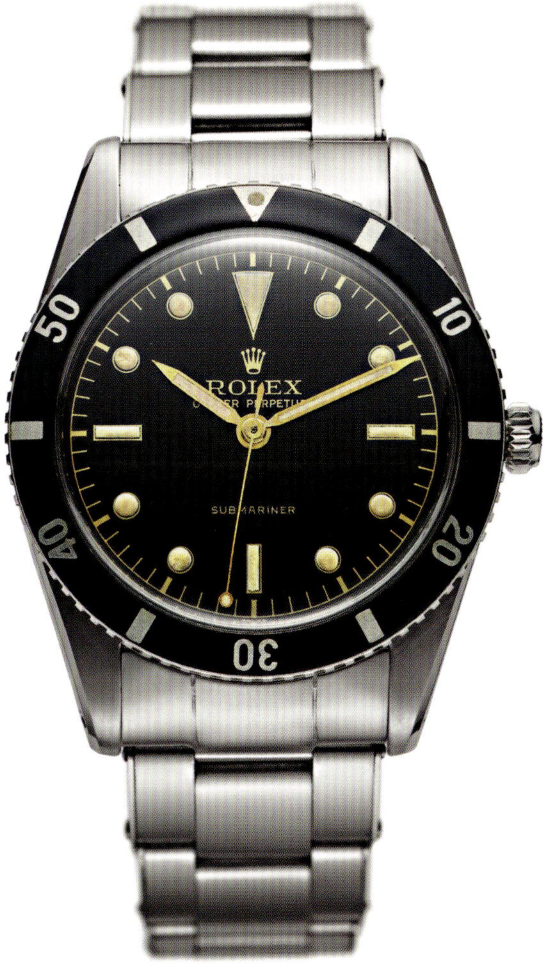

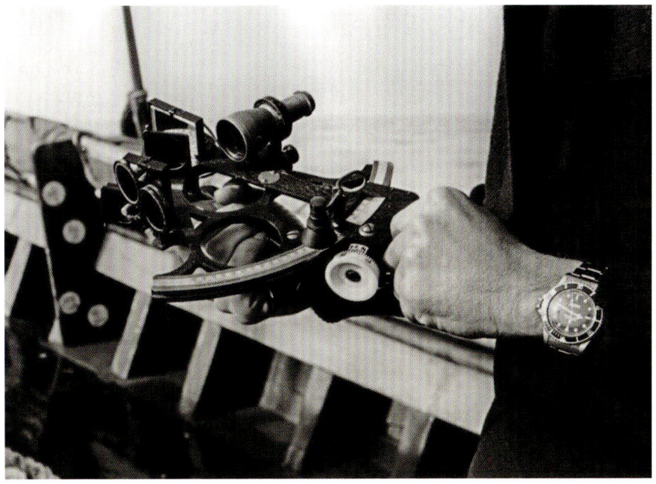

There was no GPS in 1966: navigational instruments included a sextant, the stars and a Rolex watch.

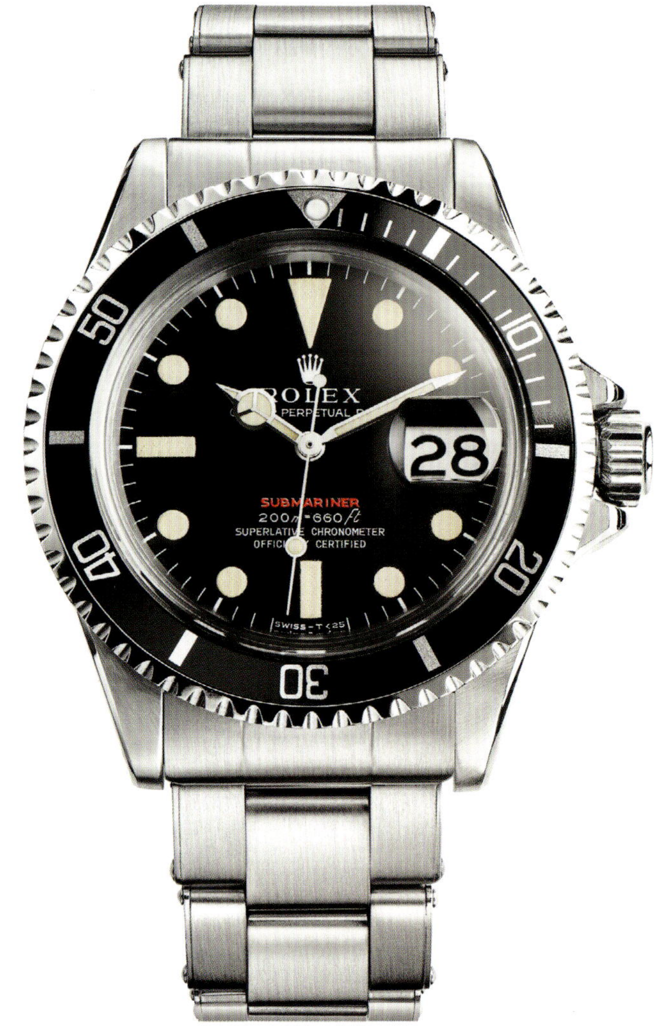

The Submariner model with a date function: the Submariner Date, 1969.

would also be brought to this depth from 1989 onward. Conceived as a professional tool for underwater exploration and diving, the Submariner soon became a symbol of practical and versatile masculine elegance, soon rising to the level of status symbol. Two characters contributed to this transformation, completely different from each other yet equally relevant in terms of media impact. One case was that of Sean Connery as Agent 007. In the opening sequence of the film *Goldfinger* (1964), he dives underwater, equipped with scuba gear, to enter the enemy base, planting some explosive charges. Then, arriving at a bar, he removes his scuba gear to reveal a flawless evening ensemble. James Bond, slyly sipping his drink, then measures the amount of time remaining before the bombs detonate via the rotating bezel on his Rolex Submariner. The explosion upsets the entire bar as he walks away, expressionless and unnoticed. The second character was Ernesto 'Che' Guevara, immortalised in a photo wearing an olive green uniform and enjoying the first puff of one of his beloved Montecristos, with a Submariner on his wrist. The timepiece thus became the archetype of the action watch.

Over the years, the Submariner and Submariner Date have freed themselves of their 'water' associations, coming to also conquer terra firma.

American anthropologist Johan Reinhard wore a Submariner during his explorations in various environments and climates: 'I have worn a Rolex Submariner for 17 years now and used it throughout the world on mountain climbing expeditions (including Mount Everest), underwater archaeological research (including the world's highest dive at 19,300 feet in Northern Chile), desert crossing, sky diving, jungle exploration and during numerous anthropological investigations. I doubt that many people have subjected a Rolex watch to worse punishment over such a period of time. Yet it still functions perfectly.'

In 2020, Rolex presented the new gen-

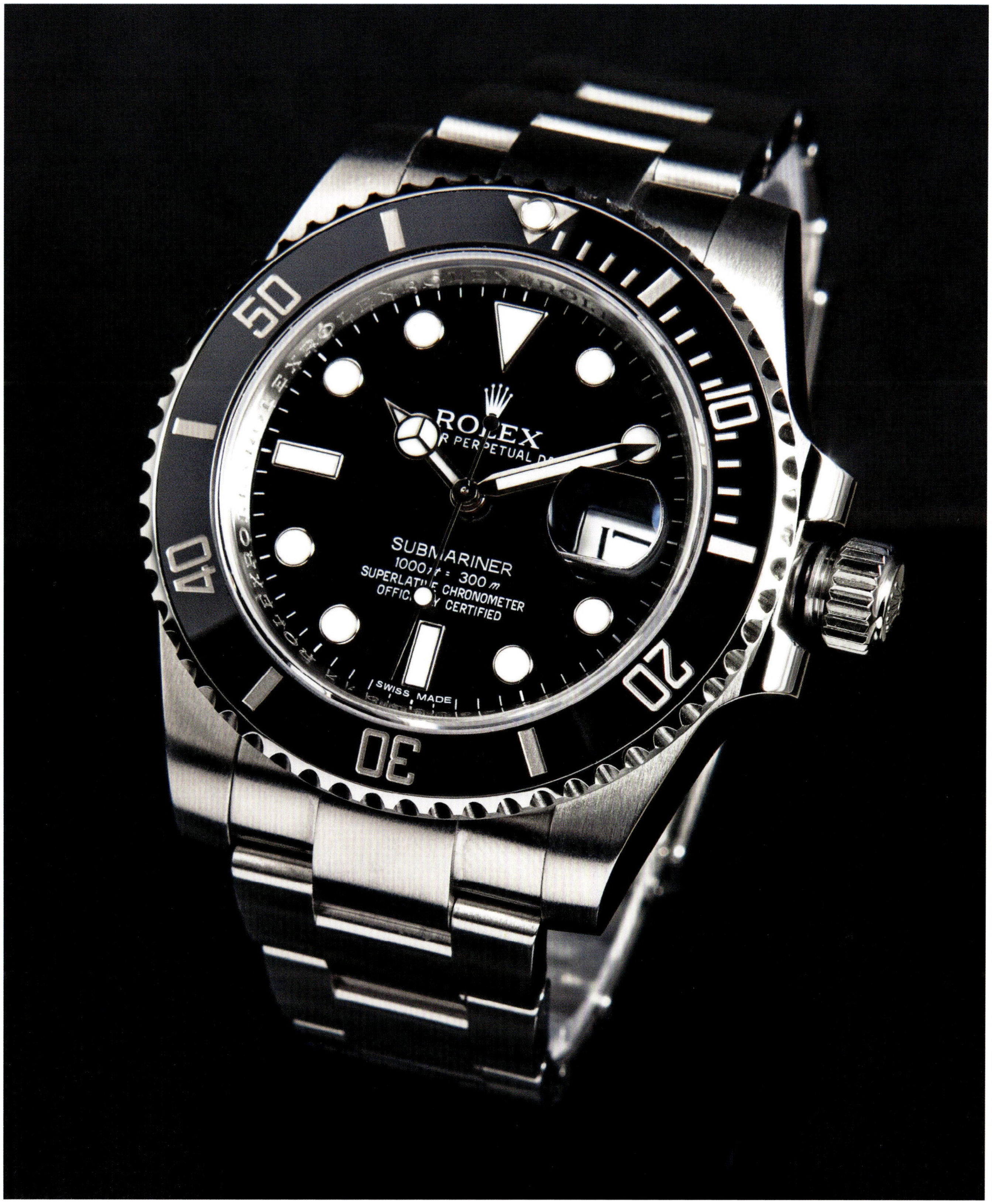

The Submariner Date model, black version with white markers.

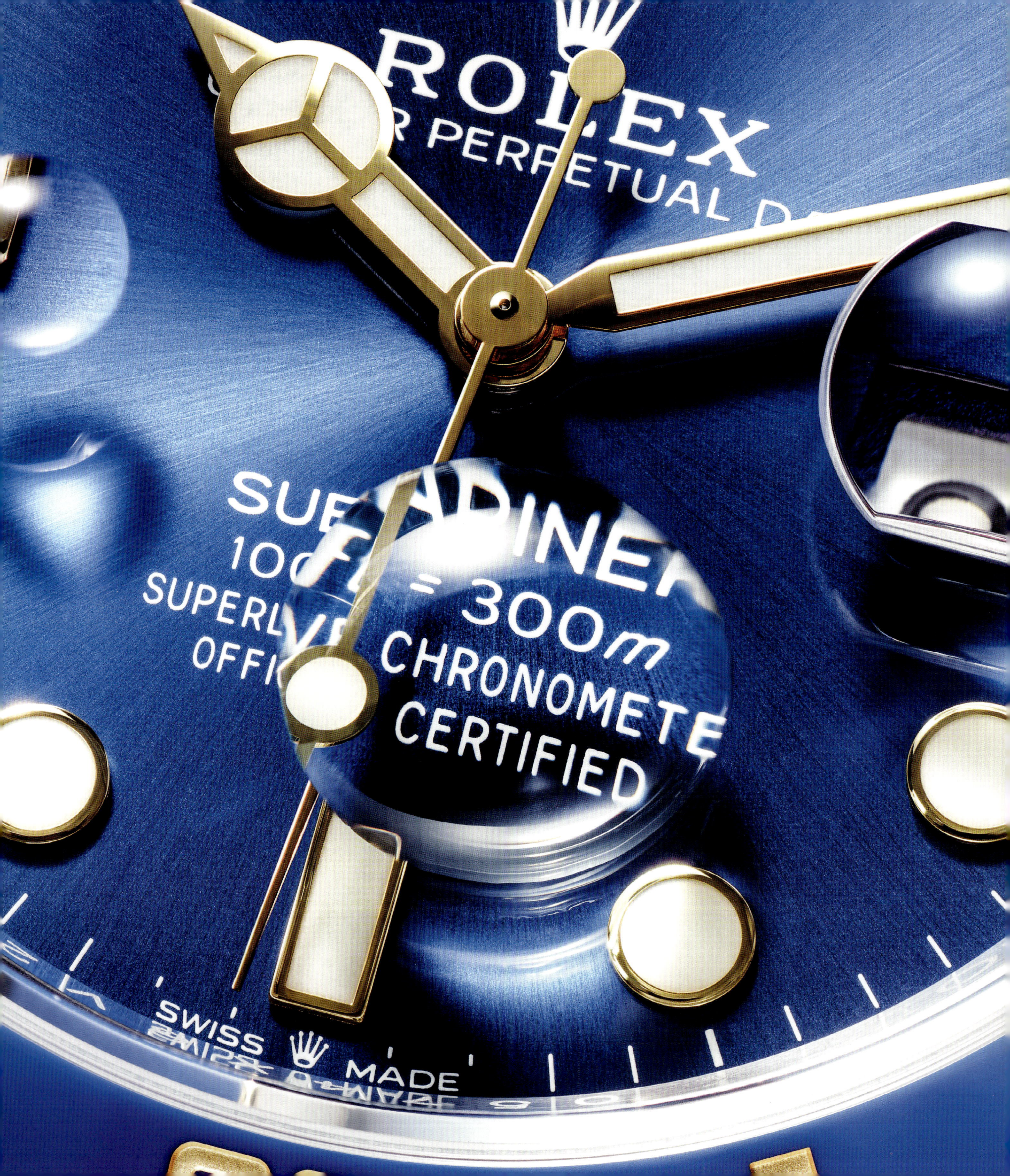
ROLEX
300m
CERTIFIED
SWISS MADE

eration of the Oyster Perpetual Submariner and the Oyster Perpetual Submariner Date, models representing the historic link between Rolex and the world of scuba diving. Today, both watches feature a case enlarged to 41 millimetres in diameter with a bracelet of modified proportions. The timepieces are respectively equipped with calibres 3230 and 3235, with the latter displaying the date as well as the time.

Faithful to the historical aesthetics of the model, the Submariner in Oystersteel exhibits a black dial and rotating bezel with Cerachrom insert. The Submariner Date's edition in yellow Rolesor (a combination of Oystersteel and yellow gold) combines a royal blue dial and rotatable bezel with a blue Cerachrom insert. Two versions of the Submariner Date, the first in Oystersteel and the second in white gold, feature a dial and rotatable bezel with a Cerachrom insert in two different colours. The first combines a black dial and green bezel, the second a black dial and blue bezel. The Submariner and Submariner Date feature a Chromalight display: the hands and indices are coated or filled with a long-lasting luminescent substance that generates a blue light when the watch is in a dark environment. The zero on the bezel's graduated scale, represented by a triangle, can also be seen in the dark, thanks to a capsule that contains the same luminescent material.

Like all Rolex watches, these both have a

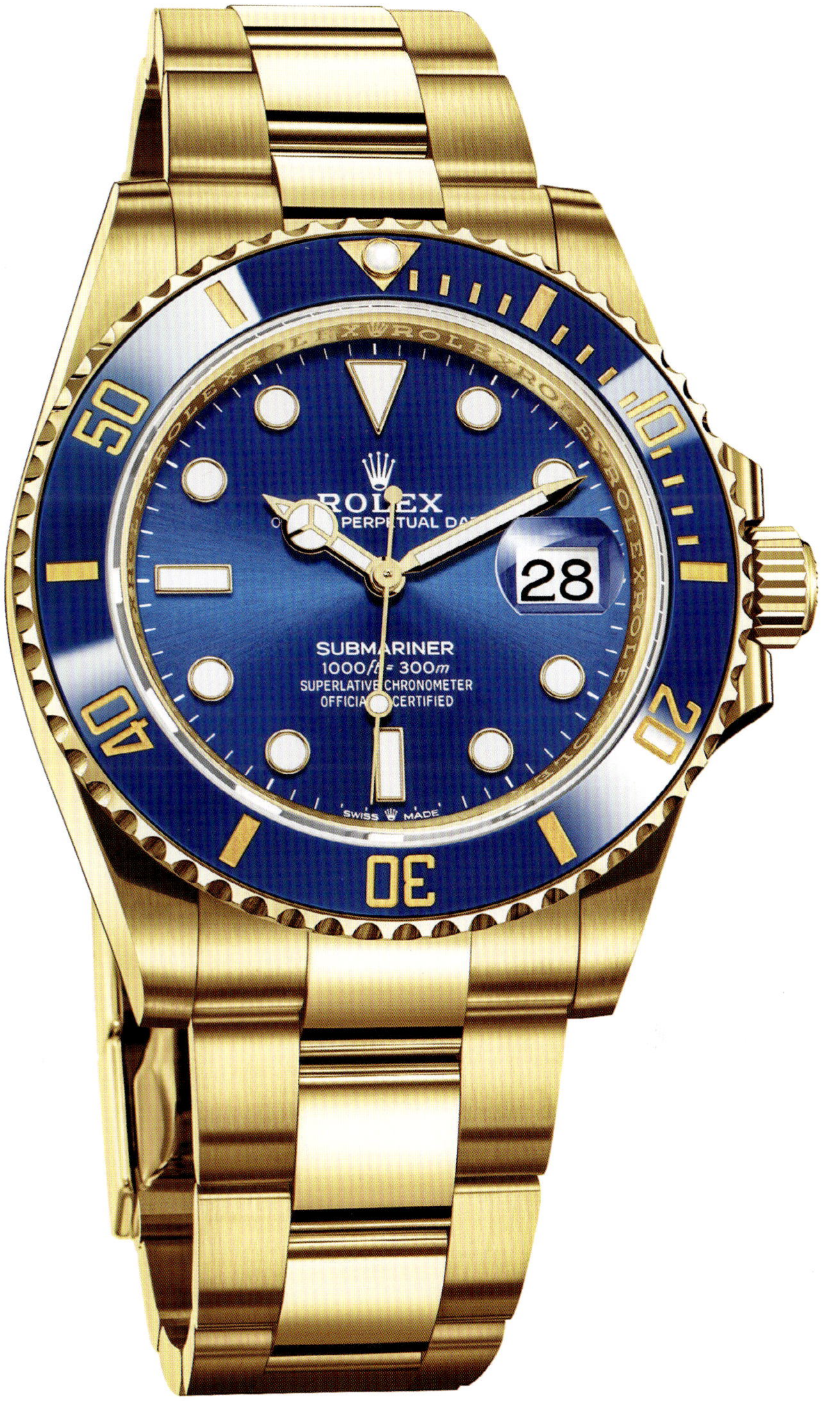

The Oyster Perpetual Submariner Date model, yellow Rolesor version with a blue Cerachrom insert.

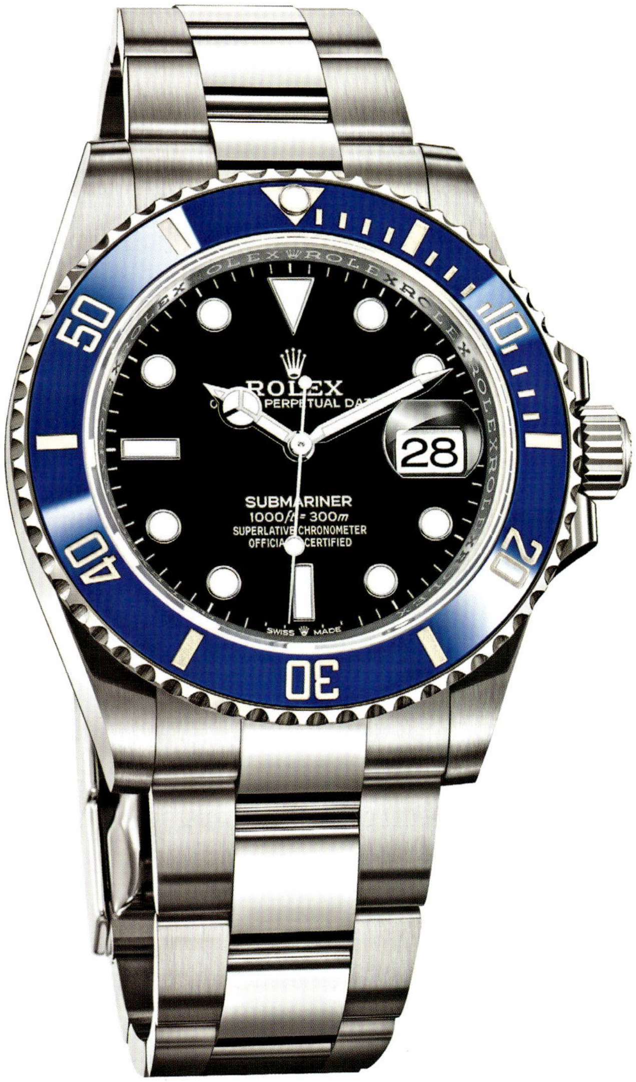

The Oyster Perpetual Submariner Date model, steel version with a black dial and blue bezel.

Superlative Chronometer certification. The most recent Submariner and Submariner Date are equipped with calibres 3230 and 3235, respectively, which were entirely developed and manufactured by Rolex. Both calibres are powered by a self-winding module with Perpetual rotor and, thanks to their barrel architecture and the superior efficiency of their escapements, boast a power reserve of approximately seventy hours.

DIAL: During a dive, being able to read the time with no possibility of error is a factor of survival. For this reason, the dial of the Submariner and Submariner Date has a simple, clean design. The hour and minute hands, clearly differentiated in shape and size and the indices using simple geometric shapes – a triangle, circles and rectangles – help the wearer to instantly and reliably consult the time, preventing any risk of confusion to divers while underwater.

CASE: The Submariner and Submariner Date feature a rotatable bezel with a graduated insert that allows divers to track their time spent underwater, helping them to manage their breathing gas reserves. The Oyster case's security is enhanced by its winding crown. This screw-down crown, called 'Twinlock', has been perfected over time, featuring two waterproof zones.

BRACELET: The Submariner and the Submariner Date editions are combined with an oyster bracelet. Developed in the late 1930s, this three-piece-link bracelet is distinguished for its robustness. The Oyster bracelet on these new watches is fitted with a Rolex-designed and -patented Oysterlock folding safety clasp, which prevents any inadvertent opening. It also features the Rolex Glidelock extension system, also developed and patented by the brand. This particularly ingenious device, with a rack integrated under its clasp cover, can be adjusted using a toothed sliding link, which locks into a chosen notch. The Rolex Glidelock on the Oyster bracelet has ten notches about 2 millimetres apart, allowing its length to be easily adjusted by about 20 millimetres without the aid of any tools.

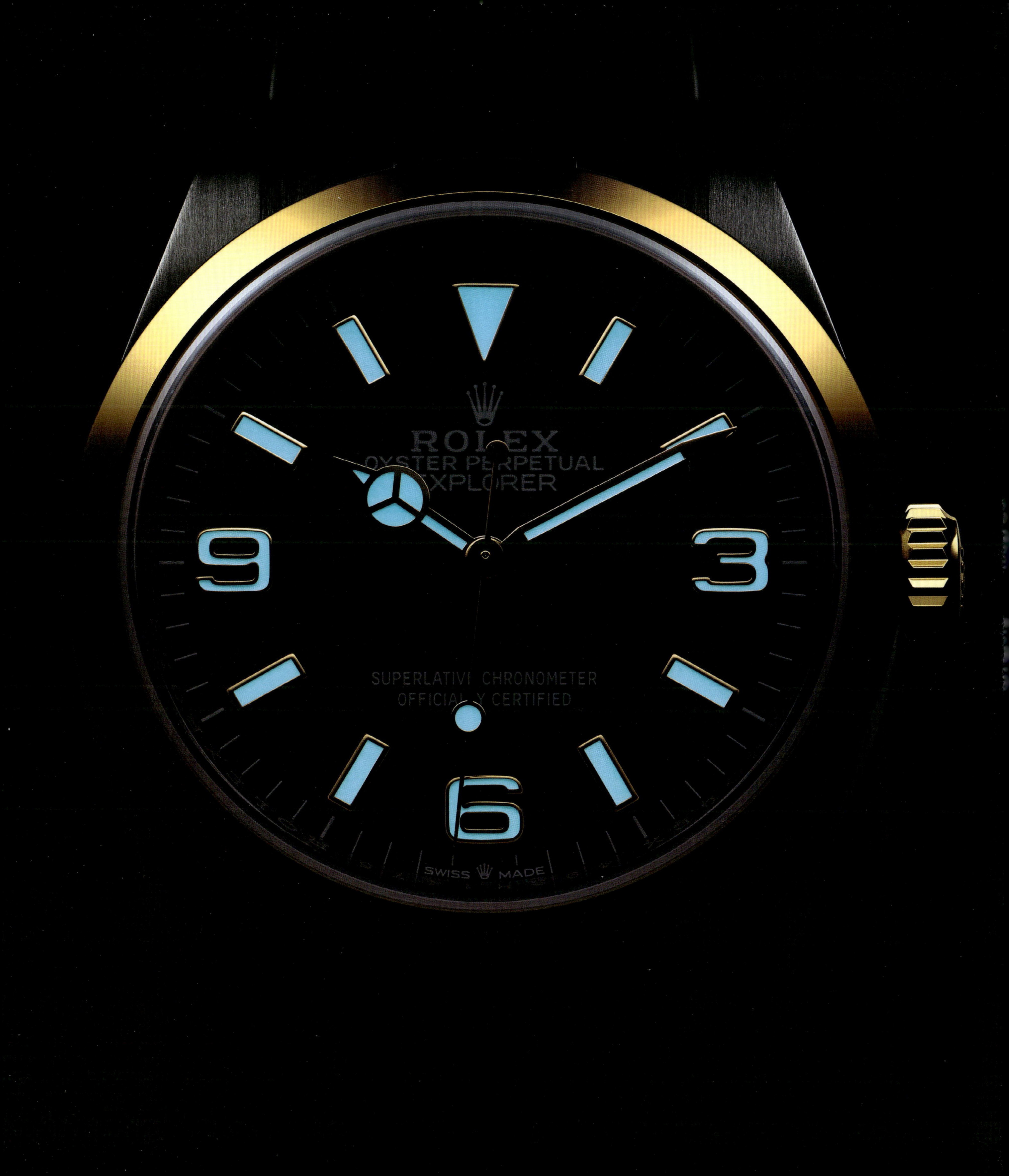
ROLEX
SWISS MADE

GMT-Master

Developed in the mid-1950s, the GMT-Master was the pioneer of the 'dual-time' function.

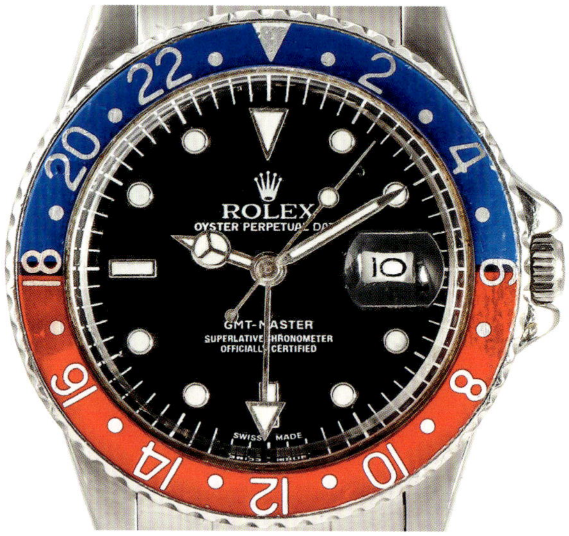

The 1950s were a time of economic recovery following the devastation brought on by the Second World War. With the development of the aeronautic industry, leading airlines in this period were able for the first time to make long-distance journeys without stopping, crossing multiple time zones in a short span of time. While the development of intercontinental flights signaled a turning point in travel on the one hand, on the other a new issue emerged for airline crews: jetlag. It was the subject of multiple scientific studies. These aimed to mitigate the effect of the phenomenon, which was causing fatigue and confusion for pilots and onboard staff.

One method to counteract jetlag that emerged from the various tests was to own a tool that could simultaneously show one's home time zone and that of the flight's destination. Management at Pan American World Airways, then a leader in world air transport, invited Rolex to devise a watch model that could alleviate the effects of jetlag. Following a period of research, the 'crown' company met the challenge and developed the GMT-Master, an Oyster watch that simultaneously told the time in two different time zones, and it soon became the official watch of various airlines. 'GMT' stands for Greenwich Mean Time, or the time at the prime meridian, in relation to which the time in every part of the world is calculated. In this model, the second time zone was displayed by means of an additional triangle-tipped hand that completed a full circle every 24 hours, indicating the time on its unique bezel, graduated from 0 to 24. This bezel could be rotated according to the time zone to be displayed, thus coinciding with the classic time read on its dial.

One thing that further bonded the Rolex GMT-Master to the world of aviation at the end of the 1960s was the testing of the new Aérospatiale/BAC Concorde supersonic aircraft. With great pride, the Swiss company announced that both test pilots on the futuristic aircraft were wearing Rolex GMT-Masters.

Though it had been created for professional pilots, the GMT-Master soon attracted the frequent-flyer class, becoming the watch for jetsetters. It was a symbol of sophistication for an elite group that moved above the clouds, traveling for work or leisure. No longer just reserved for teams at Pan Am, the GMT-Master could be spotted on the wrists of various famous figures, from Pablo Picasso to Che Guevara.

The GMT-Master made countless appearances on the big screen. One of the first to wear it was British actress Honor Blackman, who played the famous 'Bond Girl' Pussy Galore, sporting the model in various scenes in the film *Goldfinger* (1964). A GMT-Master appeared on Dustin Hoffman's wrist in *Straw Dogs* (1971) and *Marathon Man* (1976). The movie *In the Line of Fire* (1993) showed Clint Eastwood wearing an edition with a two-tone steel and yellow-gold bracelet, a brown dial and bezel, and gold applied indices; it was nicknamed 'Root Beer', since the American soft drink was advertised in similar colours to the model. The version that Marlon Brando chose

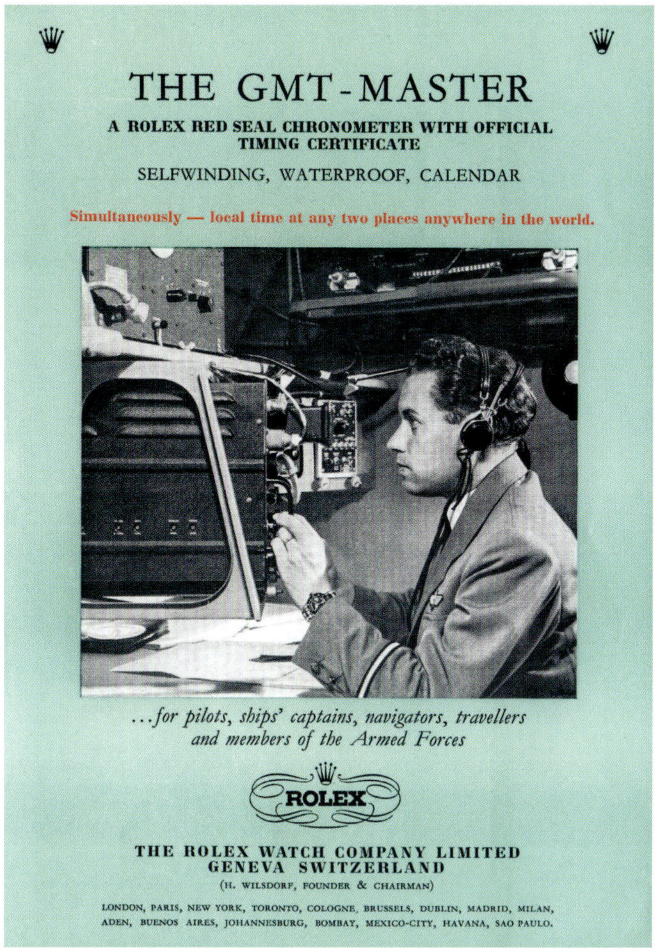

Original Rolex GMT-Master brochure.

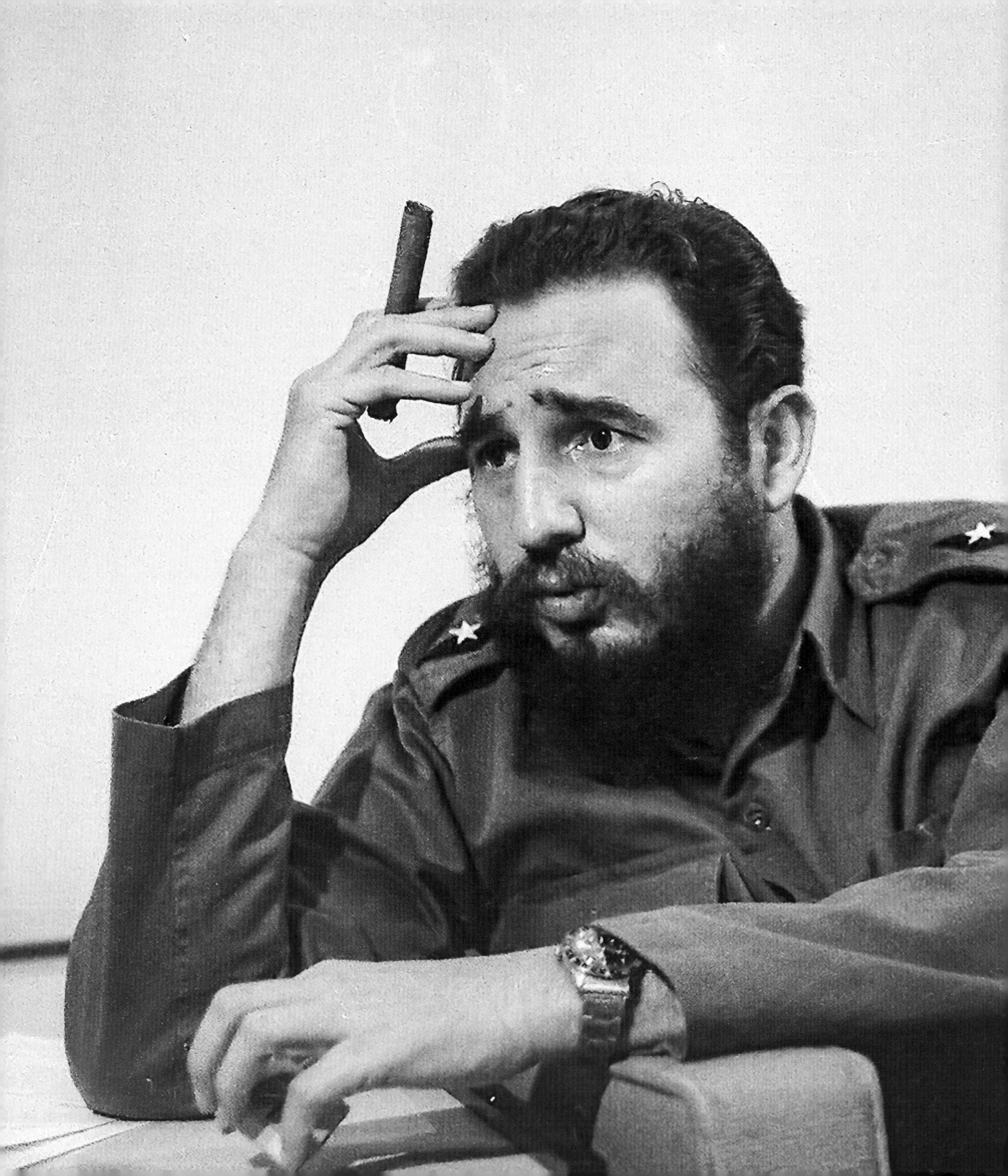

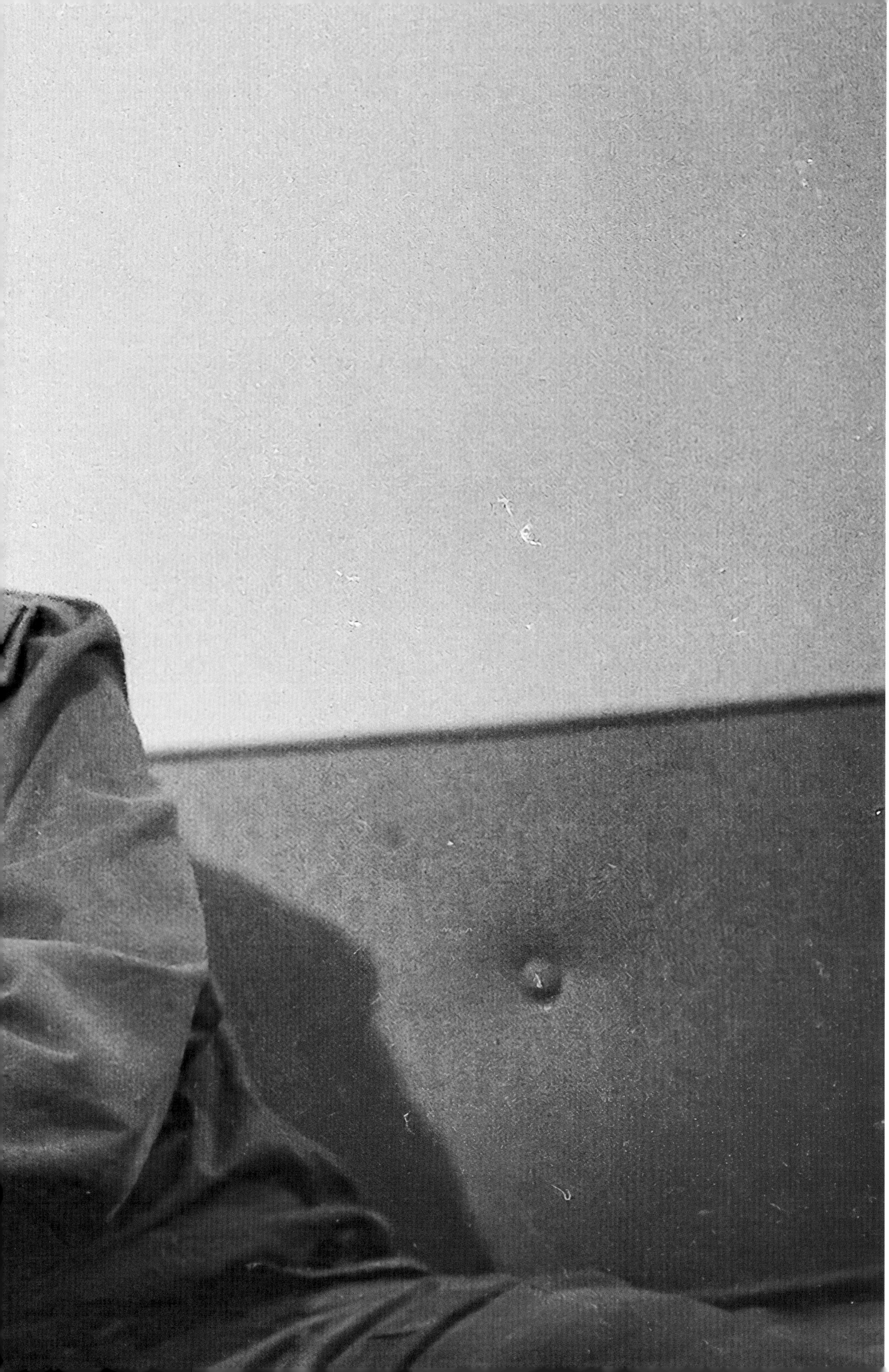

Fidel Castro wearing a GMT-Master.

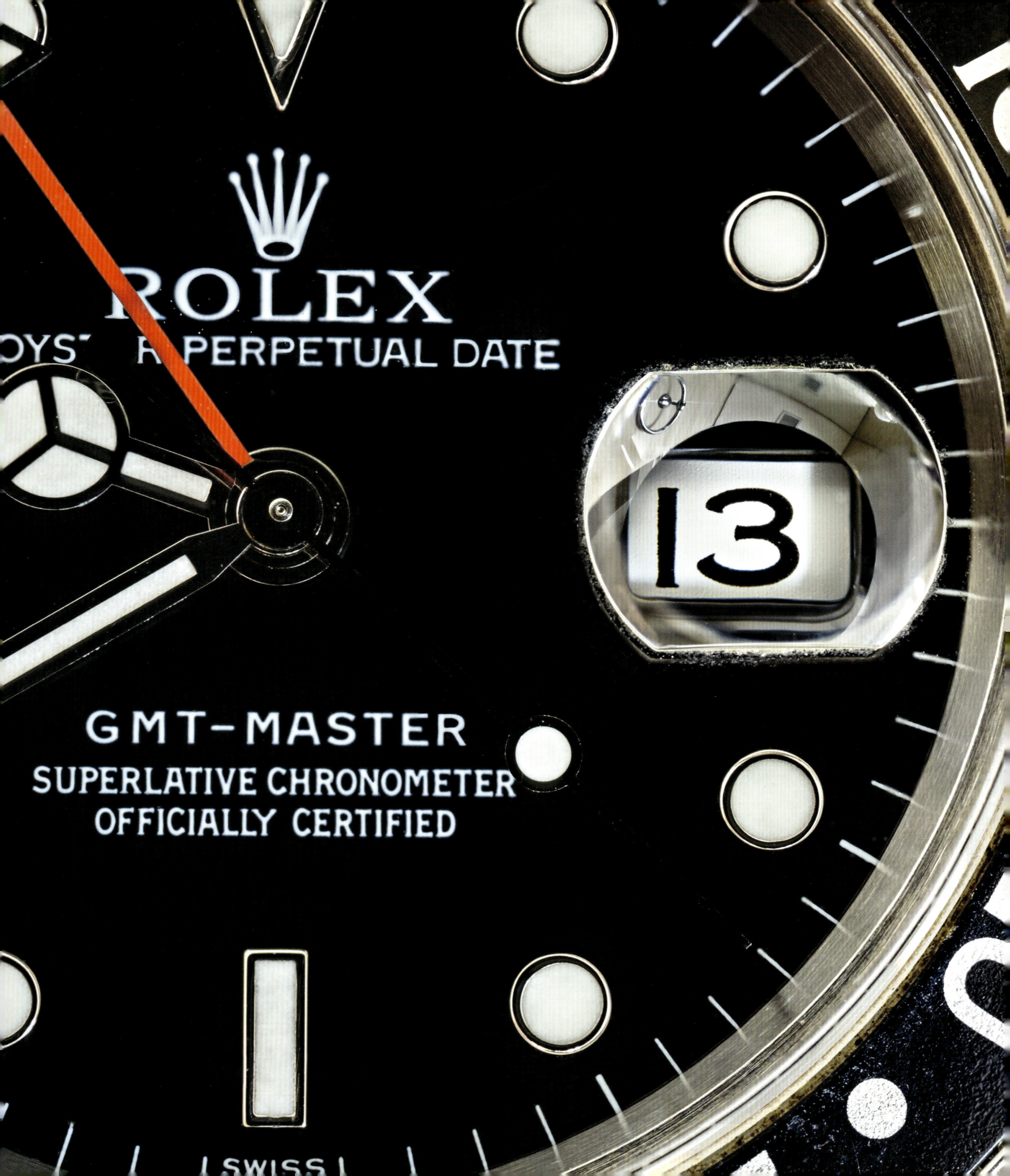
ROLEX
OYST R PERPETUAL DATE
13
GMT-MASTER
SUPERLATIVE CHRONOMETER
OFFICIALLY CERTIFIED
SWISS

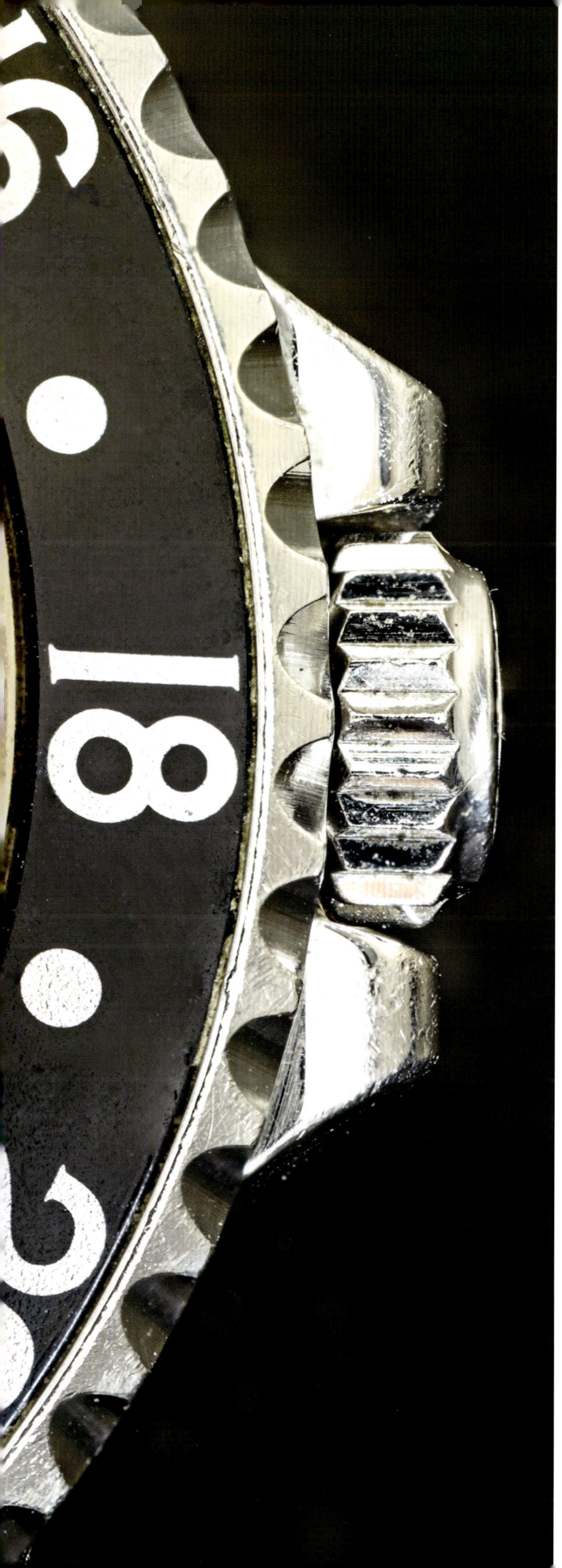

while portraying Colonel Walter Kurz in *Apocalypse Now* (1979) is a famous one. In the movie, the character of Kurz wears a GMT-Master without its rotatable bezel; the watch became famous, bringing in a high price at auction.

The GMT-Master has gone down in history for its unmistakable two-toned bezel. To create a more pleasant experience for crewmembers while consulting the watch, Rolex decided to adopt a bezel with two colours: red indicated daytime hours and blue nighttime.

The first version of the GMT-Master was represented by reference number 6542, whose Bakelite bezel rendered the model easily recognisable, yet extremely fragile and thus it is incredibly difficult to find in its original condition. Its two colours very soon became emblematic of the GMT-Master, even if the Geneva *maison* would later offer other colour pairings to choose from. The original blue-and-red bezel, affectionately known as 'Pepsi' among collectors, was joined by other models earning nicknames like 'Coca-Cola' (for its red-and-black bezel), 'Root Beer' (with a brown-and-gold bezel), and 'Batman' (which was blue and black).

In 1959, reference 6542 was replaced by reference 1675, which would remain in production up until the 1970s. The chief difference in this new version was its transition from the iconic (and problematic) Bakelite insert to a two-toned aluminium insert.

Version of the GMT model with the reference number 6542, with a Bakelite bezel.

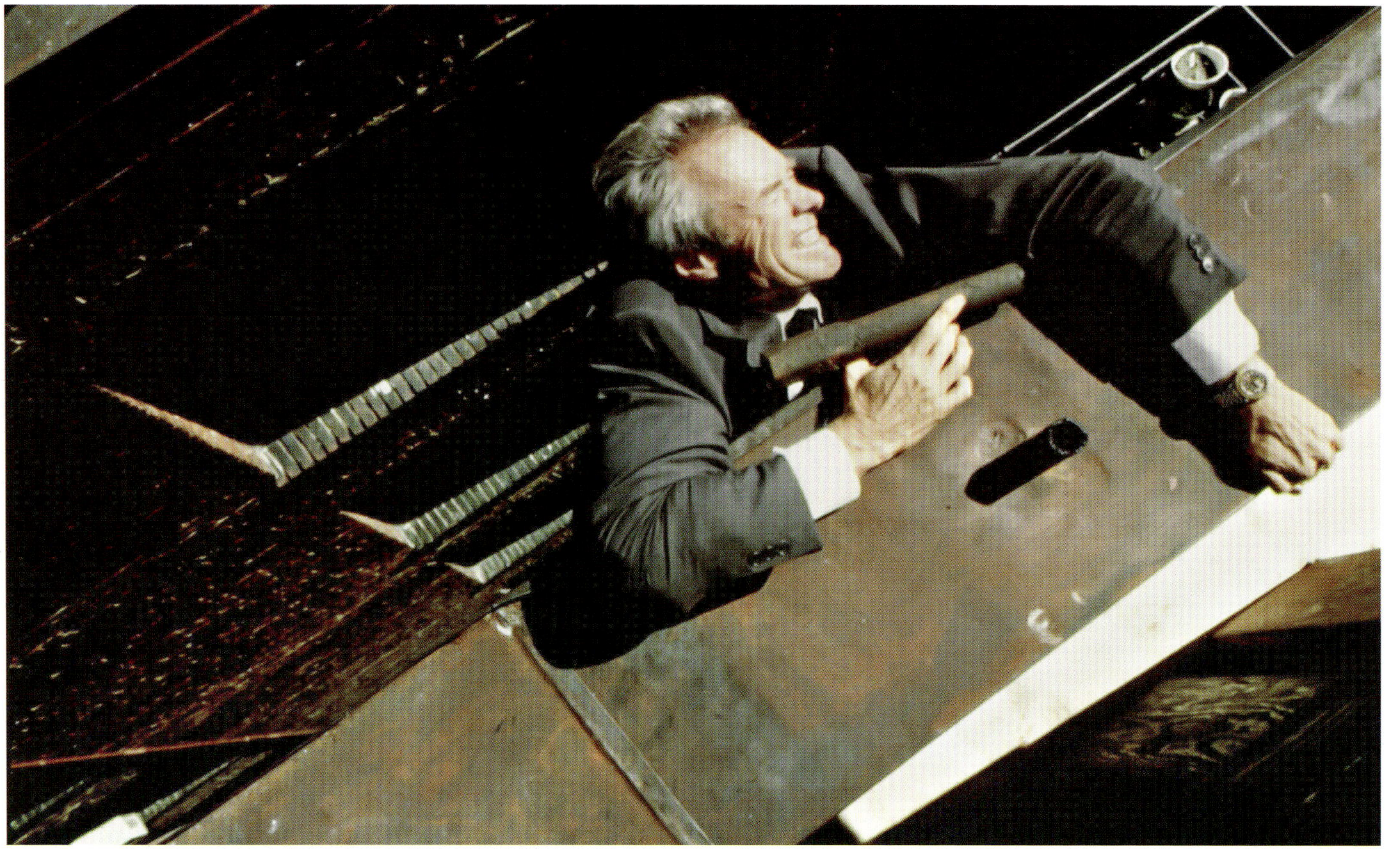

Clint Eastwood wears a GMT-Master in the 1993 film *In the Line of Fire.*

The solution was especially long-lasting, enduring from the insert's replacement up until it was switched to ceramic as late as 2005. The case would also be given a crown guard. In 1981, the reference 16750 was introduced. Its movement featured a quick-set date but with no substantial aesthetic variations: it still had the retiring generation's matte 'buckshot' dial; this would later change to a glossy dial (nicknamed 'bicchierini' or 'little glasses' by Italians) with indices made of tritium with white-gold surrounds. In 1988, the latest version of the GMT-Master was launched, the reference 16700, introducing a sapphire crystal.

DIAL: A highly distinctive rotatable bezel in red and blue with a 24-hour graduation, allowing for easy discernment between hours of the day and night.

CASE: Reference 6542 features no crown guard.

BRACELET: The watch was initially mounted on an Oyster bracelet. Starting in 1960, however, the version following the first model, reference 1675, was the only professional model to be sold by default with the elegant and less sporty Jubilee bracelet. During the long production period, the watch was equipped with various steel, gold and steel-gold bracelets.

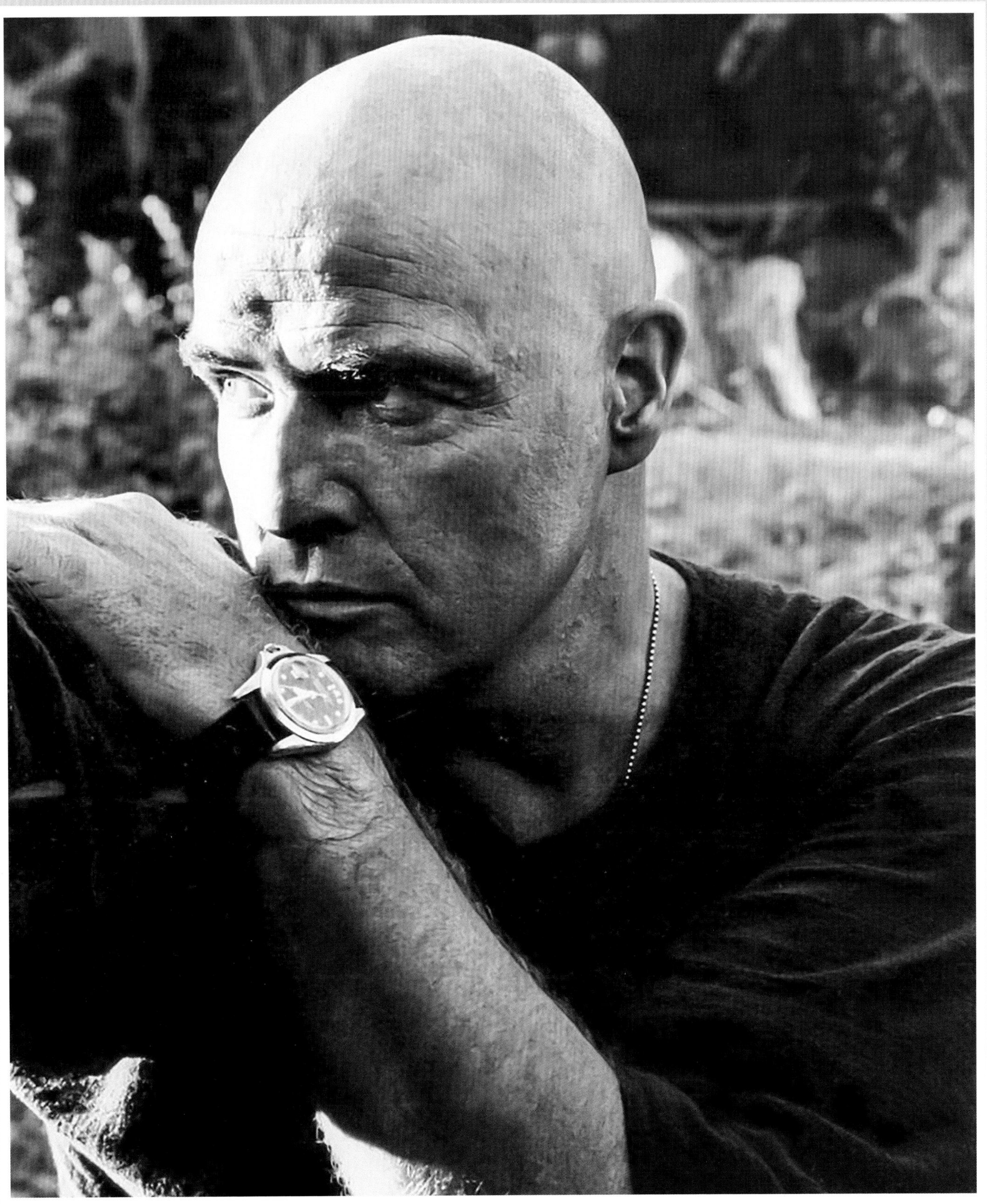

Marlon Brando's Rolex in *Apocalypse Now* (1979), a GMT-Master reference 1675, customised with 'M. Brando' engraved on its case back.

Day-Date

Introduced in 1956, the Day-Date was the first watch to display, in addition to the date, the full name of the weekday inside a small window on its dial.

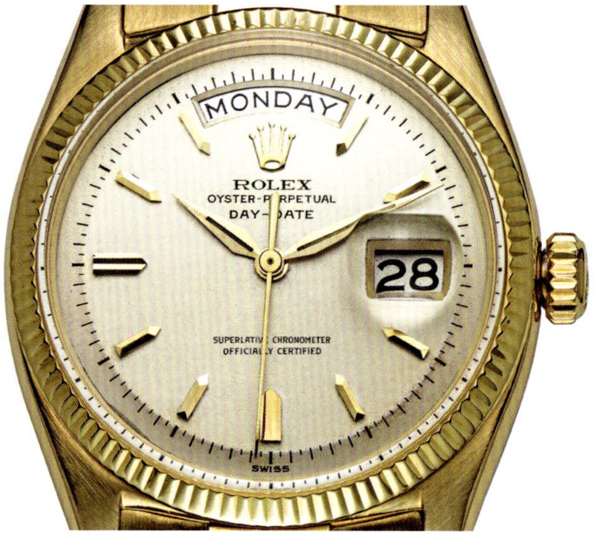

Innovative and revolutionary, the Oyster Perpetual Day-Date is perhaps the *maison*'s most prestigious model. Worn by international elites, so much so that it was re-dubbed 'President', it features a semicircular window at 12 o'clock indicating the day of the week, available in a wide range of languages. Since its launch, it has been produced exclusively in precious metals: in yellow, white or Everose gold, or in platinum. Worn over the years by politicians, executives and visionaries from around the globe, the Day-Date is instantly recognisable on one's wrist, especially for its characteristic President bracelet. Destiny was in its name; given the stature of the men who wore it. It is also said, however, that Hans Wilsdorf began paying homage to American presidents with certain watches after World War Two: first the Datejust, which was given to Dwight D. Eisenhower, then this model.

John F. Kennedy went down in history as the president who did not wear a Rolex because he declined a gift from the *maison*. A gold Day-Date engraved with the words 'Jack, with love as always from Marilyn' was auctioned off in 2005. Legend has it that Marilyn Monroe gave it to him on his birthday in 1962, the day she gave a public performance of 'Happy Birthday, Mr. President'. According to the serial number, however, the watch was made in 1965, two years after Kennedy was assassinated. It could be a mystery, or speculation. Kennedy's successor Lyndon B. Johnson was the first president to officially wear a yellow-gold Day-Date with the President bracelet.

Presented as something of precious value, the Day-Date features a calendar indicating the day of the week, spelled out in a semicircular window at 12 o'clock, with the date in another window at 3 o'clock. This display is based on a disc mechanism whereby the day of the week and the date change simultaneously and instantaneously. A 1963 Rolex catalogue stated that the calendar existed in eleven languages; today, the day-of-the-week display is available in a choice of twenty-six languages.

The first Day-Date watches available were the reference 6510 with a smooth bezel and the reference 6511 with a fluted, so-called 'coin-edge' bezel. Both were powered by the calibre 1055. The mechanism's thickness gave the case a slightly domed appearance. A year later, these references were replaced by the reference 6611 model with a fluted bezel, reference 6612 with a smooth bezel, and reference 6613 with a diamond-set bezel. The substantial difference was on a mechanical level, with the introduction of the new balance wheel with Microstella screws for regulation. This modification of the movement brought the reference 6611 Rolex Day-Date to bear the prestigious Superlative

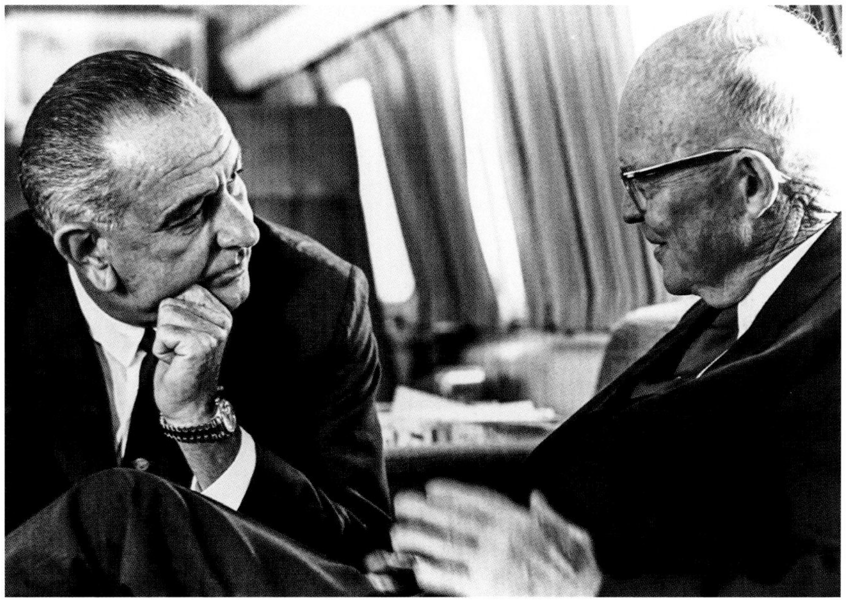

Presidents Dwight D. Eisenhower and Lyndon B. Johnson. Johnson is wearing a Day-Date.

Chronometer certification, attesting to its precision.

In 1959, the reference 18XX series was introduced in a wide range of variations, characterised by a slimmer case housing calibre 1555 (with a frequency of 18,000 VPH). After 1965, the movement gave way to its successor, calibre 1556 (19,800 VPH). The only difference between reference numbers was decoration: 1802 had a smooth bezel, 1803 a fluted one, and 1804 a diamond-set one. References

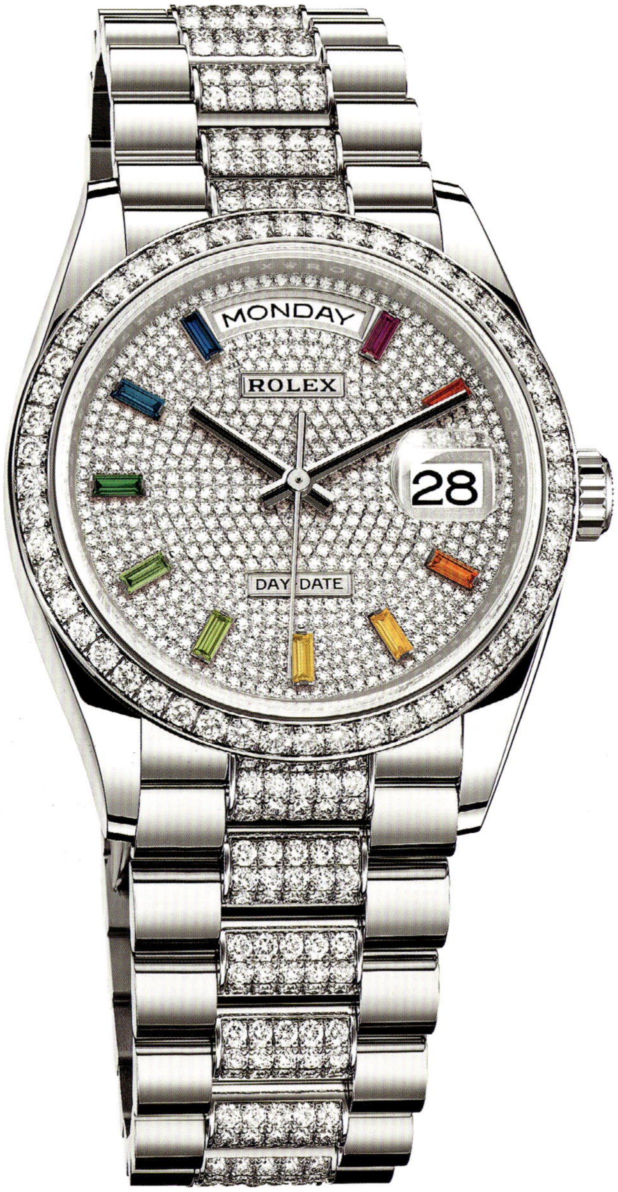

A version of the Day-Date model embellished with diamonds and coloured stone baguettes.

1806, 1807 and 1811 respectively had 'Florentine', 'Bark' and 'Morellis' finishes.

In 1972, Rolex introduced a hacking seconds feature to the movement, useful for perfectly synchronising the watch. Reference 1803 remained in production until 1977, becoming the classic, iconic Day-Date. Like its predecessors, it was only produced in gold – in yellow, rose and white variants – in addition to a rare platinum version.

One distinctive feature of the 1970s was its highly sought-after enamel dials, which came in a wide range of brilliant colours – including mauve, yellow, coral, turquoise, blue, green, orange and oxblood – which were called 'Lacquered Stella', from the name of the Geneva company that produced them.

A few innovations coincided in 1977: a sapphire crystal and new movement, number 3055 with a quick-set date, accompanied the introduction of new, five-digit reference numbers. Their last digit indicated the material of the case, following a precise standard that assigned a number to each metal.

It was specifically with this series that one of the most luxurious and exclusive Rolex Day-Dates was born: the Tridor, featuring the extraordinary union of all three shades of gold.

Calibre 3155 was introduced in 1988, characterised by a double quickset function that allowed the date and day to be set independently. The Day-Date series fitted with this movement was produced from 1988 until 2000.

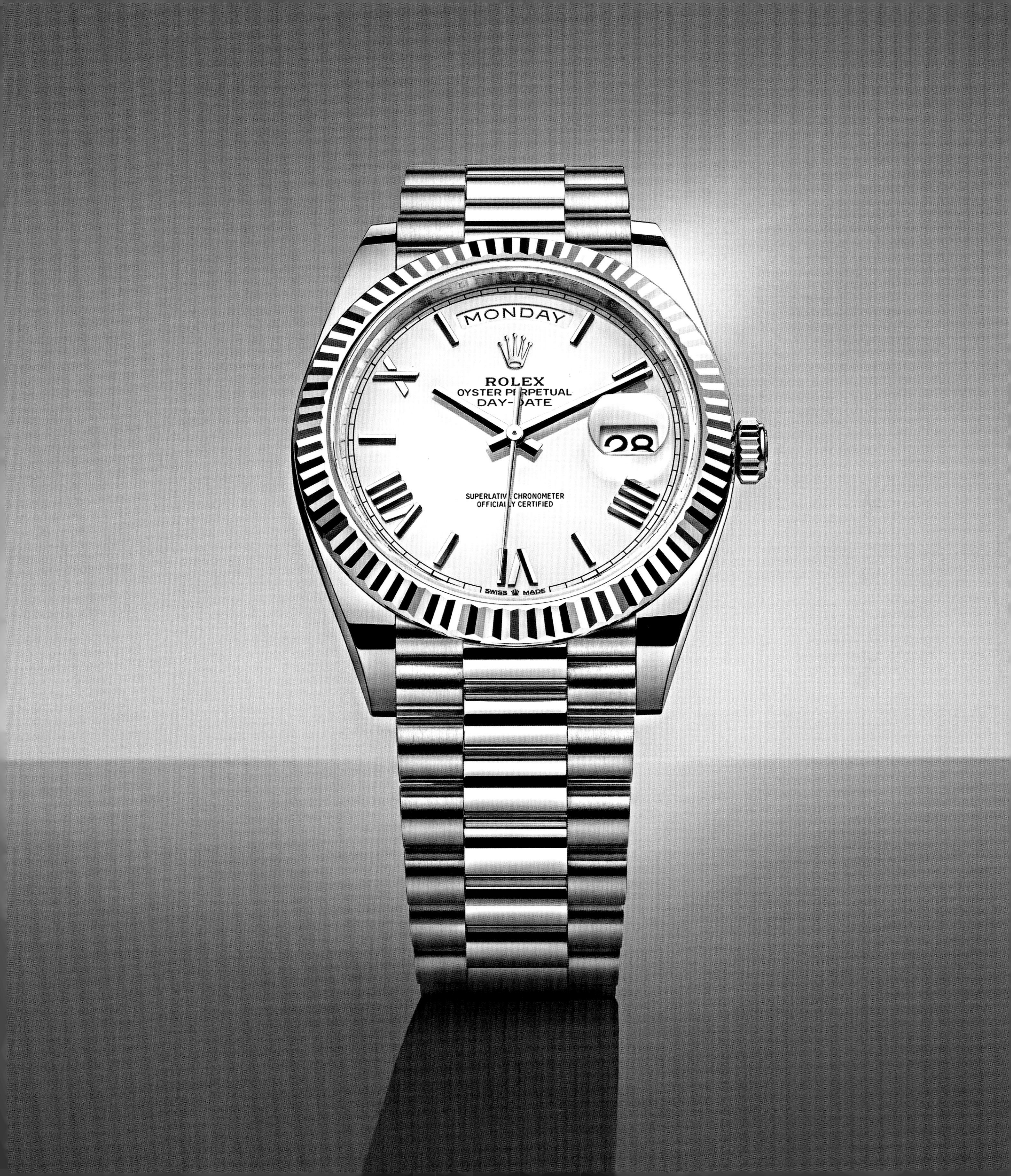
MONDAY
ROLEX
OYSTER PERPETUAL
DAY-DATE
28
SUPERLATIVE CHRONOMETER
OFFICIALLY CERTIFIED
SWISS MADE

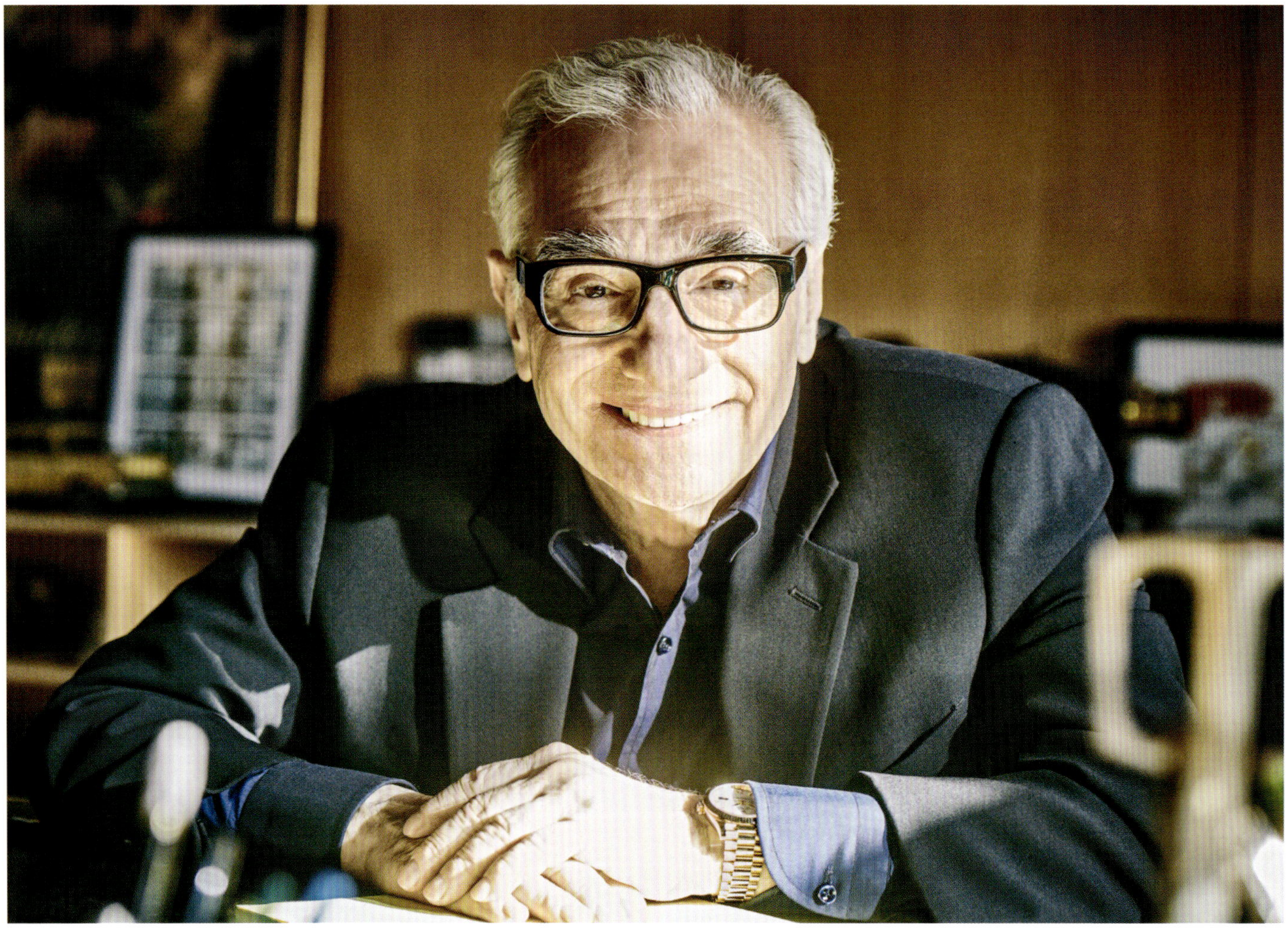

Director Martin Scorsese, Rolex ambassador.

A new version of the Day-Date was released in 2000, the 118XXX series with six-digit numbering. The most significant innovation came, however, in 2008 with the introduction of the Day-Date II, which had a 41-millimetre case and was waterproof to 100 metres, in keeping with Rolex's modern scheme.

The Day-Date 40 was presented in 2015 with a new movement, numbered 3255. The model is still on Rolex's price list, along with the 36-millimetre version.

The Day-Date's Oyster case, with a diameter of either 36 or 40 millimetres, is guaranteed waterproof to a depth of 100 metres. The Twinlock winding crown, with a double waterproofing system, is firmly screwed to the case. The bezel can be fluted or set with diamonds. The crystal, with a Cyclops lens at 3 o'clock to facilitate reading of the date, is made of scratch-resistant sapphire with an anti-reflective treatment.

Since 2022, the Day-Date 36 and Day-Date 40 have both housed calibre 3255, a self-winding mechanical movement completely developed and manufactured by Rolex.

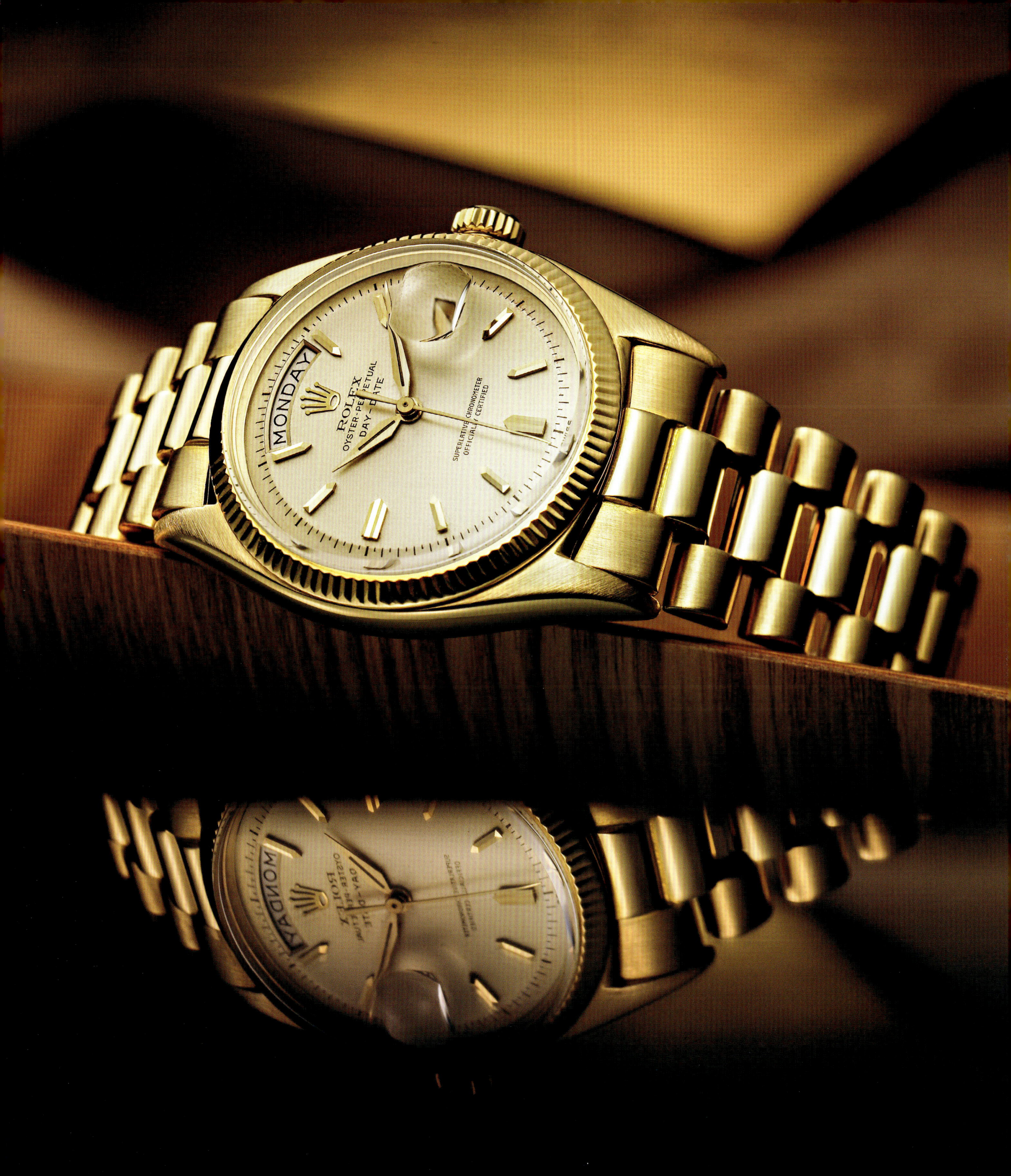
MONDAY
ROLEX
OYSTER-PERPETUAL
SUPERLATIVE CHRONOMETER
CERTIFIED

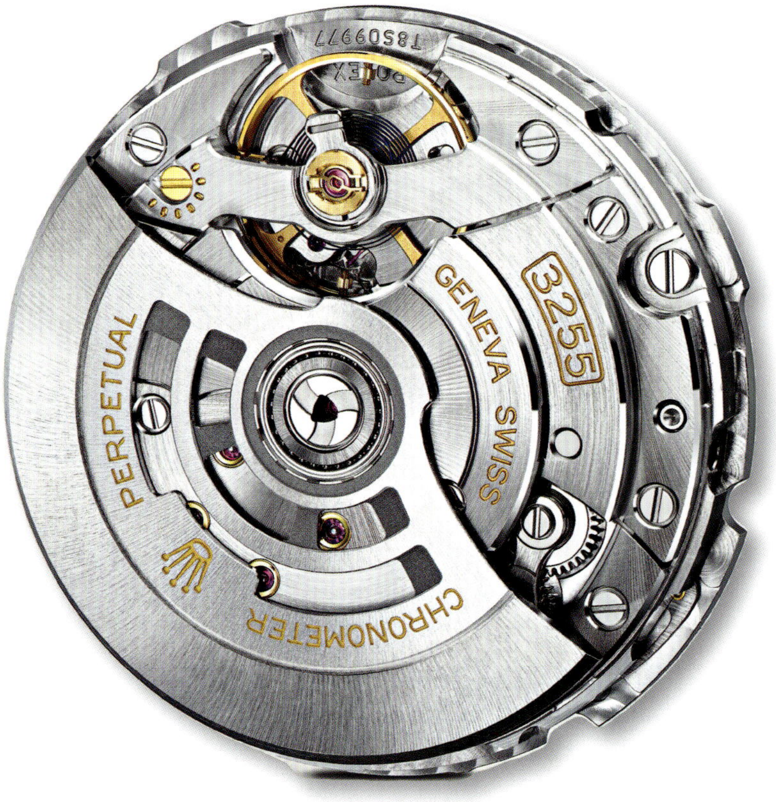

The calibre 3255 features the self-winding module with Perpetual rotor.

DIAL: Since 2022, Day-Date dials have been offered in a very wide range of colours and materials, from 'Rolex green' to olive green, onyx to meteorite. There are also diamond-paved dials.

CASE: The Day-Date is available in two diameters (36 and 40 millimetres) and is also offered in gem-set versions.

BRACELET: A characteristic feature of the Day-Date is the President bracelet, with three rounded links, created specifically for this watch. Its Crownclasp is a clasp with folding blades; its opening and closing system is concealed beneath a small lever in the shape of a Rolex crown, which renders it completely invisible.

The oscillator of calibre 3255 is equipped with a large balance wheel with variable inertia regulated with extreme precision by means of gold Microstella nuts. It is held firmly in place by a height-adjustable traversing bridge for stable placement and increased shock resistance. The oscillator, moreover, is fitted on a high-performance Paraflex shock-absorbing system patented and developed by Rolex.

Calibre 3255 contains an optimised version of the blue Parachrom hairspring, manufactured by Rolex in a paramagnetic alloy. Insensitive to magnetic fields, the hairspring boasts high stability during changes in temperature. It is equipped with a Rolex overcoil, which ensures regularity at any position. The movement includes the patented Chronergy escapement, made of nickel-phosphorus, which guarantees both high energy efficiency and great dependability. It is also unaffected by magnetic fields.

The calibre 3255 features an automatic winding module with Perpetual rotor, which allows the mainspring to be wound continuously, providing constant energy via movements of the wrist. It benefits from a power reserve of about seventy hours, thanks to its barrel architecture and its escapement's superior efficiency. The Day-Date boasts a Superlative Chronometer certification.

The President bracelet, created especially for the Day-Date and presented for its launch in 1956, contributes elegance and decisive style to the prestige of the watch.

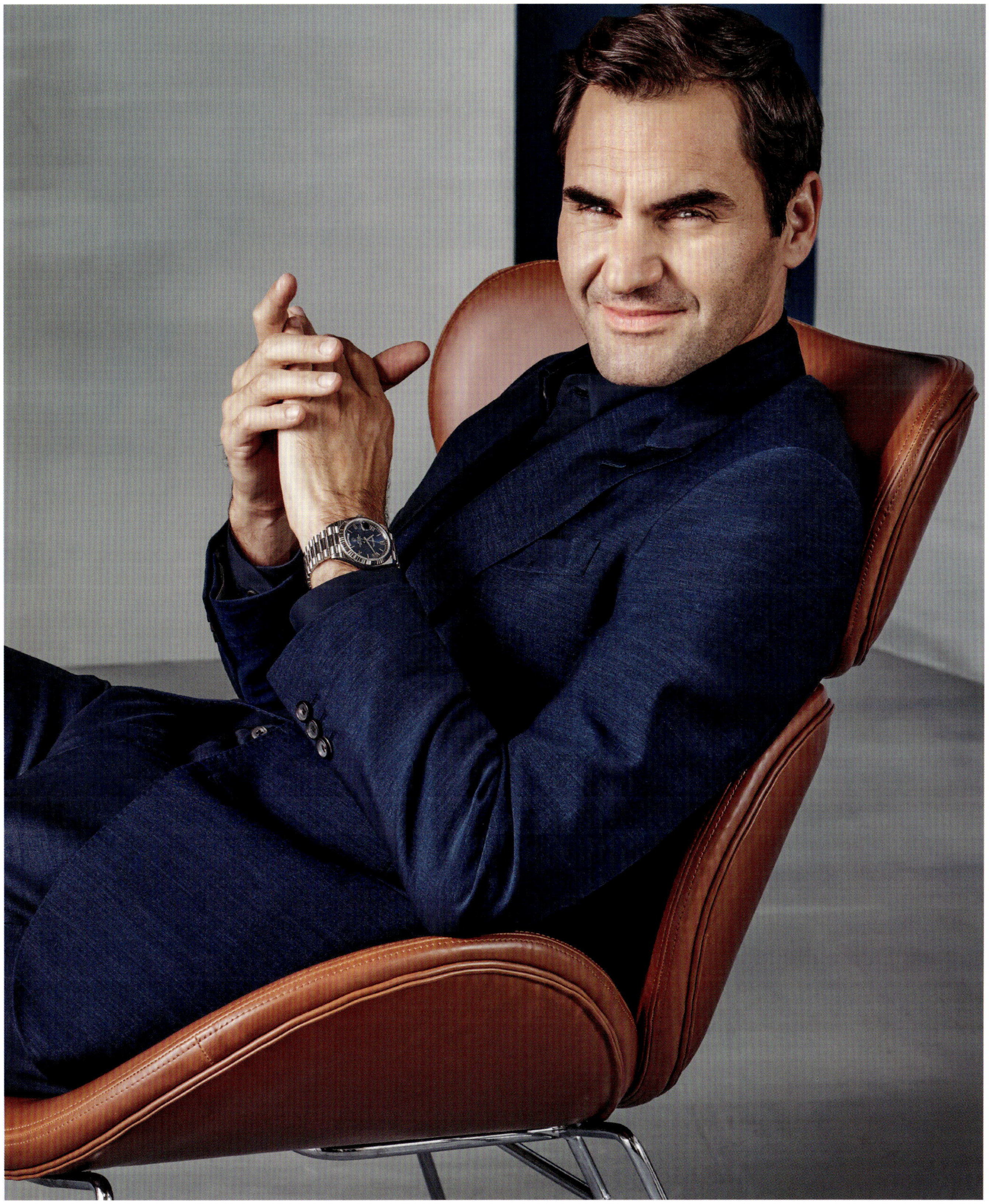

Champion Roger Federer wears Day-Date model with a President bracelet.

Milgauss

This model, introduced in 1956, adopted in particular by scientists at CERN (the European Organisation for Nuclear Research), based in Geneva, recalls historical connections between Rolex and the world of scientific research.

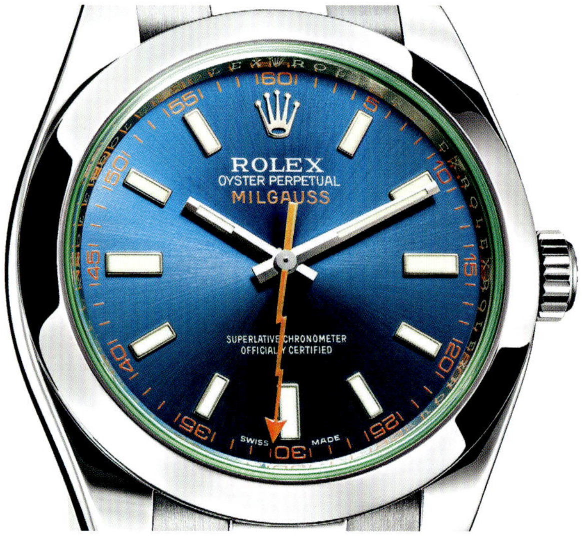

In areas like telecommunication, the space industry, electrical engineering or medical technology, the influence of magnetic fields is especially apparent. Magnetism is an element of considerable disturbance to the correct operating of a mechanical watch; the movement's components can become magnetised, especially the escapement and oscillator, custodians of a watch's precision. Depending on the intensity and length of its exposure to a magnetic field, a watch can experience significant deviations in speed, or even stop. A magnetic field of 50 to 100 gauss is enough to alter the reliability and precision of a mechanical watch. (For the sake of comparison, the magnetism of an everyday magnet approaches 200 gauss at a 5-millimetre distance.)

Rolex has managed, since 1956, to bypass the effects of magnetic fields. That year, the brand designed a watch that could withstand an interference of 1000 gauss – hence its name, combining the number one thousand in French (*mille*) with the unit of measurement for magnetic flux – while also maintaining excellent chronometric performance. What Rolex did to achieve this important aim was simple, yet ingenious: to protect the Milgauss calibre by placing it inside a 'magnetic shield' of soft iron which acts as a 'Faraday cage', isolating its contents from any external electromagnetic field. Its movement, derived from Rolex calibres already in use at the time, was also equipped with components made in a special antimagnetic alloy.

The Milgauss was a model initially intended exclusively for a professional audience. A prototype was presented as early as 1954: the reference 6543, in which antimagnetic insulation was ensured via a very thick case back, combined with a soft iron band positioned around the movement. This edition still had a straight seconds hand. In 1956, the next reference, number 6541, appeared; it was the first to employ the antimagnetic 'shield', a container made of soft iron in two parts, sealing the movement inside. This model also introduced the characteristic lightning seconds hand, which would become its distinguishing feature.

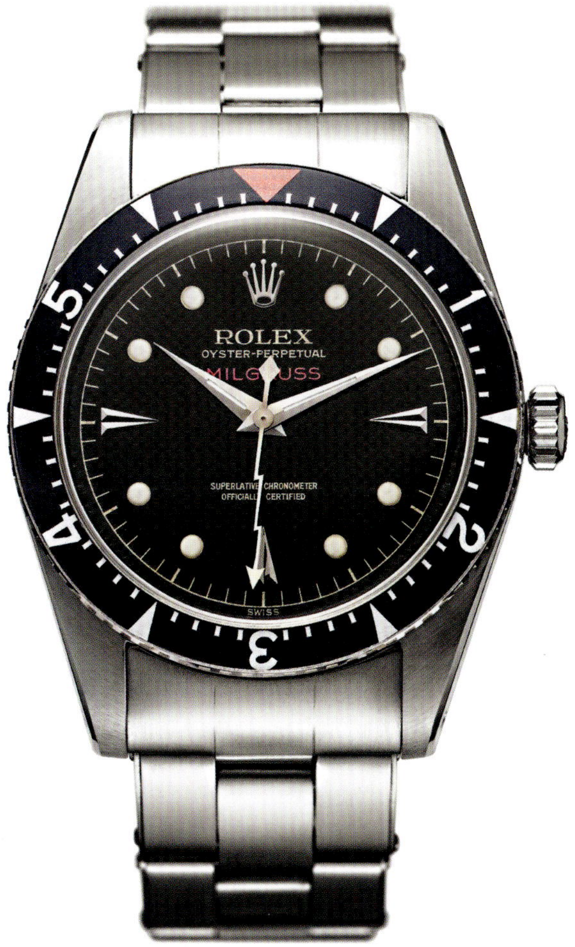

Milgauss model reference 6541 (1956), with the characteristic lightning seconds hand.

In the early 1960s, the Milgauss 1019 was launched. It featured a more sober design: the graduated bezel gave way to a simple, smooth bezel, and the movement was also enclosed inside two antimagnetic covers. Its dial was available in a silver or black version. The seconds hand returned to being upright with a small red arrow at its tip.

The 1019 continued to be produced for nearly three decades, over which minor alterations were made to the dial's finish and lettering.

While, being a hyperspecialised model made for too specific a field, its production may have been limited in the seventies and eighties, Rolex presented, in 2007, a new

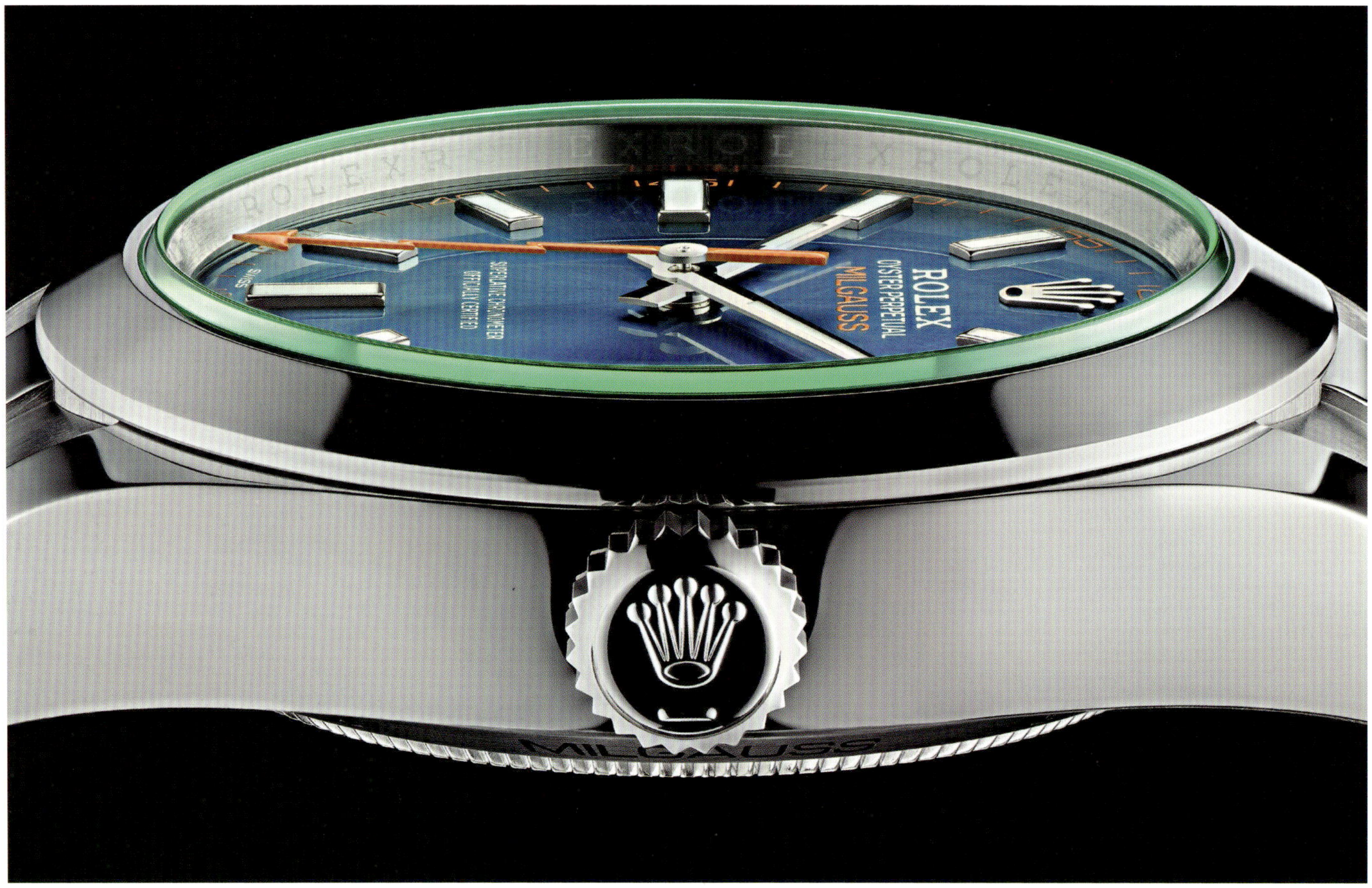

Reference 116400, introduced in 2007. Its oscillator and escapement wheel are made of paramagnetic materials.

generation of this watch connected to the world of science. New editions of the Oyster Perpetual Milgauss, with a 40-millimetre diameter, were distinguished aesthetically by their orange seconds hand shaped like a lightning bolt, taking direct inspiration from the original model. The minute and hour markers were made in the same colour. One of these models was equipped with a green-toned sapphire crystal. The brand perfected the movement's key elements: since 2007, its oscillator and escapement wheel have been manufactured from paramagnetic materials. This was the reference 116400, with an updated style and adapted for wearing on any occasion, not just scientific settings. The reference 116400 GV (for *glace verte* or 'green glass' in French) was introduced in 2014, with a gleaming Z-Blue dial with a sunray finish.

The Oyster case of the Milgauss is guaranteed waterproof to a depth of 100 metres, and its Twinlock winding crown, equipped with a double waterproofing system, screws firmly to the case. The crystal is of scratch-resistant sapphire. The watch is equipped with the calibre 3131, a self-winding mechanical move-

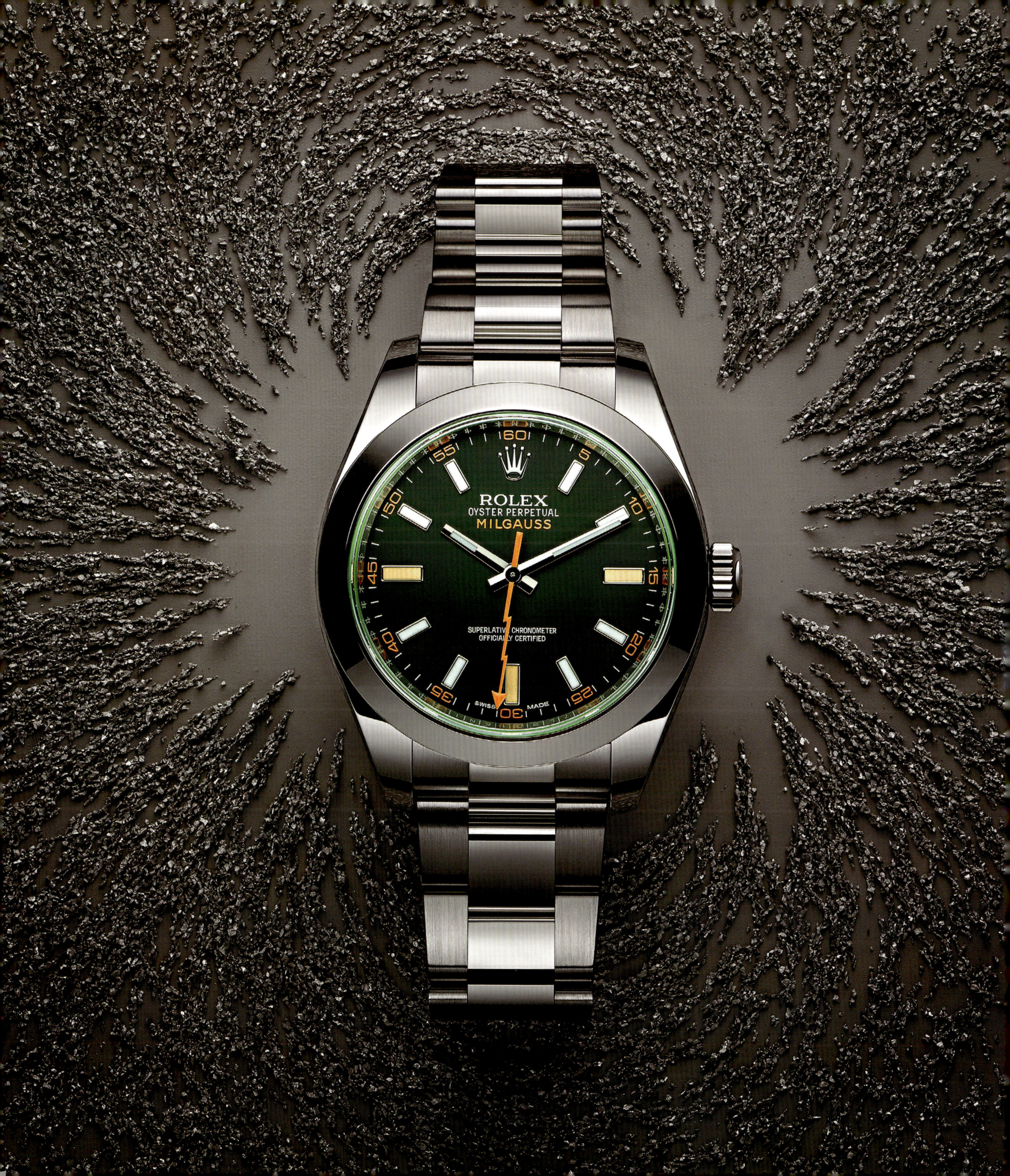
ROLEX
OYSTER PERPETUAL
MILGAUSS
SUPERLATIVE CHRONOMETER
OFFICIALLY CERTIFIED
SWISS
MADE

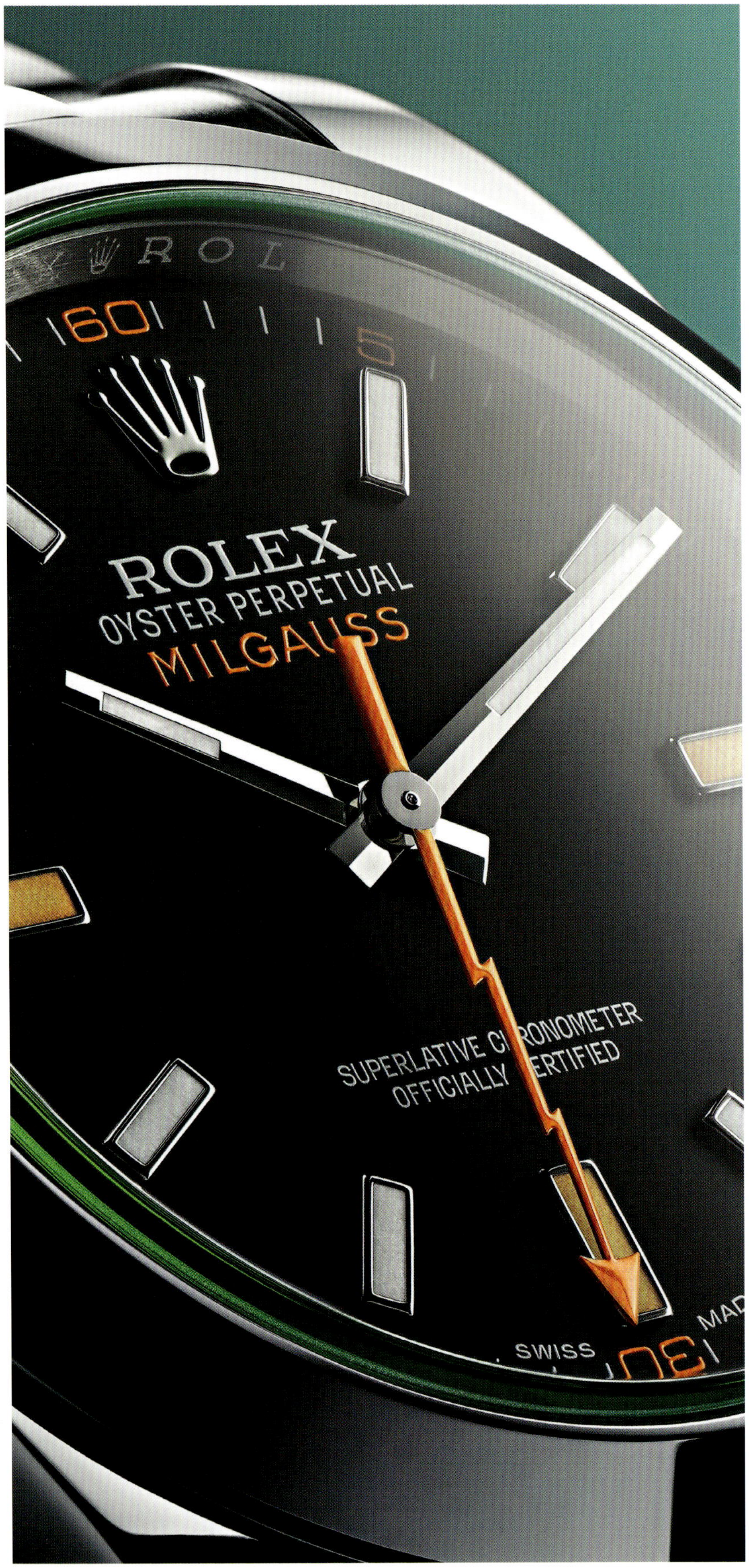

ment entirely developed and manufactured by Rolex.

It has Superlative Chronometer certification and some antimagnetic features, including a blue Parachrom hairspring, Rolex-manufactured in a paramagnetic alloy. In addition to being insensitive to magnetic fields, the hairspring boasts high stability amid fluctuations in temperature. It features a Rolex overcoil, guaranteeing regularity in any position, and a paramagnetic nickel-phosphorus escape wheel. Calibre 3131 features a self-winding module with a Perpetual rotor which ensures continuous winding of the mainspring, thus supplying constant energy through movements of the wrist. It boasts a power reserve of about forty-eight hours.

DIAL: The story of the Milgauss begins with its striking seconds hand in the shape of a lightning bolt. The appearance of this unusual shape is one of this model's most recognisable and desirable features.

CASE: The resistance of this watch to magnetic flux density is made possible via a shield made of ferromagnetic alloys and built inside the case, which is able to protect the watch movement.

BRACELET: The Milgauss is combined with a three-piece link Oyster bracelet in Oystersteel, fitted with a folding Oysterclasp designed by Rolex. Its Easylink comfort extension link was also developed by the brand, allowing for its length to be easily adjusted by about 5 millimetres.

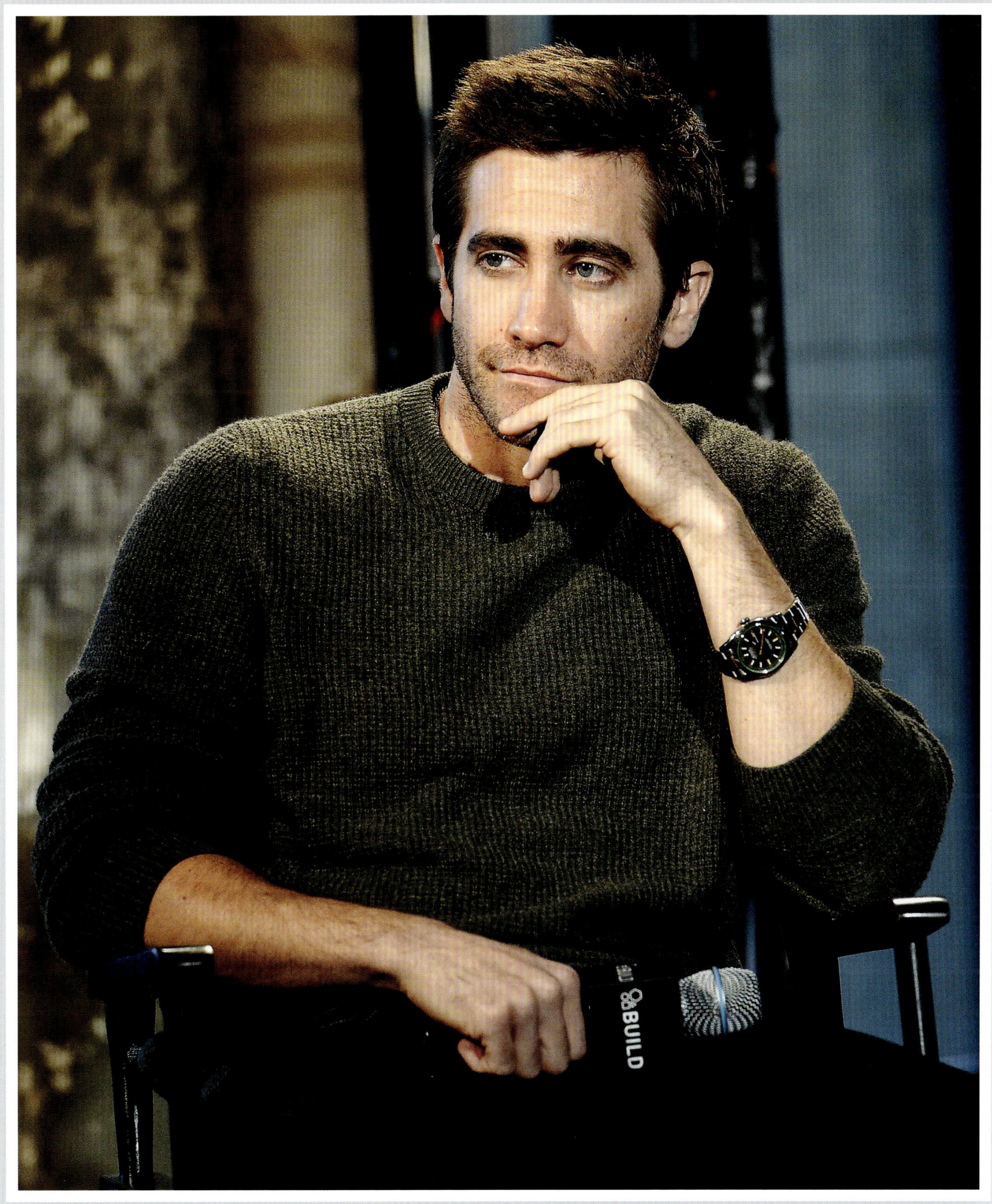

Actor Jake Gyllenhaal wears a Milgauss model.

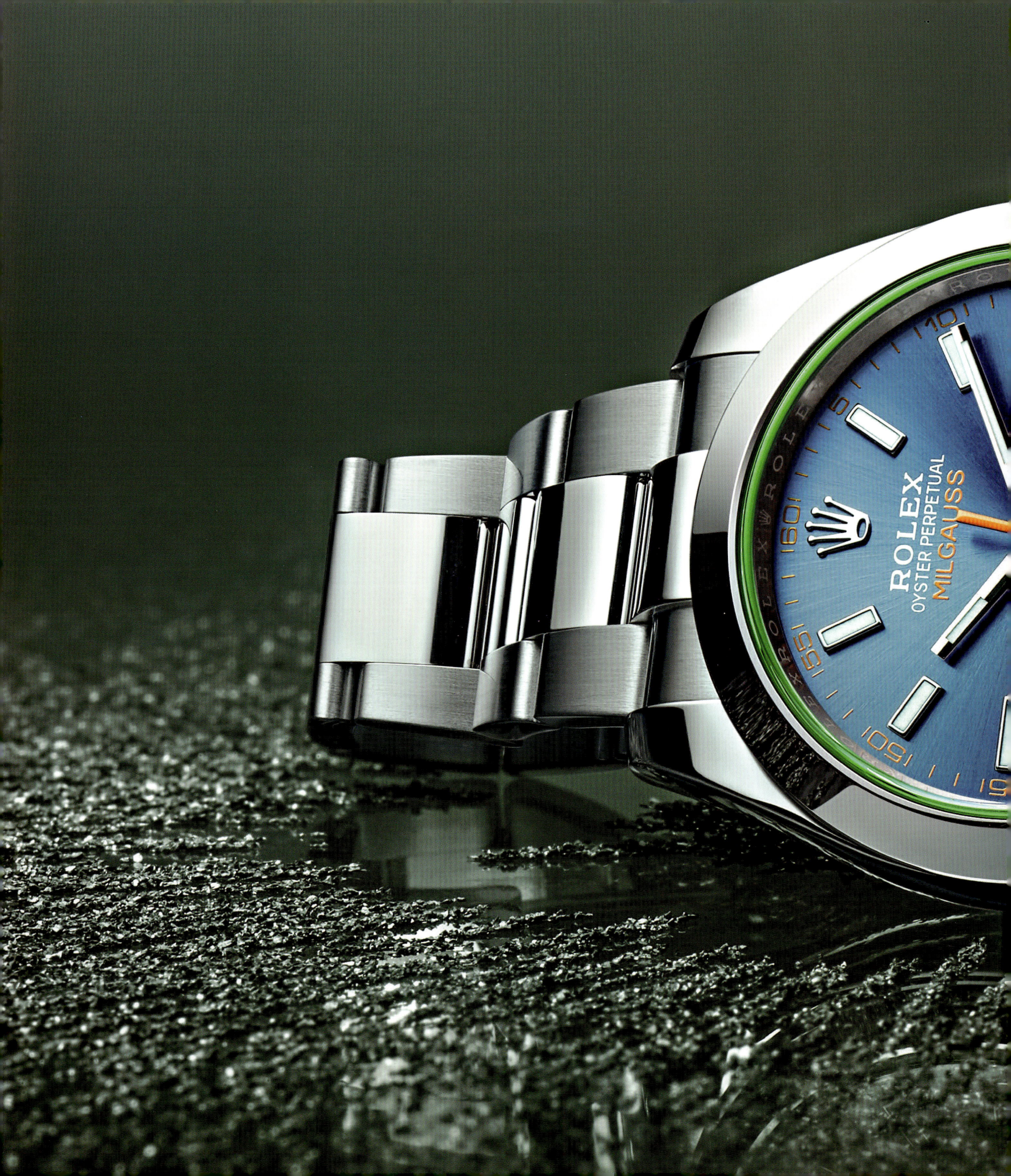
ROLEX
OYSTER PERPETUAL
MILGAUSS

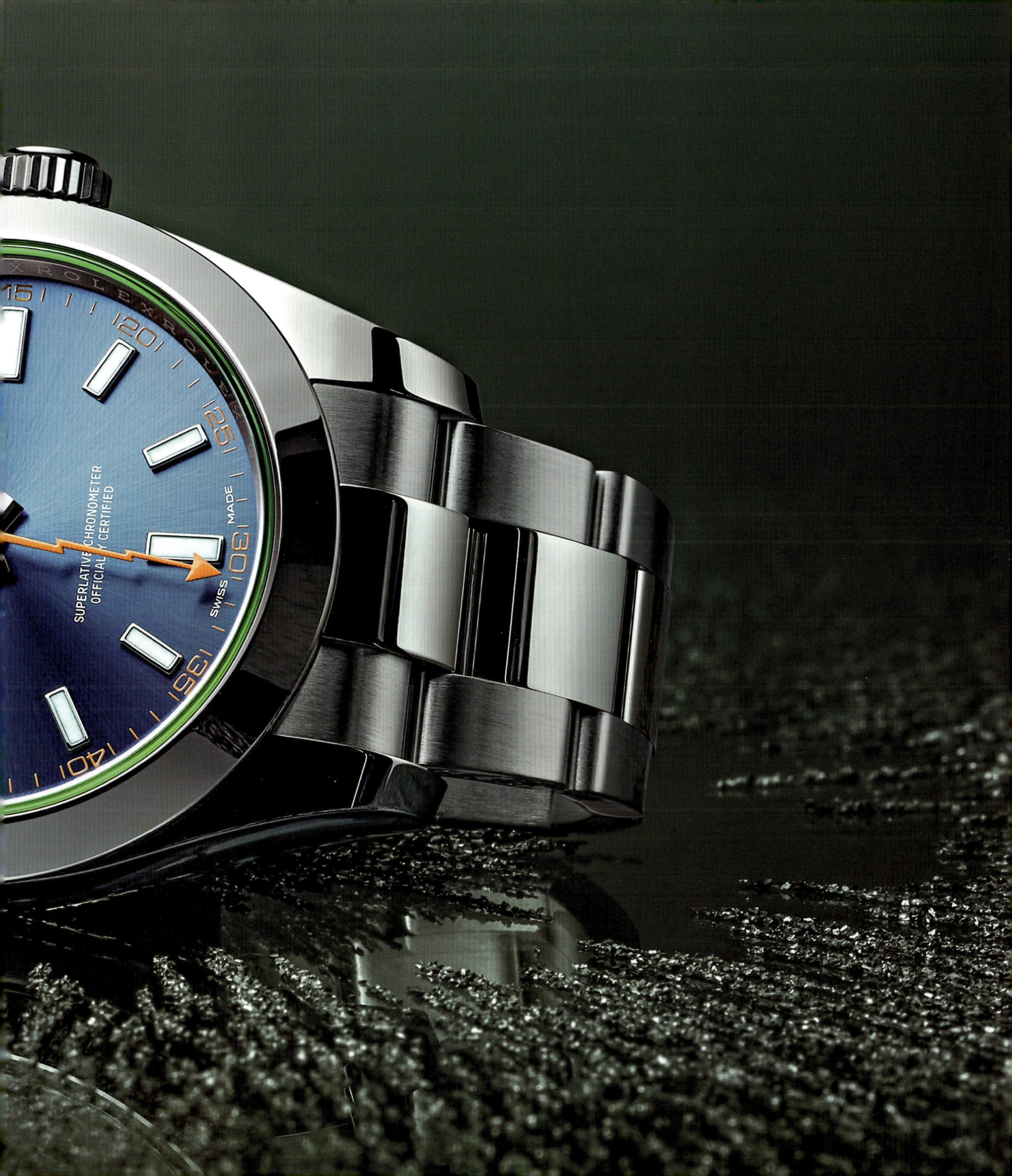
SUPERLATIVE CHRONOMETER
OFFICIALLY CERTIFIED
SWISS MADE

Lady-Datejust

Introduced in 1957, the Lady-Datejust is the crown watchmaker's first timepiece entirely dedicated to women.

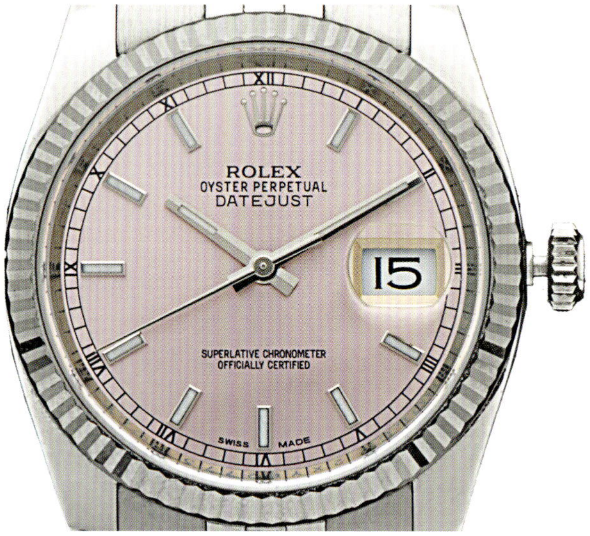

Women have played a decisive role in the history of Rolex. It was Mercedes Gleitze who swam the English Channel in 1927 wearing a Rolex Oyster, providing Rolex with their first endorsement. She indeed birthed the very concept of a Testimonee.

In the 1950s, as well-being spread among all social classes and significant engineering and industrial discoveries profoundly altered everyday life, the figure of the woman also changed, taking on a more independent role in society. In 1957, Audrey Hepburn fascinated Hollywood with her delicate elegance; Sylvia Earle obtained her Master of Science degree to later become a famous oceanographer (and the first woman to head the United States' National Oceanic and Atmospheric Administration); and Françoise Sagan, at a time when the French word for 'writer' rarely appeared in feminine form, conquered the world literary scene with her novel *Bonjour Tristesse*.

Hans Wilsdorf, ever attentive to lifestyle, understood that the modern woman sought to manage her time and therefore needed the same precision that men did. Women's watches had, to that point, been objects where aesthetics prevailed and the precision of their mechanics was secondary. A watch of small proportions posed an important engineering challenge: that of creating a mechanical movement that was relatively small, yet capable of ensuring the same high chronometric standards of the larger counterparts installed in men's watches. As Wilsdorf himself testifies: 'Ladies want the best of both worlds: a tiny watch and an accurate movement. Yet, the smaller the watch, the more difficult it is to make it accurate.'

In 1957, Rolex introduced the Lady-Datejust, succeeding in the feat of creating an extraordinarily precise, robust watch, one worn by generations of successful women around the world. The ladies' Datejust featured all characteristics that had been introduced in the Datejust of 1945. The first edition of the women's chronometer with a calendar, it continued to carry the elegance of the Datejust in a smaller format, perfectly suited to slim wrists. The watch was initially produced with a 26-millimetre diameter, a size that seamlessly followed the trends of the time and was kept for decades. In 2015, the *maison* launched the 28-millimetre Lady-Datejust.

Thanks to its iconic robust and durable Oyster case, the Lady-Datejust is water resistant to a depth of 100 metres and has a highly sophisticated mechanical movement, developed and manufactured entirely by Rolex to its Superlative Chronometer standards. With its instantaneous date, Perpetual self-winding system, silicon hairspring, Paraflex anti-shock system, and 55 hours of power reserve, the Lady-Datejust is perfectly equipped to perform its tasks with the utmost diligence at any time, in any context.

Its middle case is crafted from a solid block of Oystersteel, an alloy that is especially resistant to corrosion, or of gold. The finely fluted case back is hermetically screwed down using a special tool which allows only Rolex-approved watchmakers to gain access to its movement.

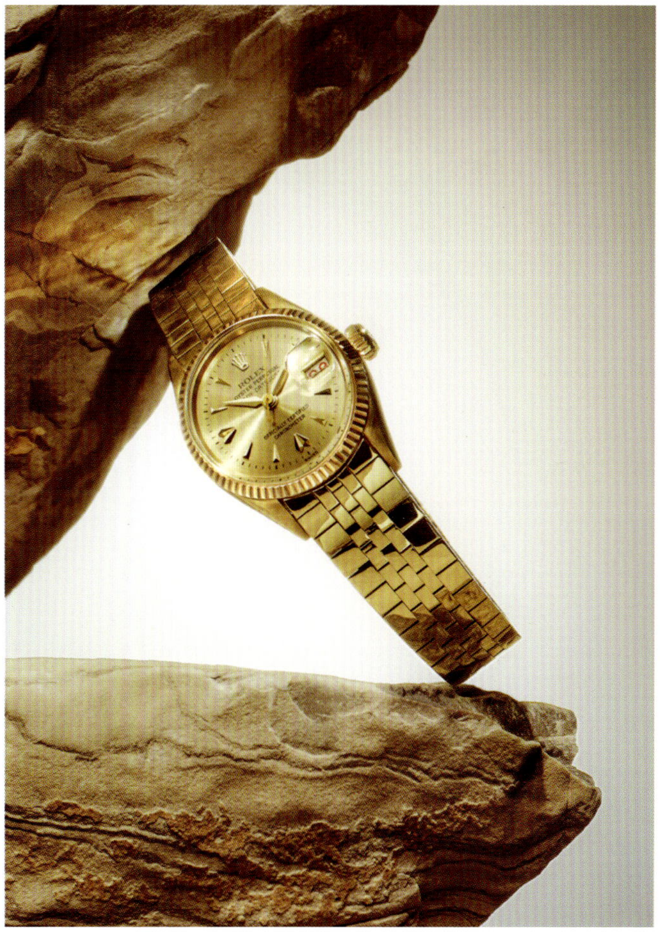

A Lady-Datejust model in gold.

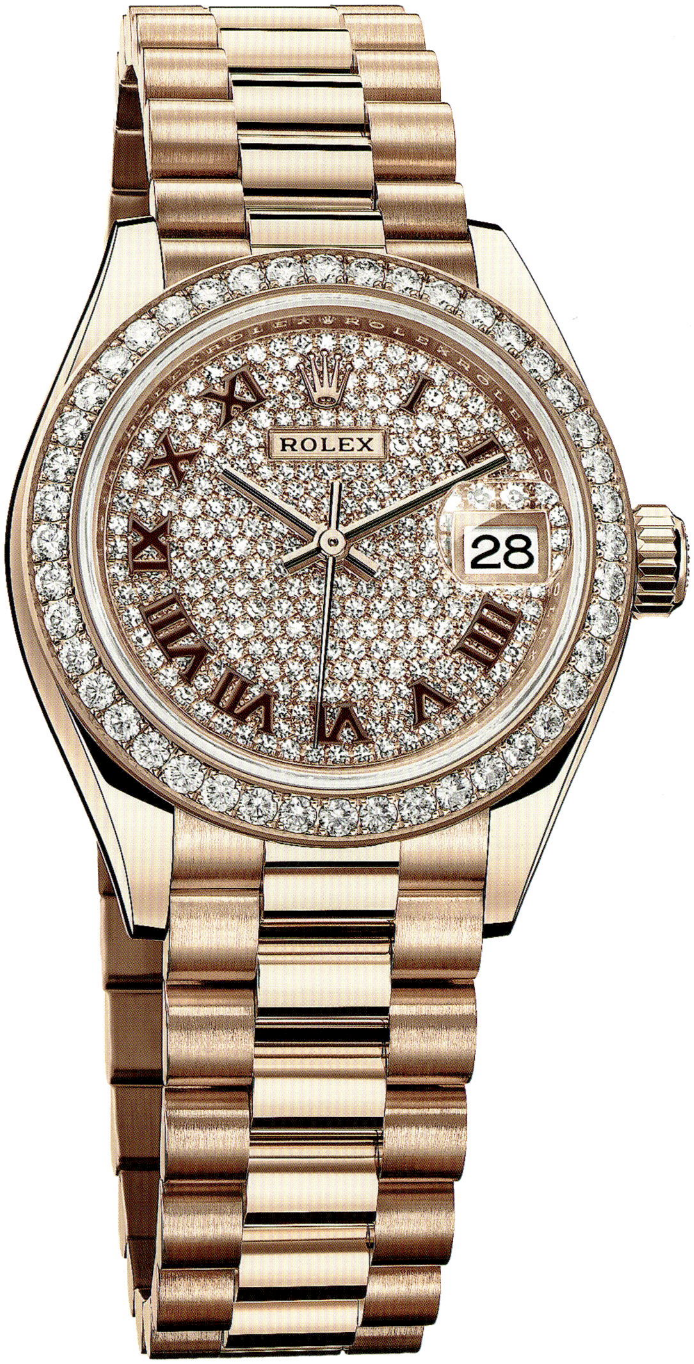

An edition of the Lady-Datejust model with diamonds.

The Twinlock winding crown, equipped with a double waterproofing system, is securely screwed against the case. The crystal, with a Cyclops lens placed above the 3 o'clock position to facilitate reading the date, is in scratch-resistant sapphire with anti-reflection coating. The completely waterproof Oyster case provides optimal protection of the movement it houses.

The Lady-Datejust is fitted with the calibre 2236, a self-winding mechanical movement entirely developed and manufactured by Rolex. Its construction gives it a degree of precision and reliability second to none. Calibre 2236 incorporates a Syloxi silicon hairspring, patented and produced by the *maison*. Resistant to magnetic fields, this hairspring boasts high stability with fluctuations in temperature and, in the event of an impact, remains up to ten times more precise than a traditional hairspring. Its patented geometry ensures accuracy in all positions.

The Lady-Datejust is a model with some of the widest range of presentations in the Oyster Perpetual collection, including diamond-set versions. The Lady-Datejust's diversity of materials is equaled only by the variety of its bracelets and its innumerable dials in delicate colours, available with guilloché (engine-turned) decoration, in mother-of-pearl, or with diamonds set in place of indices or around the bezel.

Depending on the version, the Lady-Datejust is available with a domed, fluted or diamond-set bezel. It can be paired with

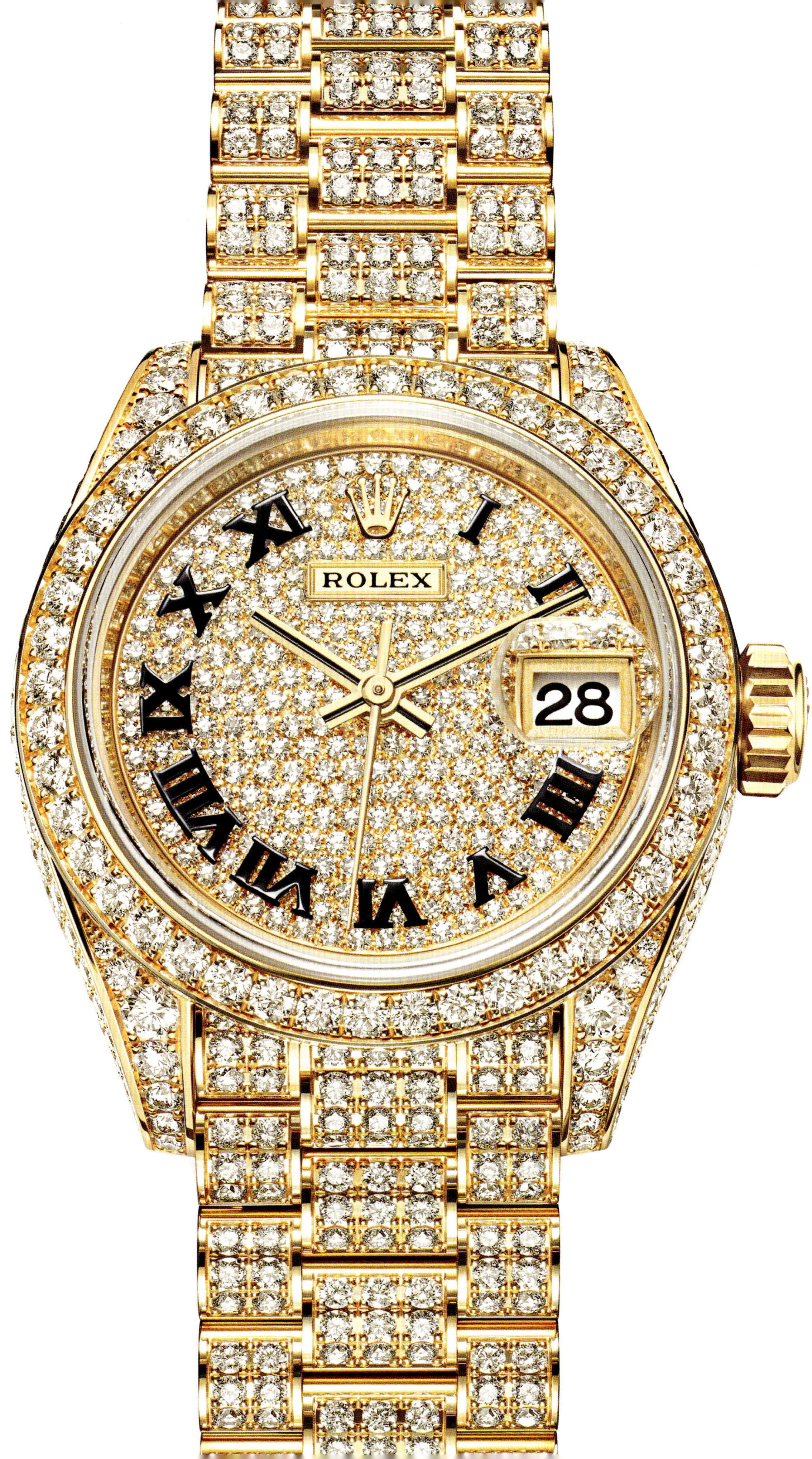
ROLEX
28

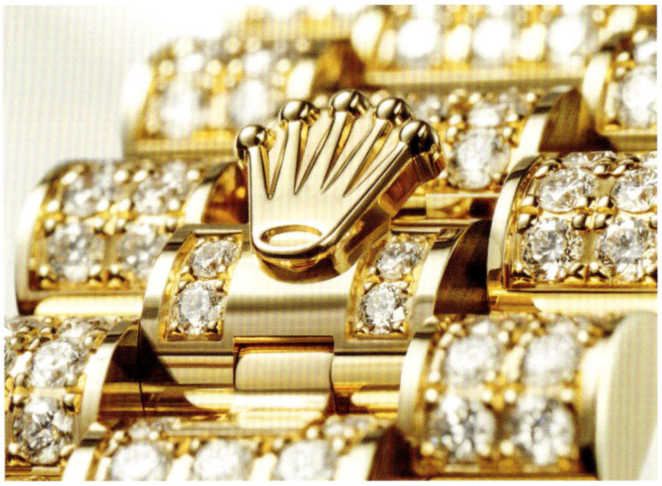

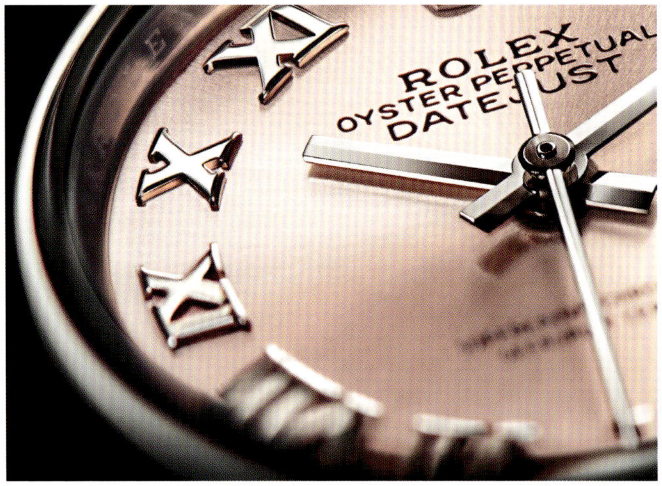

DIAL: The dial's character is expressed through features like the Rolex crown at 12 o'clock, Arabic and Roman numerals, hour markers with classic or professional geometric shapes and brilliant precious stones.

CASE: The gem-set version of the Lady-Datejust presented in 2021 is an eye-catcher with its glittering diamond paving, a testament to Rolex's skill in the art of jewelry.

BRACELET: The Lady-Datejust can be combined with a three-piece-link Oyster bracelet with an Oysterclasp, a five-piece-link Jubilee bracelet – created specifically for the Datejust in 1945 or a three-piece-link President bracelet.

a three-piece-link Oyster bracelet with an Oysterclasp, a five-piece-link Jubilee bracelet (created especially for the Datejust in 1945), or a three-piece-link President bracelet. The Jubilee and President bracelets feature an elegant concealed Crownclasp and a hidden attachment system ensures complete visual continuity between the bracelet and the case. The Lady-Datejust's Oyster bracelet also features the Easylink comfort extension link developed by the brand, allowing the wearer to easily adjust its length by about five millimetres for better comfort in all circumstances. Some gold versions are equipped with a President bracelet whose links contain internal ceramic inserts to improve its flexibility when worn and its durability over time.

In keeping with its founder's dynamic spirit, the company has chosen certain Testimonees over the years who have, each in their own way, demonstrated an ability to approach life with a vision and a good dose of determination. Whether athletes, artists, great explorers or scientists, each has a story to tell of passion and audacity in taking up challenges, as well as perseverance in realising a dream. Through their exploits and their pioneering spirit, their endorsements embody the values so closely held by the brand.

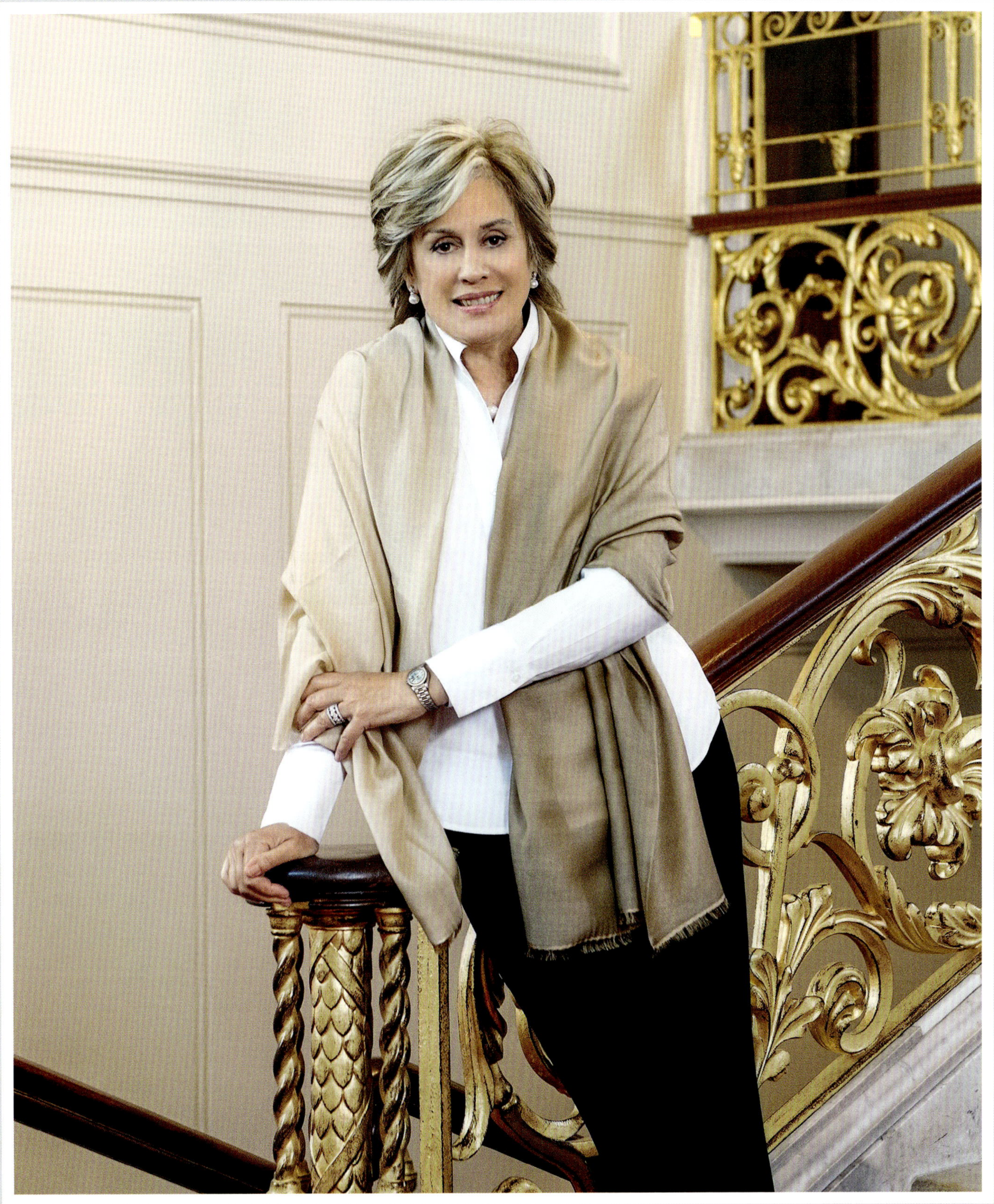

Soprano Kiri Te Kanawa is a Rolex ambassador.

Air-King

This 'king of the skies' was developed in honour of British World War II fighter pilots.

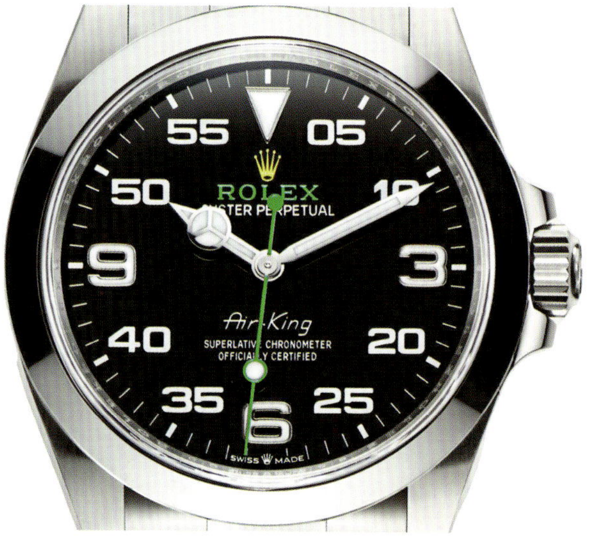

Rolex has been linked closely to aviation since the 1930s. In that decade, aircraft performance was making giant strides, allowing for ever-growing progress in the conquest of the skies and ushering in long-haul flights. Several pilots set their records with an Oyster on their wrists. Among these explorers was the English pilot Charles Douglas Barnard, who declared of the Oyster 'The peculiar qualities of this Rolex watch render it eminently suitable for flying purposes and I propose to use it on all my long-distance flights in the future.' In 1933, Oyster watches helped equip the Houston expedition, the first to fly over Mt. Everest in extreme climate conditions at an altitude of over 10,000 metres (33,000 feet). In 1934, Owen Cathcart-Jones and Ken Waller completed a round-trip flight from London, England to Melbourne, Australia in record time aboard the de Havilland

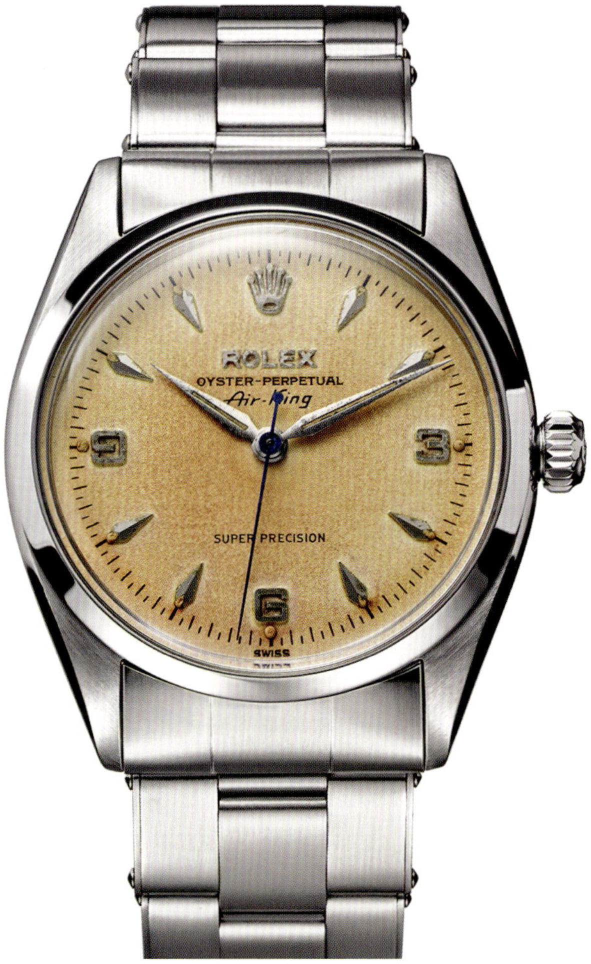

Model Air-King, Super Precision edition.

DH.88 Comet, using a Rolex Oyster as an on-board chronometer. In a letter to Rolex in 1935, Waller wrote: 'I relied faithfully upon my Oyster watch, and am glad to say that it has continued to behave with absolute precision, in spite of the great differences in climate and temperature, air pressure and many other causes which might have stopped such a tiny mechanism.'

The Air-King, however, would be created some years later. Rolex began production of a series of models with a diameter of 34 millimetres, which appears rather small by today's standards but was considered large during and immediately after the Second World War – thus its name, the 'King'.

With its innovative diameter and large dial, offering improved legibility, it became a favourite model of pilots in Britain's Royal Air Force. The watches became so popular that pilots began to put their military watches aside to purchase a Rolex with their own earnings.

When Rolex founder Hans Wilsdorf understood the appetite for this model, he dedicated a special line of watches to the courageous pilots, which was then put on the market toward the end of the 1950s. When it was released, the model was part of a collection of watches with various 'Air' names, each referring to the pilots' valour: there were the Air-Lion, Air-Tiger and Air-Giant, but the Air-King prevailed.

From its launch in 1958, the 'Air-King' has had its name written on the dial in rounded lettering. Featuring a simple design, the reference 5500 had a 34-millimetre steel Oyster case and matching Oyster bracelet. The watch indicated the hour by way of two central hands, with indices and numerals.

Various hearts have beaten within the Rolex Air-King since its introduction: a 'Precision' inscription meant that Calibre 1520 moved the watch hands, while a movement numbered 1530 was the mechanism inside 'Super Precision editions. Over the years, the Air-King's dial would be occupied by stick indices with Roman or Arabic numerals and its colour varied from cream to blue to white. The bezel, meanwhile, has been made in three different variations: smooth, fluted and engine-turned. Amid all these adaptations, however, the design of the Rolex Air-King has always kept one constant, namely the steel Oyster case, both perfectly recognisable and highly representative of the brand.

The Air-King is known for its dials, many of them personalised with company logos. The well-known American pizza chain Domino's, for example, encouraged its affiliates to hit their sales targets by rewarding their success with an Air-King. This unique collaboration brought together two brands that were typically unaffiliated with each other, apart perhaps from a shared obsession with time (given the pizza chain's '30 minutes or it's free' delivery guarantee).

In the early 1990s, Rolex modified the Air-King's dial graphic to display Arabic numerals at the hours of 3, 6 and 9, combined with stick indices. The model was redesigned in 2007 with wider lugs and bezel; a sapphire crystal was introduced; and its bracelet was substituted by a steel Oyster model equipped with a folding clasp. Its movement was also COSC-certified (by the Contrôle Officiel Suisse des Chronomètres) for the first time.

Production of the Air-King was suspended in 2014, then taken up again in 2016 when a new model, the reference 116900, was introduced, this time with a 40-millimetre diameter case.

Today's Oyster Perpetual Air-King features a black dial with large Arabic numerals at the hours of 3, 6 and 9 and an especially legible minute scale with two digits for an ideal reading of navigation times. The union of yellow in the crown with the green seconds hand and 'Rolex' name reflects the emblematic colouring of the brand's logo.

Since 2022, the case has incorporated crown guards and features a middle case with straight sides to match the cases of most models in the professional category.

The current version of the Air-King boasts the 'Superlative Chronometer' certification. The model is equipped with the Calibre 3230, a self-winding mechanical movement developed and manufactured entirely by Rolex. This movement is the quintessence of the technology, boasting several new patents and guaranteeing high performance in terms of precision, autonomy, shock resistance, ease of use and reliability.

Calibre 3230's oscillator is equipped with a large balance wheel with variable inertia, which is regulated with extreme precision via gold Microstella nuts. It is held firmly in place by a height-adjustable traversing bridge for stable positioning and better shock resistance. It is also fitted on high-performance Paraflex shock absorbers, which were developed and patented by Rolex.

Calibre 3230 contains an optimised version of the blue Parachrom hairspring, produced by Rolex in a paramagnetic alloy. In addition to being unaffected by magnetic fields, the hairspring is highly stable during sudden temperature changes. It is equipped with a Rolex overcoil which guarantees regularity in any position.

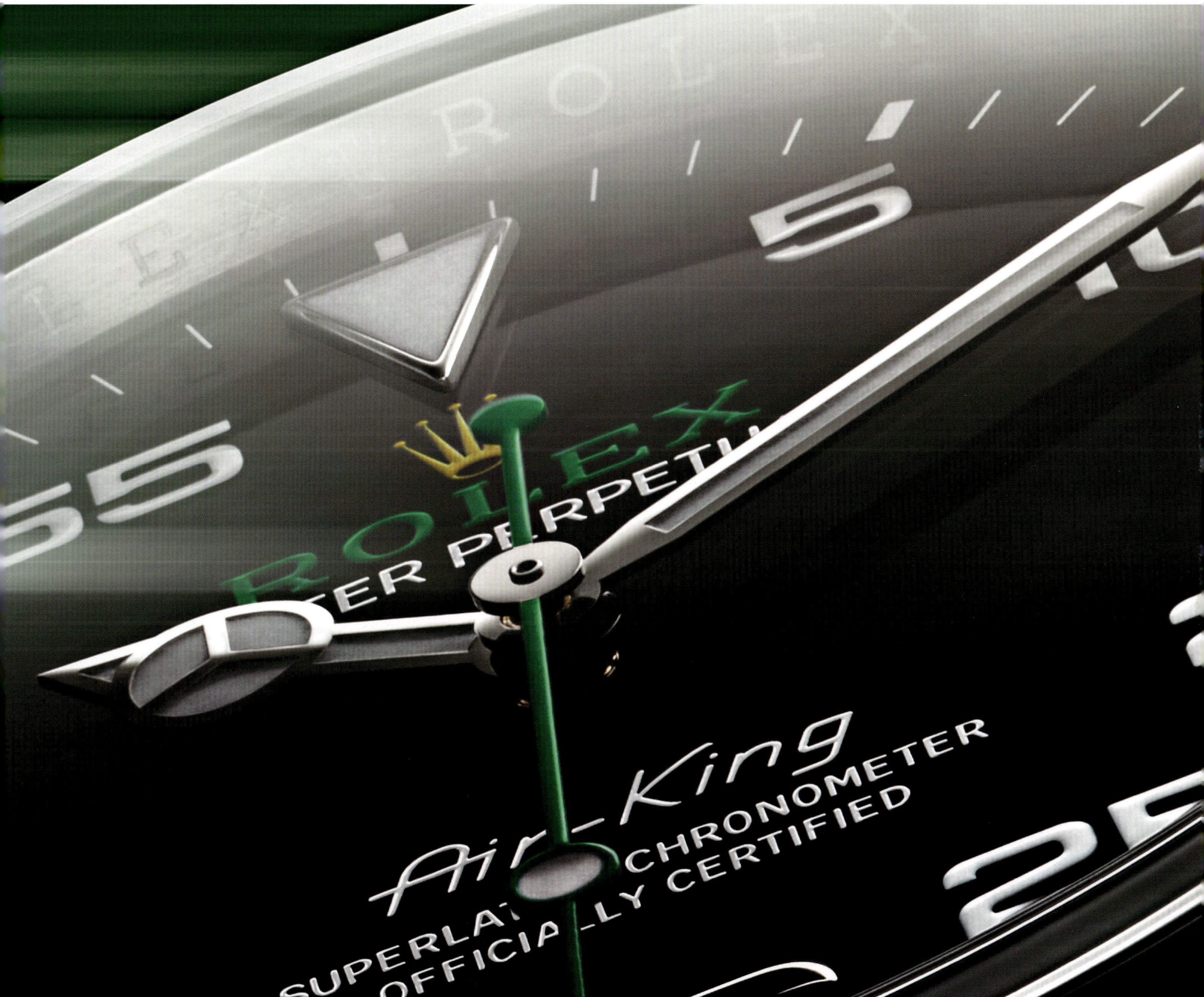
ROLEX
Air-King
SUPERLATIVE CHRONOMETER
OFFICIALLY CERTIFIED

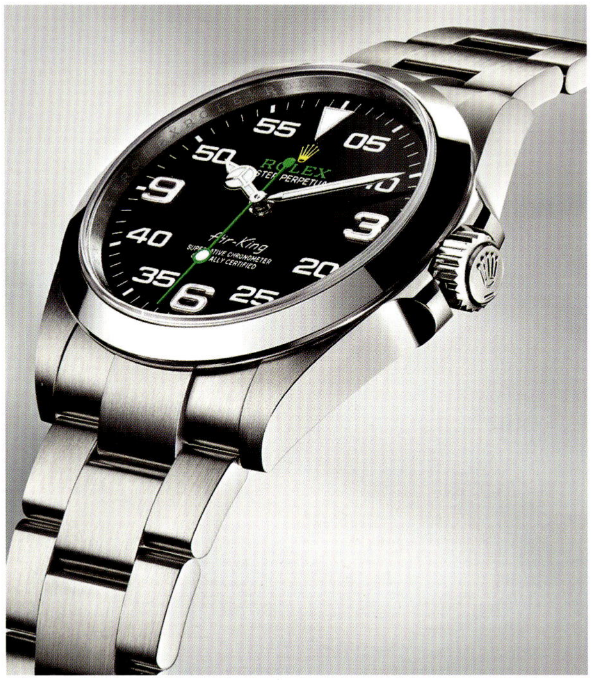

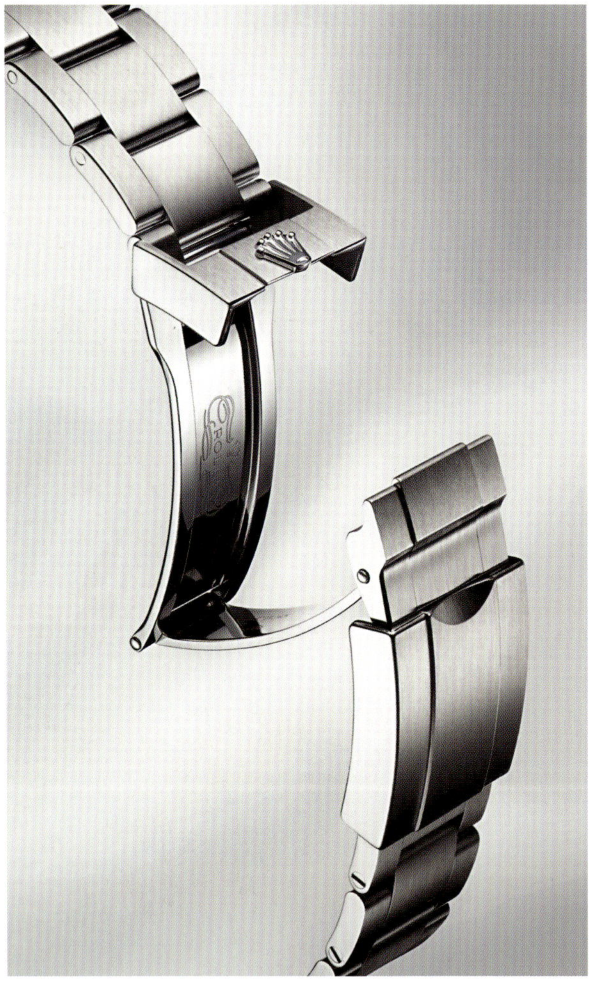

The movement includes the company's patented Chronergy escapement made of nickel-phosphorus, which simultaneously guarantees high energy efficiency and extreme dependability while also being insensitive to magnetic fields.

The Calibre 3230 has a self-winding module with Perpetual rotor, which allows the mainspring to be wound continuously, thus providing constant energy through movement of the wrist. It boasts a power reserve of approximately seventy hours due to its barrel architecture and the superior efficiency of its escapement. The Air-King's movement, accessible only to Rolex-qualified watch technicians, is exceptionally finished, in accordance with the brand's quality criteria.

The Air-King's Oyster case, with a diameter of 40 millimetres, is guaranteed waterproof to a depth of one hundred metres.

DIAL: The Air-King guarantees excellent legibility in all circumstances; this includes the dark, thanks to its unique Chromalight display: the numerals at 3, 6 and 9 o'clock and the hands are respectively filled and coated with long-lasting luminescent material (in the colour blue), emitting a glow up to two times longer than that of traditional phosphorescent materials.

CASE: The Air-King's Oyster case is completely waterproof, protecting the movement it houses perfectly.

BRACELET: The Air-King is paired with a three-piece-link Oystersteel Oyster bracelet. This bracelet is fitted with a Rolex-designed and -patented Oysterlock safety clasp which prevents any inadvertent opening. It also features the Easylink comfort extension link developed by the brand, which allows the wearer to easily adjust its length by about 5 millimetres, for greater comfort in any circumstance. A hidden attachment system provides for seamless visual cohesion between bracelet and case.

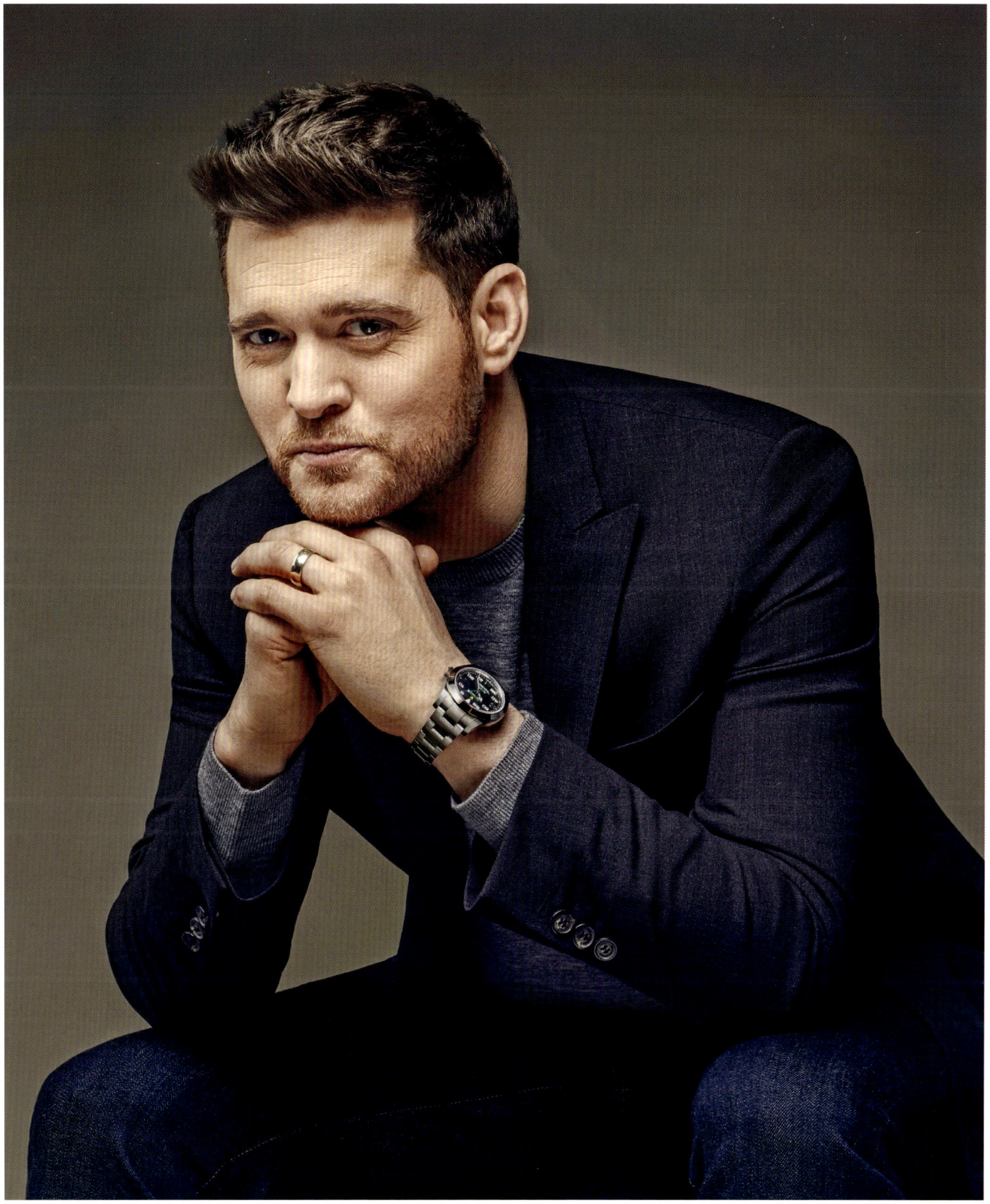

Singer Michael Bublé wearing an Oyster Perpetual Air-King model.

Cosmograph Daytona

A chronograph produced for auto-racing professionals, this watch became an icon.

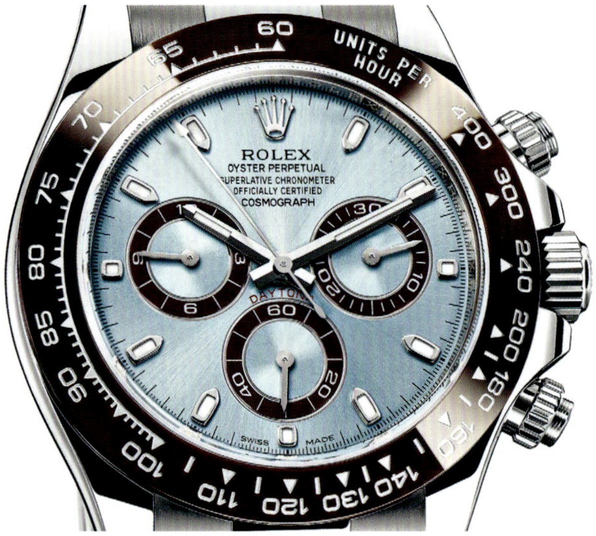

Produced in 1963, this chronograph's two-fold name would become legendary; it alludes to the concept of exploring the cosmos, linking it to the great undertakings and discoveries among the clouds that took place in the 1960s. On the other hand, it also recalls the Daytona International Speedway in Florida, where the historic 24-hour race is held, already sponsored by the Geneva house at the time. As early as the 1930s, Daytona Beach provided a stage for some of the most memorable speed records and for exciting competitions. It was on this hard-sand beach hosting high-speed race-cars that Sir Malcolm Campbell, at the wheel of his racecar Bluebird with an Oyster on his wrist, was the first to reach the landmark land speed of three hundred miles per hour in 1935. Nick-named the 'Speed King', the racer would break the world speed record nine times. Campbell was already wearing a Rolex Oyster in 1930, and he testified to its extraordinary resistance to shock and vibrations in the era's advertising. Following his last endeavour, he sent Rolex a telegram: 'The Rolex watch is still keeping perfect time – I was wearing it yesterday when Bluebird exceeded 300 MPH – Campbell'.

The Cosmograph Daytona therefore succeeded a long series of chronographs produced from the 1930s, and Rolex concentrated all its technical expertise into the model: it was thus designed to be the finest chronometric instrument for racecar drivers for its precision and reliability.

The production of this watch can be divided into three phases:

The first, from 1963 to 1988, concerned the manually wound Cosmograph Daytona. This model, reference 6239, featured a metal bezel bearing an engraved tachymeter scale, had pump pushers, and was available in gold and steel. Other manually wound versions were introduced over the years with a red 'Daytona' inscription; the very first examples did not have one, but it would become a characteristic feature of the model.

The chief distinctive feature of this watch, however, was its bicolour graphic scheme: the chronograph's counters stood out clearly due to their strong colour contrast with the dial: black on a light-toned dial, for instance, or a light colour on a black dial. The background of the sexagesimal scale was coloured to match the numerals and indices on the counters, contrasting with the counters' backgrounds. This was realised in two layers, creating an invisible step between these elements and the central dial.

In the early years producing this watch, Rolex modified details in certain batches, including colour combinations, index styles and numeral positions: this is how nonstandard dials – which the *maison* termed 'exotic' – were born; collectors called them 'Paul Newman' dials, as the actor wore a watch with this configuration.

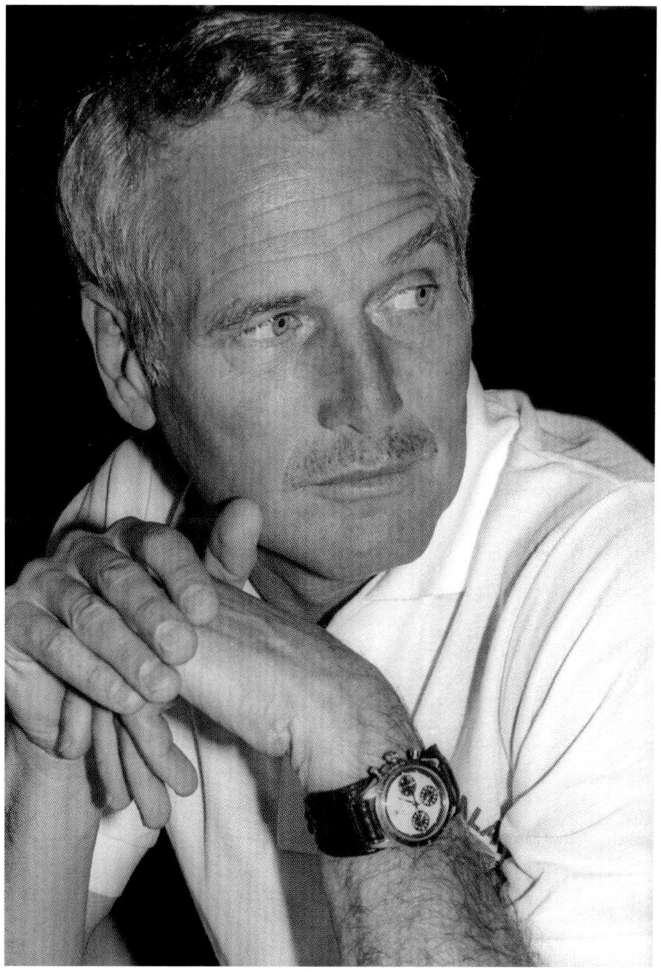

Actor Paul Newman wearing the Cosmograph Daytona.

Its second phase was of the Cosmograph Daytona from 1988, when it became a self-winding watch. The movement animating it was the calibre 4030, which was based on the Zenith El Primero calibre 400.

In addition to its 'Oyster Perpetual Cosmograph Daytona' branding, the model started to bear the inscription reading 'Superlative Chronometer, Officially Certified'. In terms of aesthetic, the bezel also featured a tachymetric scale (200 units/hour in the first series, 400 in the second), with screw-down crown and pushers, and the dial, which kept its 'chronograph' label and 'Daytona' in red, was available in just two variations: white with a black circular field surrounding each counter, or black with white fields. This reference was available in three versions: steel, a combination of steel with gold, and gold. This model was discontinued in 2000, when the third era of this watch began, with the introduction of reference 116520; it was equipped with calibre 4131, the first entirely manufactured by Rolex, with a COSC certification.

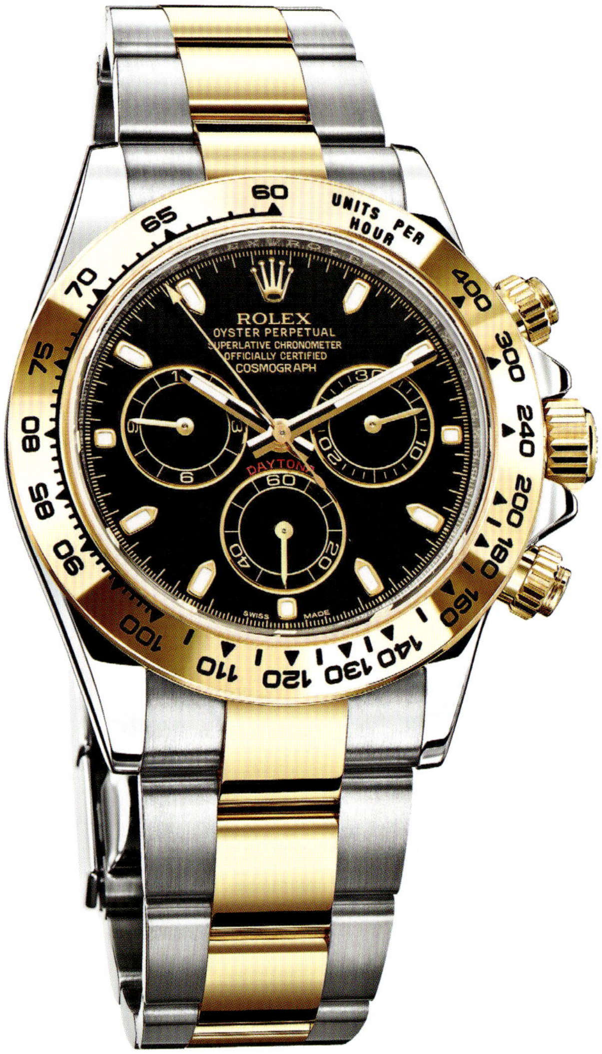

The Cosmograph Daytona model, reference 116520.

Beyond the steel version of reference 116520, available only with a black and white dial, the collection included various versions in precious metals, paired with a variety of dials and bands. They were also made in special editions, among them the Daytona 'Leopard', which had an ornate yellow-gold case, lug decorations paved with forty-eight diamonds, a bezel embellished with thirty-six baguette-cut cognac sapphires, and leopard-patterned dial and straps. There was a successful mini collection known as Daytona 'Beach' in white gold, combined with a coloured strap; four pastel dial colours were available: blue turquoise, yellow or pink mother-of-pearl and green chrysoprase.

The Daytona's dial decoration has been an element of focus in creating new editions: there is the precious Daytona Rainbow, with a blend of multicoloured sapphires recreating the colours of the rainbow, as well as the

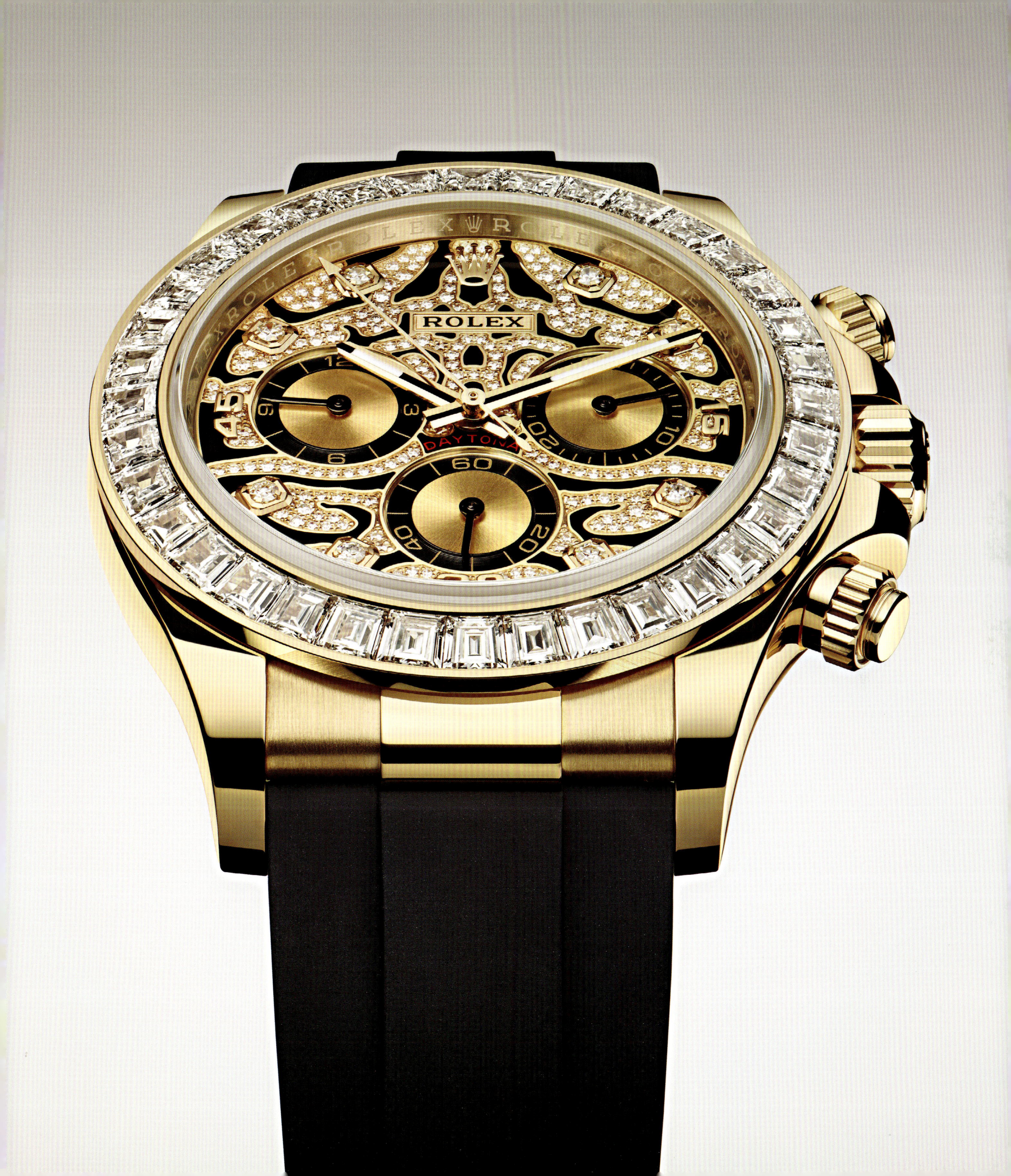
ROLEX
DAYTONA

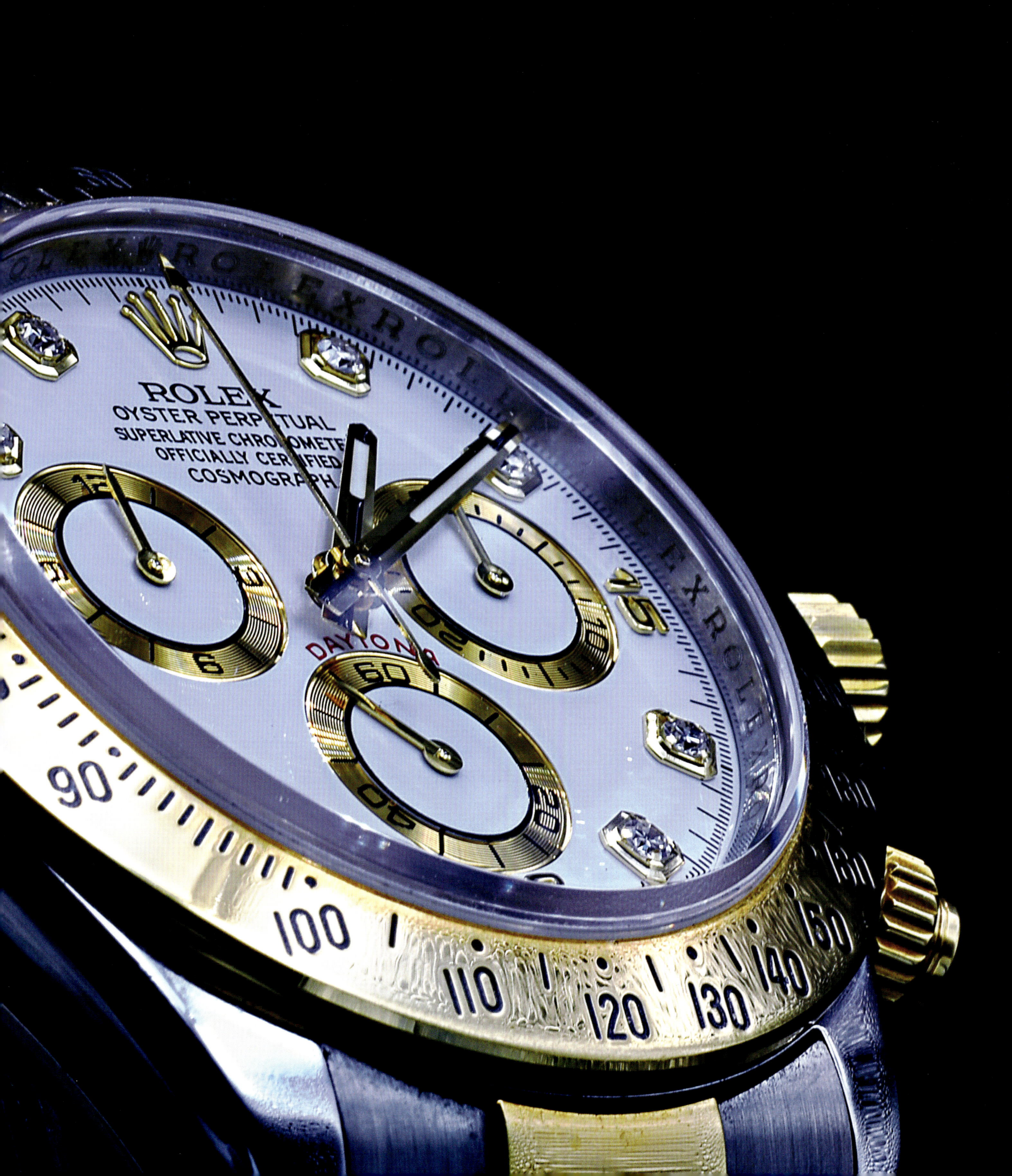
ROLEX
DAYTONA
90
100
110
120
130
140

reference 116515 in ceramic, admired for its harmonised contrast between its Everose case, the black Cerachrom bezel with a tachymeter scale engraved in pink gold, and the dial's warm ivory and chocolate tones.

The application of a ceramic bezel came in 2013 and was even used in the first platinum version of the Daytona. This was reference 116506, created to celebrate the model's fiftieth anniversary, which paired a brown bezel with an ice-blue dial. In 2014, a version was presented with a jewelled bezel, embellished with baguette diamonds, with a pave diamond dial.

In 2016, the reference 116500LN made its debut, differing from the previous version only in its ceramic bezel, paired with a black-and-white dial.

The most recent version of the Cosmograph Daytona is equipped with the calibre 4130, a self-winding chronographic movement entirely developed and manufactured by Rolex. It has a self-winding module with Perpetual rotor, guaranteeing that the mainspring is kept under continuous tension, providing a power reserve of about seventy-two hours. It also boasts a Superlative Chronometer certification.

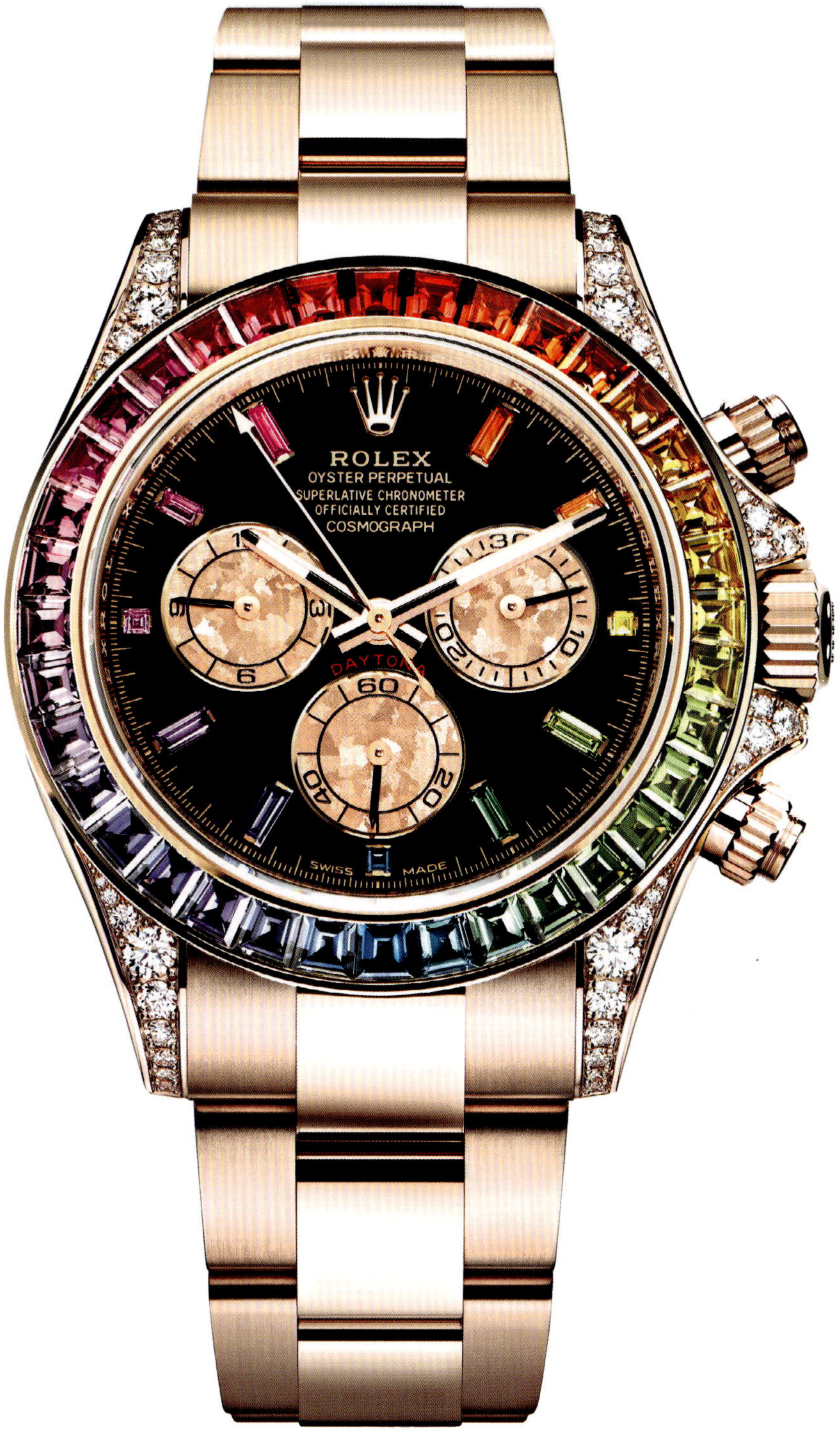

Daytona model in platinum, dubbed reference 116506.

The Daytona Rainbow model.

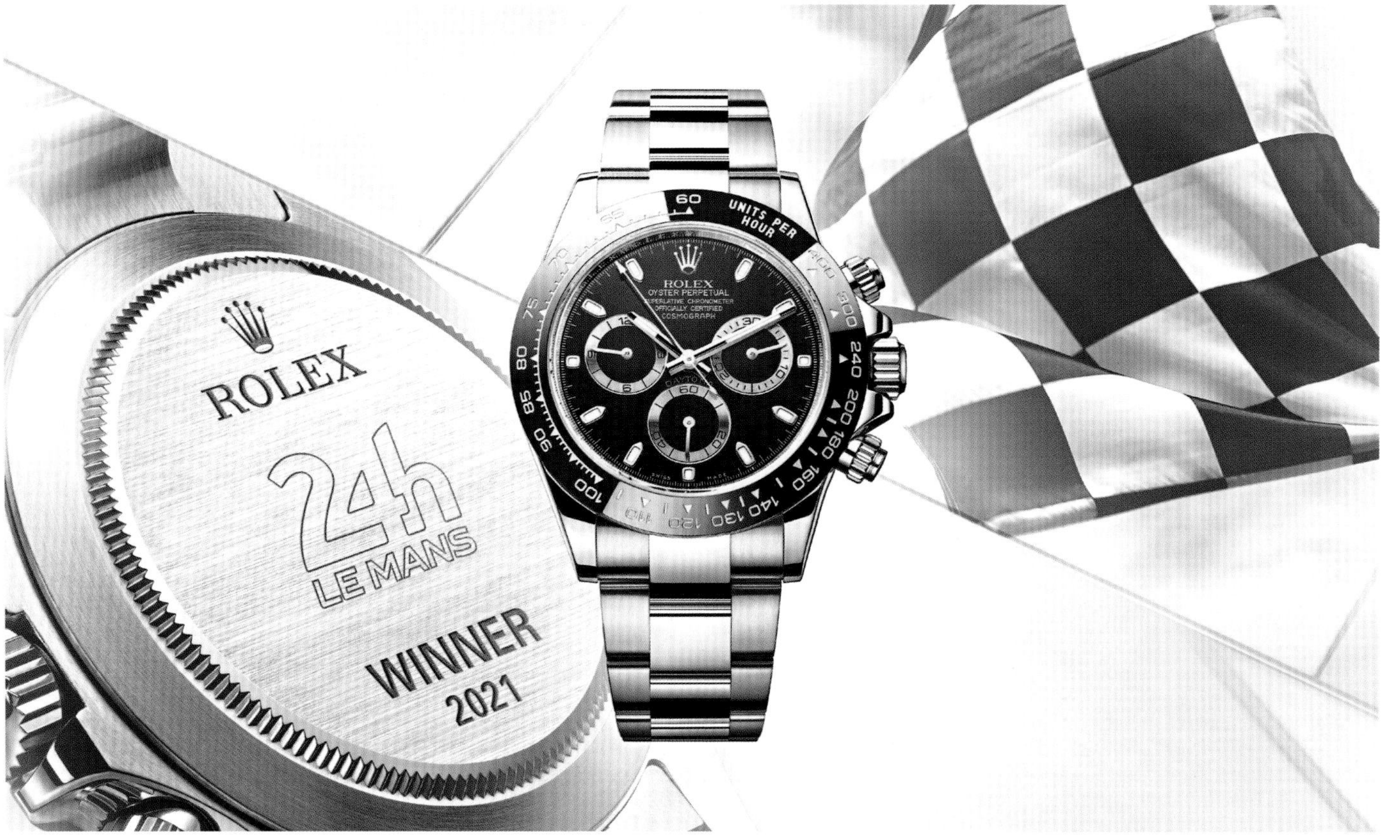

DIAL: The chronograph counters are vividly distinguished on the dial due to the strong contrast, between black on a light-coloured dial or a light tone on a black dial, for example. The tachymeter scale, or the graduated scale used to determine an average speed over a given distance using the chronograph's seconds hand, can be found along the bezel's circumference.

CASE: The Oyster case, with a 40-millimetre diameter, houses a high-performance mechanical chronograph movement that allows for measurement of a time interval or determination of average speeds up to four hundred miles or kilometres per hour. These additional functions are activated via its pushers, which, like the winding crown, screw down to the Oyster case.

BRACELET: Depending on the edition, the Cosmograph Daytona is combined with a three-piece-link Oyster bracelet or an Oysterflex bracelet, both fitted with an Oysterlock safety clasp. The two bracelets also feature the Easylink comfort extension link to easily adjust the length (by about five millimetres).

The Cosmograph Daytona's tachymeter scale on its metal or high-tech ceramic bezel allows for an average speed to be read over a given distance based on the measured time.

Thanks to this easily legible scale, the Cosmograph Daytona is the ideal instrument for determining speeds up to four hundred units per hour, whether it be kilometres, miles or any other unit of distance. Equipped with an additional mechanism that can measure and display a time interval via its secondary hands, the watch offers optimal legibility for all its functions. The chronometer's central seconds hand allows for a time to be measured to within one-eighth of a second. Its counters at 3 and 9 o'clock respectively indicate the time elapsed in minutes and hours.

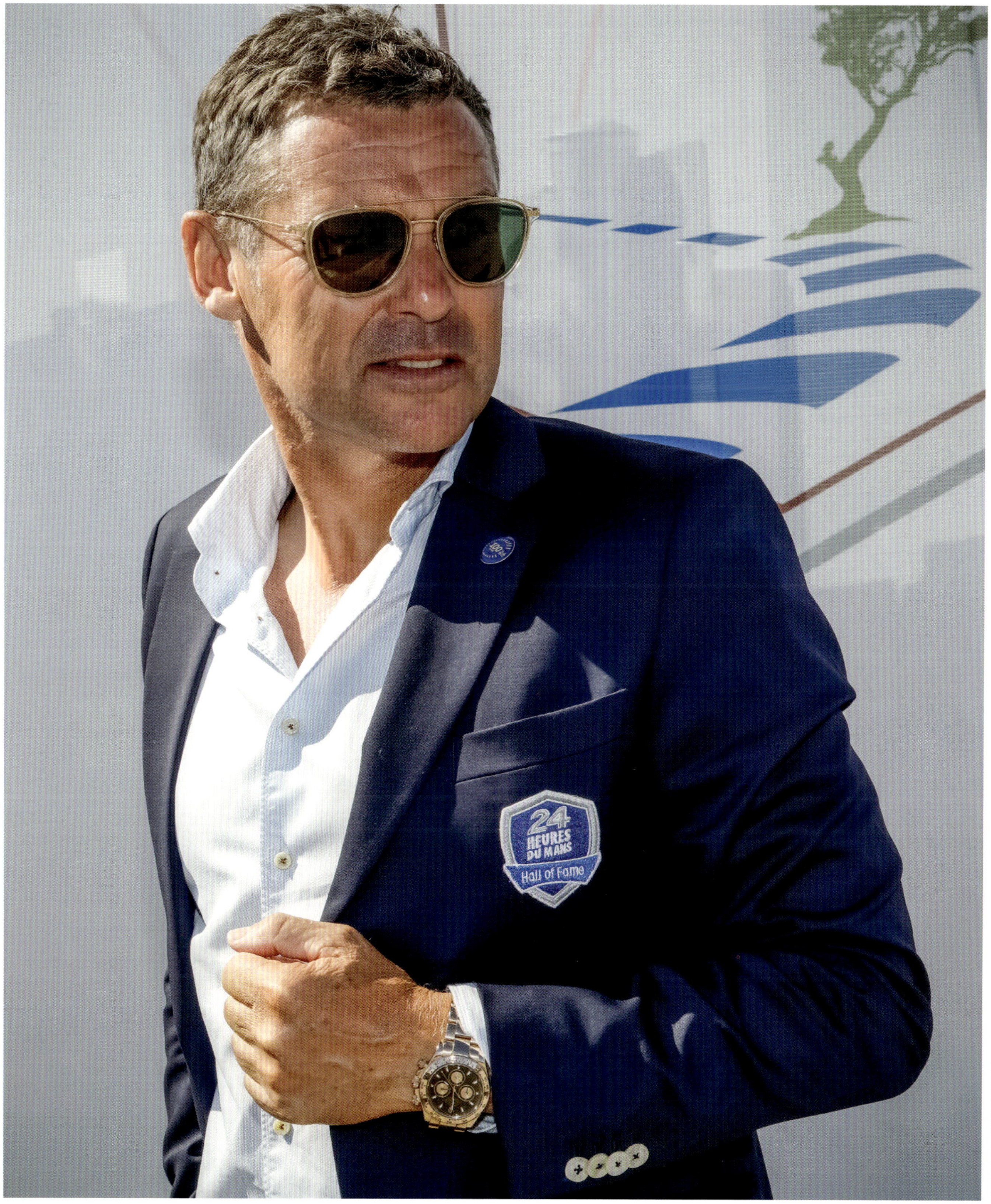

Former racecar driver Tom Kristensen, nine-time 24 Hours of Le Mans winner, wears a Rolex Cosmograph Daytona.

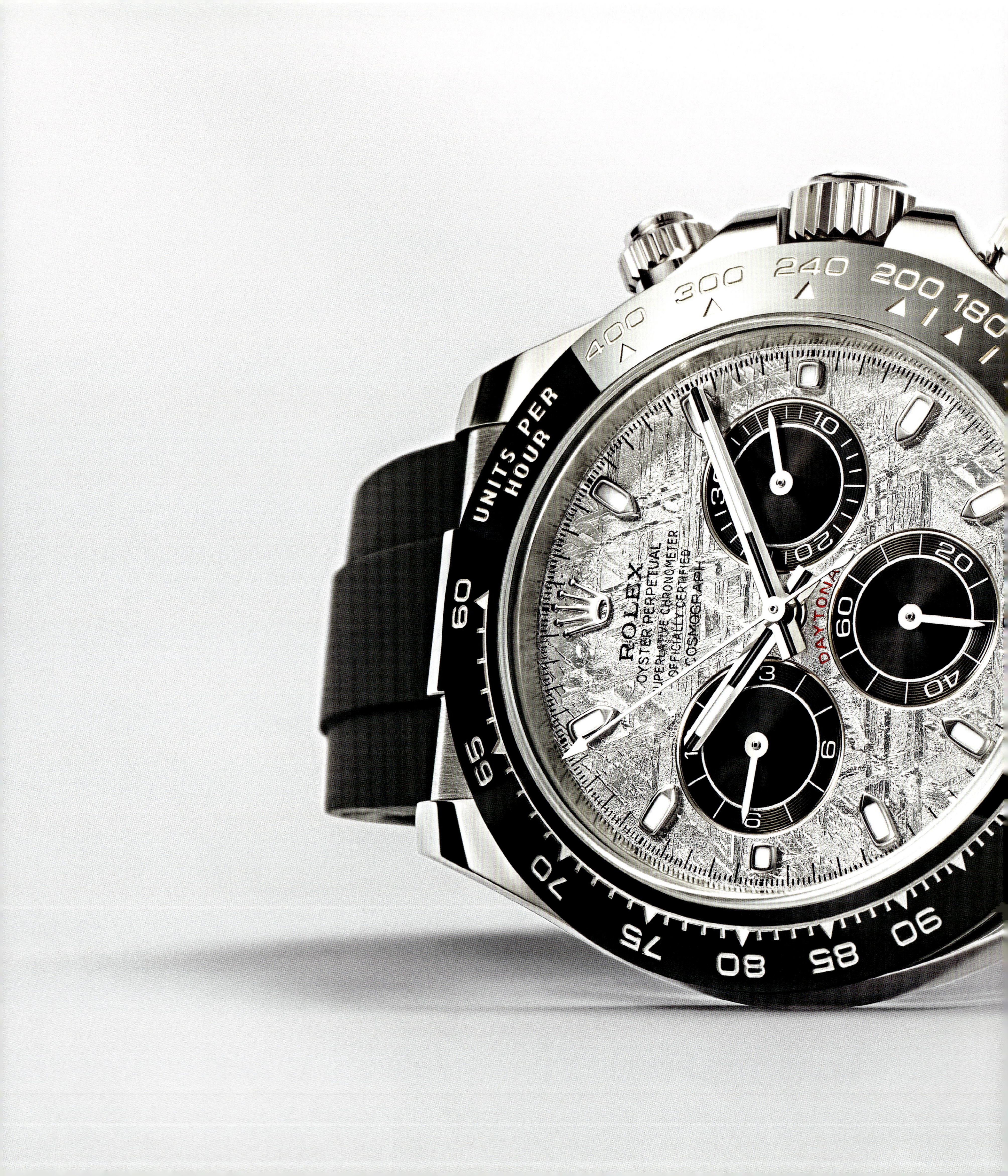

UNITS PER HOUR
ROLEX
OYSTER PERPETUAL
OFFICIALLY CERTIFIED
DAYTONA

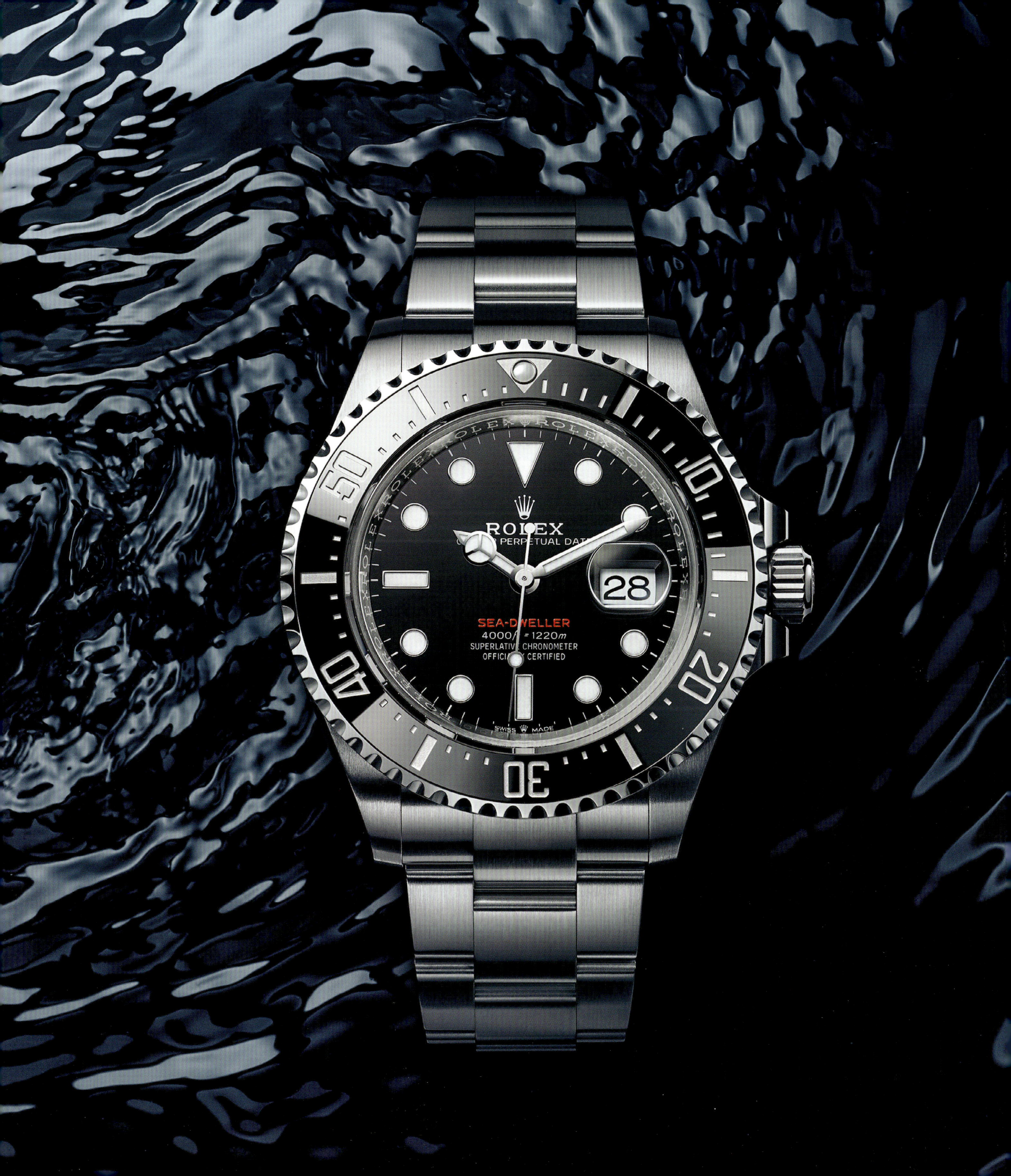
ROLEX
SEA-DWELLER
28
SWISS MADE

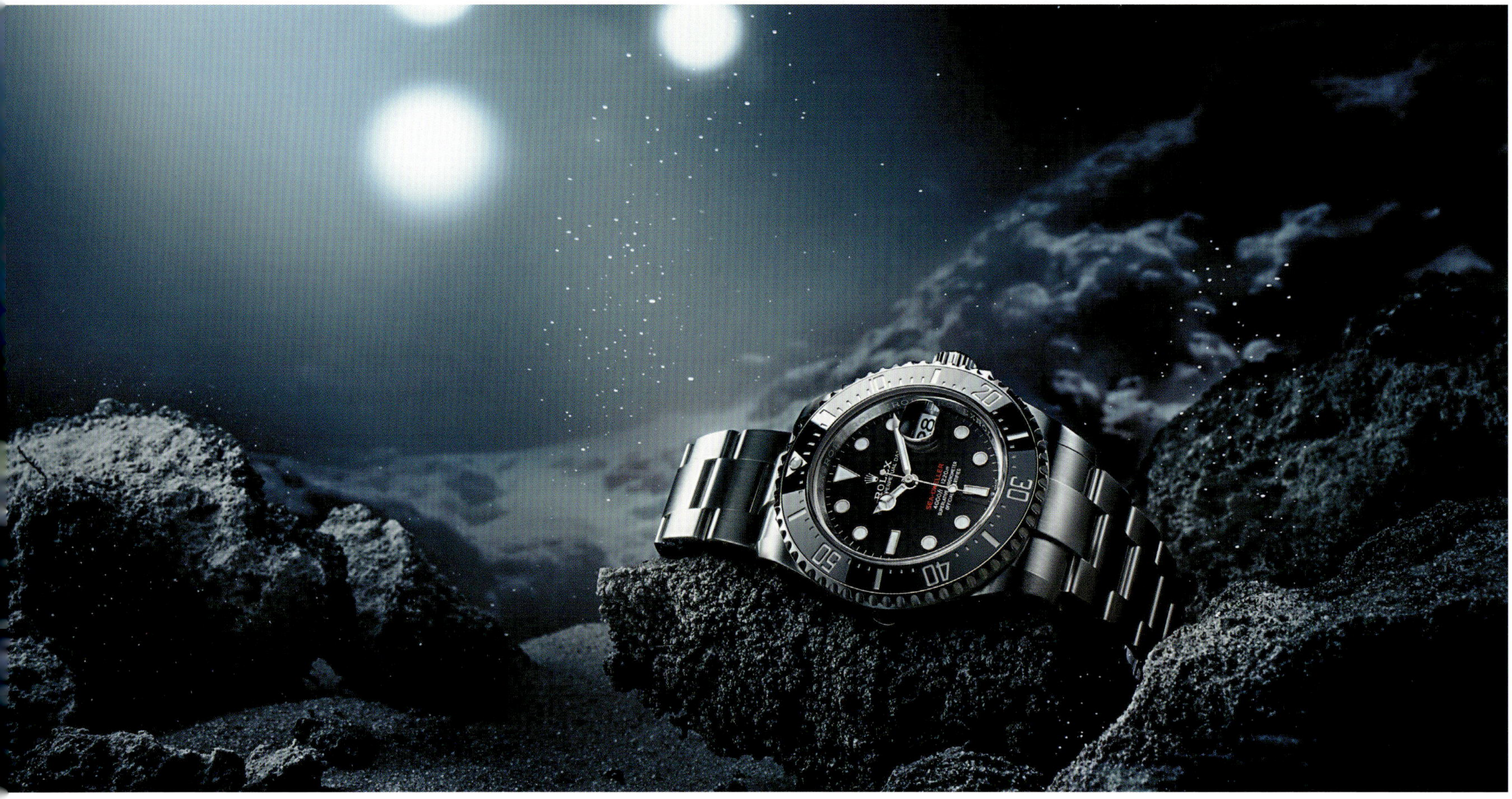

in temperature and is unaffected by magnetic fields. It is equipped with a Rolex overcoil, which ensures regularity in any position. This movement includes the patented Chronergy escapement, which is made of nickel-phosphorus, simultaneously guaranteeing high energy efficiency and extreme dependability, and is insensitive to magnetic fields. Calibre 3235 features the self-winding module with Perpetual rotor, which guarantees a power reserve of approximately seventy hours.

The Sea-Dweller is paired with a three-piece link Oyster bracelet fitted with a Rolex-designed and -patented Oysterlock safety clasp, which prevents any accidental opening and features the Rolex Glidelock extension system.

DIAL: The dial of the Sea-Dweller is glossy black, with indices in yellow and gold on the steel-and-gold reference 126603 and in white gold on the reference 126600. The former watch has 'Sea-Dweller' inscribed in gold, and the latter in red. The 1967 model had two lines of red text, reading 'Submariner 2000' underneath 'Sea-Dweller'. Some rare examples of the reference 1665 feature only one row of red lettering – dubbed 'Single Red' Sea-Dwellers. These are especially rare and sought after.

CASE: The Sea-Dweller is equipped with a valve on the left side of the case which allows helium to be expelled during the decompression phase of saturation diving, thus guaranteeing the integrity of the watch in even the most extreme of situations.

BRACELET: The Glidelock extension system on the Oyster bracelet has ten notches approximately two millimetres apart, allowing the length to be adjusted by approximately twenty millimetres. This way, the watch can be worn over a diving suit up to three millimetres thick.

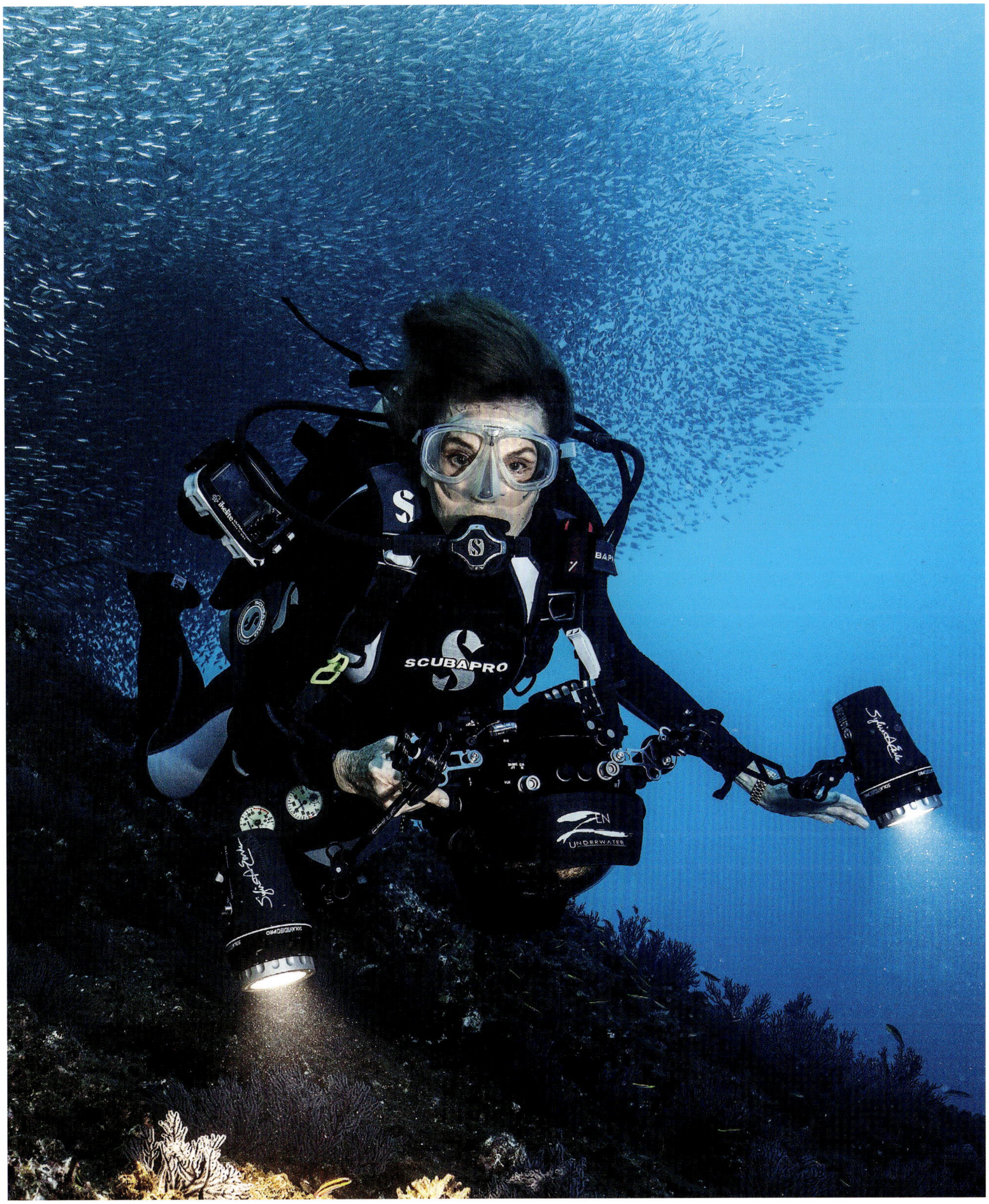

Sylvia Earle, marine biologist, a Rolex Testimonee since 1982 and an Explorer-in-Residence at the National Geographic Society since 1999. She wore a Rolex watch while on the 'Tektite II' mission, for which a group of scientists stayed in an undersea habitat for about fifteen days.

Explorer II

The Oyster Perpetual Explorer II reintroduces the Explorer's spirit of adventure and extreme durability as it integrates other functions especially useful for subterranean and polar exploration.

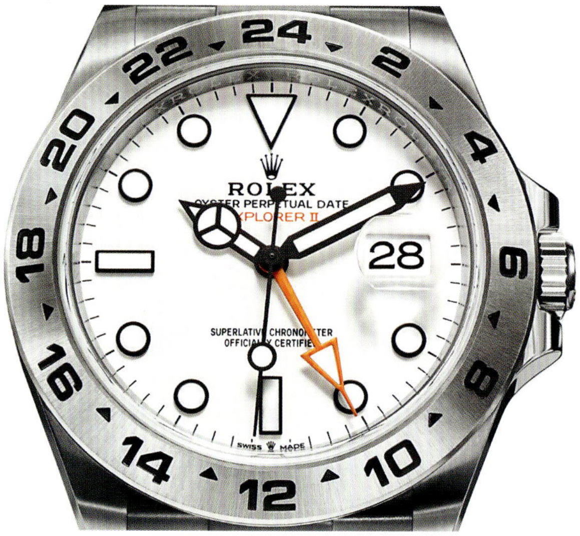

From the earliest decades of the last century, Rolex has accompanied explorers in their most daring exploits to discover the edges of the earth, individuals testing their own strength, contributing to the planet's preservation through their observations. Increasingly concerned about humans' impact on the balance between various ecosystems, these explorers have embarked on new expeditions, not only dictated by a thirst for new discovery but seeking to raise awareness surrounding the fragility of the planet as well. Driven by their desire to face ever more extreme challenges, these figures' stories made their mark in the record books. Many explorers, mountaineers and scientists have donned an Explorer or an Explorer II and become a 'Rolex Testimonee'.

Introduced in 1971, the Explorer II stood out in the world of exploration for its functions, as well as its ability to withstand more extreme conditions. It thus became the watch of polar explorers, speleologists and vulcanologists. The watch dial hosts a new orange hand. Pointing to a fixed graduated bezel in 24-hour format, this hand allows for daytime and nighttime hours to be clearly distinguished from one another, a function essential for cave explorers, who spend entire days without seeing sunlight, as well as for members of polar expeditions, who have to contend with the midnight sun or polar night. When it is not being used to tell between day and night, the 24-hour hand can be used with the graduated bezel to read the time in another time zone (a reference time). The Oyster Perpetual Explorer II is also equipped with a date display, magnified by a Cyclops lens on the crystal.

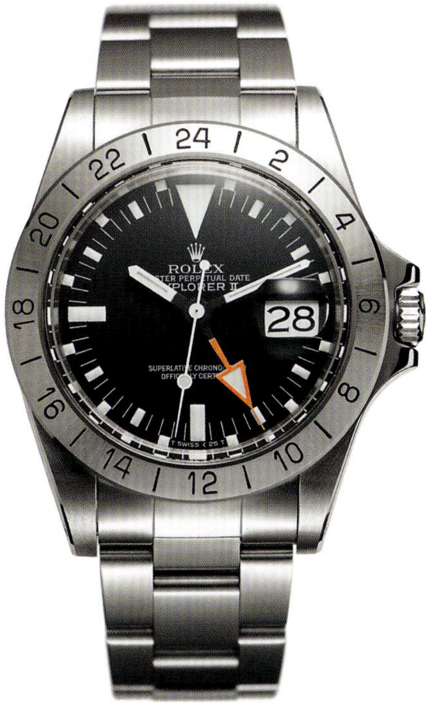

The Explorer II model, version with a black dial.

One of the figures that used an Explorer II in the course of their ventures in exploration or scientific research was Kenyan paleoanthropologist and biologist Richard Leakey. In the 1980s, Leakey and his team unearthed about four hundred pieces of hominid bone in the Koobi Fora region in northern Kenya; this included the Turkana Boy, whose skeleton is one of the most complete ever reconstructed, a discovery that proved to be of capital importance. Leakey also distinguished himself for his commitment to responsible environmental stewardship in East Africa. 'Saving our wildlife is not only an environmental necessity, it's a cultural one as well', the paleoanthropologist stated in a 1991 Rolex ad.

An Explorer II accompanied the Norwegian explorer Erling Kagge in his conquest of the 'three poles', crossing both poles on foot and climbing Mount Everest. The words 'North Pole 1990, South Pole 1992–3, Mt Everest 1994' are engraved on his Explorer II case back.

Since the 1980s, Belgian polar explorer Alain Hubert has taken part in various mountaineering and polar expeditions to the North and South poles, and in 2002 he created the International Polar Foundation to support polar science as key to understanding major changes in climate.

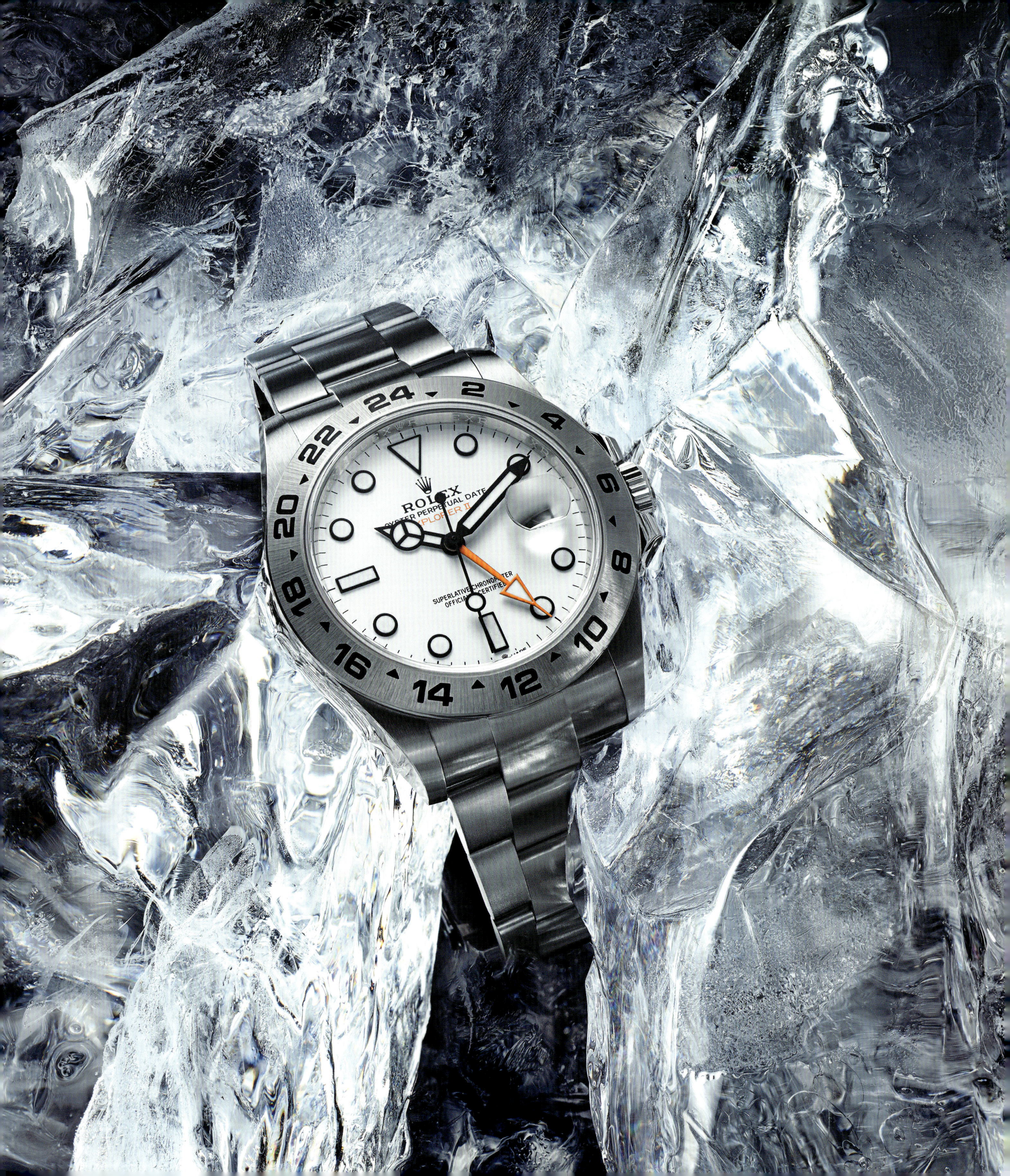
ROLEX

He never sets out on an expedition without a satellite phone, a topographic map and an Explorer II. His 2007 Arctic Arc expedition was the challenge of a lifetime: a strenuous trek, from Siberia to Greenland via the North Pole, to study the effects of climate change.

As part of his Endeavor 8000 project, completed in 2005, American Ed Viesturs climbed all fourteen of the planet's 'eight-thousanders' without supplemental oxygen. To Viesturs, the wristwatch is a central part of his equipment because of its significant contribution to the success and safety of an ascent – when time management directly influences not only one's ability to complete the climb but also, in many cases, one's chance of survival. Ed Viesturs wears the same Explorer II with white dial for all his ascents. 'It has never failed me and includes all the features I need for mountaineering: it's self-winding, robust, and the hands are easy to read against the dial, even in the dark. A durable crystal is also an advantage, as it may hit rock and ice on an ascent. I have to admit it's probably the most important piece of equipment I have with me. I am a serious clock-watcher during my climbs. My watch – and the time it tells – is the key to my safety.'

The Explorer model 1655, introduced with a 40-millimetre case, was the first reference created in this family of Rolex watches. It was often referred to by the nickname 'Big Arrow', thanks to the large orange arrow-shaped hand telling the time according to a 24-hour system.

The reference 16550, introduced in the mid-1980s, was only available in steel but with a dial in two colour variations, black or white. The indices of the watch had white-gold surrounds and its Mercedes hands were of the same material. The fixed bezel featured an updated graphic design with square, squat numerals alternating with small triangular indices. The 24-hour hand, meanwhile, exchanged its 'big arrow' for a shape already introduced in the GMT-Master.

In 1989, Rolex introduced the Explorer II reference 16570. It inherited many of the changes made in the reference 16550 and was equipped with the calibre 3185. Its dial was available in black or white. The white dial's indices and hands changed from silvered to

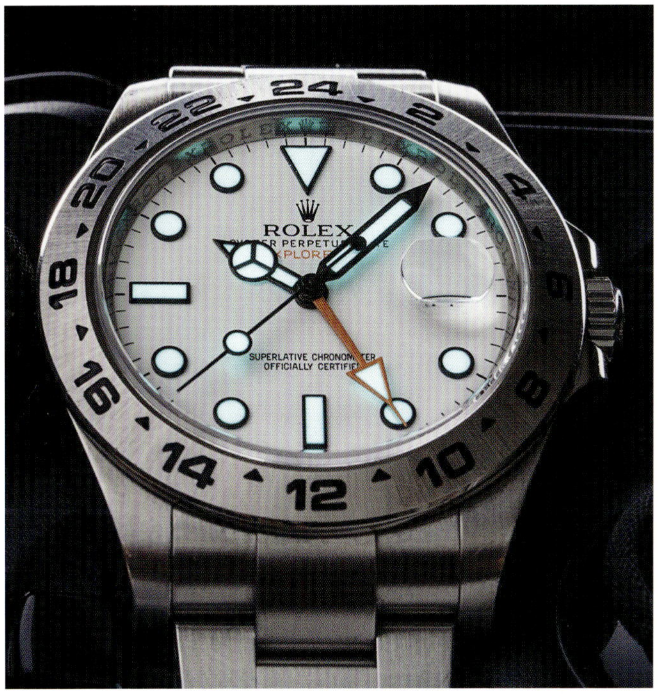

The dial of the Explorer II model, with a fixed bezel and renewed graphics: square numerals alternating with small triangular indices and a 24-hour hand in the shape of a 'big arrow'.

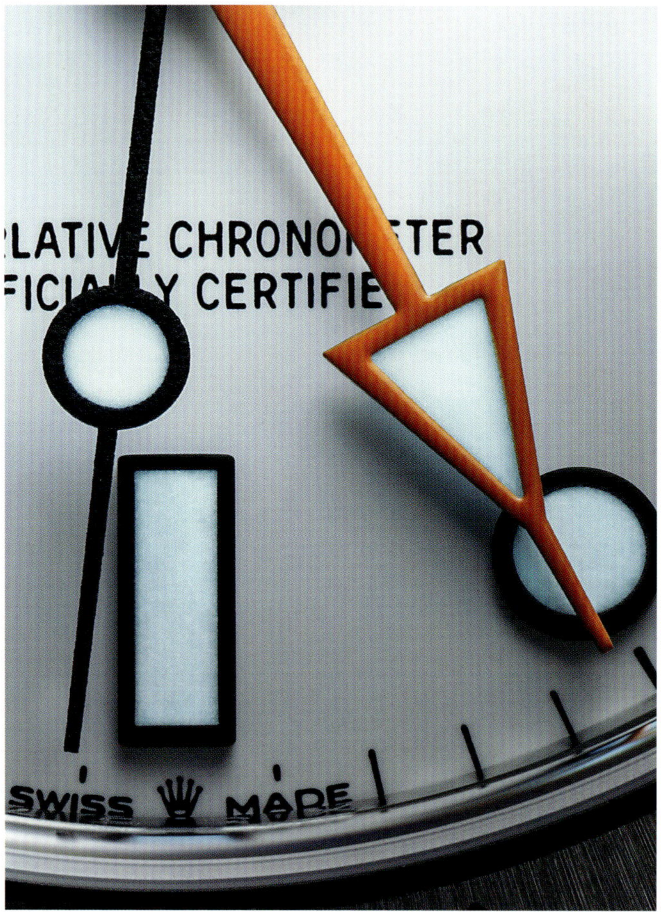

DIAL: The Explorer II has a date display and is differentiated by the extra orange 24-hour hand and fixed 24-hour graduated bezel, allowing for daytime to be clearly distinguished from nighttime. Its indices and hands feature a highly legible Chromalight display with long-lasting blue luminescent material. When the Explorer II's additional hand is not used together with the graduated bezel for its 24-hour time display, it can be used to reference another time zone.

CASE: The Explorer II's 42-millimetre fully waterproof Oyster case provides optimum protection for its movement.

BRACELET: The Oyster bracelet features an Oysterlock folding safety clasp, which prevents accidental opening, and the Easylink comfort extension link, another Rolex exclusive. The ingenious system allows the wearer to easily adjust the bracelet's length by approximately 5 millimetres for greater comfort in all circumstances.

black, heightening the contrast and improving visibility.

In 2011, Rolex produced the reference 216570 with a 42-millimetre case, which re-introduced the orange 'big arrow' 24-hour hand in honour of the reference 1655.

The most recent version of the Explorer II boasts a Superlative Chronometer certification. The Explorer II's Oyster case, with a 42-millimetre diameter, is guaranteed waterproof to a depth of 100 metres. The middle case is produced from a solid block of Oystersteel, an alloy especially resistant to corrosion. The finely fluted case back is hermetically screwed down using a dedicated instrument, and its Twinlock winding crown, fitted with a double waterproofing system and protected by an integrated crown guard, screws down firmly against the case. The crystal, with a Cyclops lens above 3 o'clock for easy reading of the date, is made of scratch-resistant sapphire, benefiting from an anti-reflective coating. The Explorer II is equipped with the calibre 3285 with a Chronergy escapement, which lengthens the power reserve from forty hours to seventy. This mechanical self-winding movement is entirely developed and manufactured by Rolex, boasting numerous patents and ensuring high performance in precision, autonomy, shock resistance and reliability.

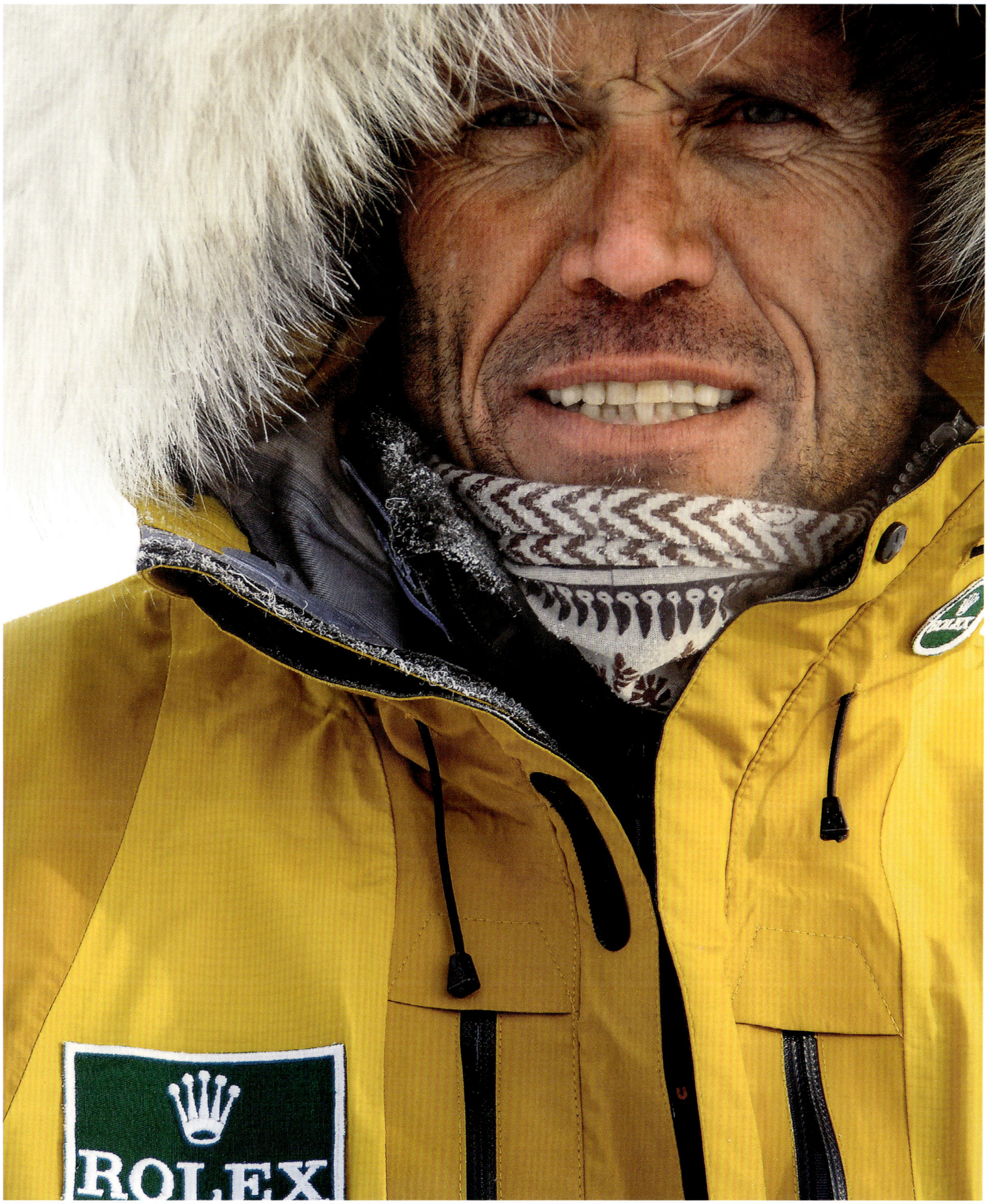

Alain Hubert is a Belgian explorer and Rolex Testimonee. His many achievements include the longest crossing of Antarctica, a journey of more than 4,000km. Throughout his various explorations of the poles, Hubert has appealed to science to bring attention to the devastating effects of global warming.

GMT-Master II

A cosmopolitan watch, the Oyster Perpetual GMT-Master II has an hour hand that can be adjusted independently of its minute and 24-hour hands.

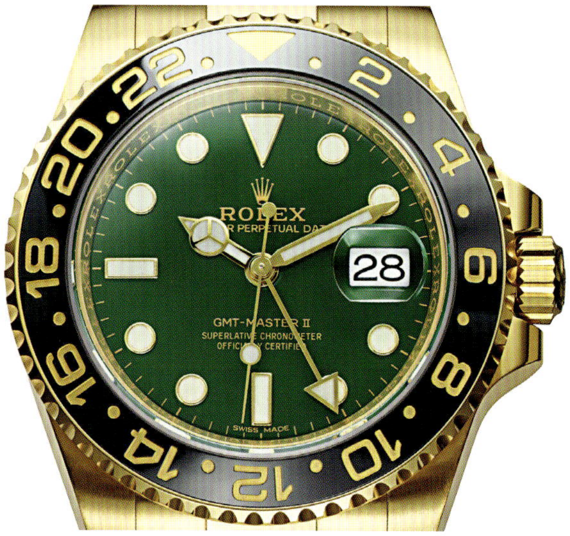

In 1982, Rolex introduced a new calibre which allowed the hour hands to be regulated independently from the 24-hour hand. To emphasise this development and prevent any confusion with the GMT-Master, the model equipped with the optimised movement was dubbed the GMT-Master II.

Complementing traditional hours, minutes and seconds hands displaying the local time, the GMT-Master II exhibits an additional hand with a triangular tip that travels around the dial over the course of 24 hours and a bi-directional rotatable bezel with a two-tone 24-hour graduated monobloc Cerachrom ceramic insert. The supplementary hand indicates the reference time (at home, for example), which is consulted via the graduated bezel when at a neutral position, with the triangle positioned at 12 o'clock.

The local time can be easily adjusted while traveling, jumping from hour to hour, using a mechanism controlled via the winding crown.

The hour hand can be adjusted independently of the minute and seconds hands without impacting the 24-hour hand (or reference time display). This way, a traveller would be able to tell the local time and the reference time at any moment, simultaneously.

The time in an additional time zone could also be told with a turn of the rotatable bezel. In this case, the reference time is no longer indicated and the 24-hour hand is used to read the new time zone on the bezel's graduation.

So that the bezel would perform perfectly, Rolex developed an innovative rotation system equipped with a spring which allows for crisp, symmetrical shifts in either direction, from one position on the 24-hour graduation to another. Thanks to its knurled edge, providing excellent grip, the bezel can be turned with ease.

The GMT-Master II presents one of the few complications that still has practical relevance in the modern era. For those doing business on an international level or communicating with loved ones far from home, navigating time zones is a highly tangible aspect of everyday life.

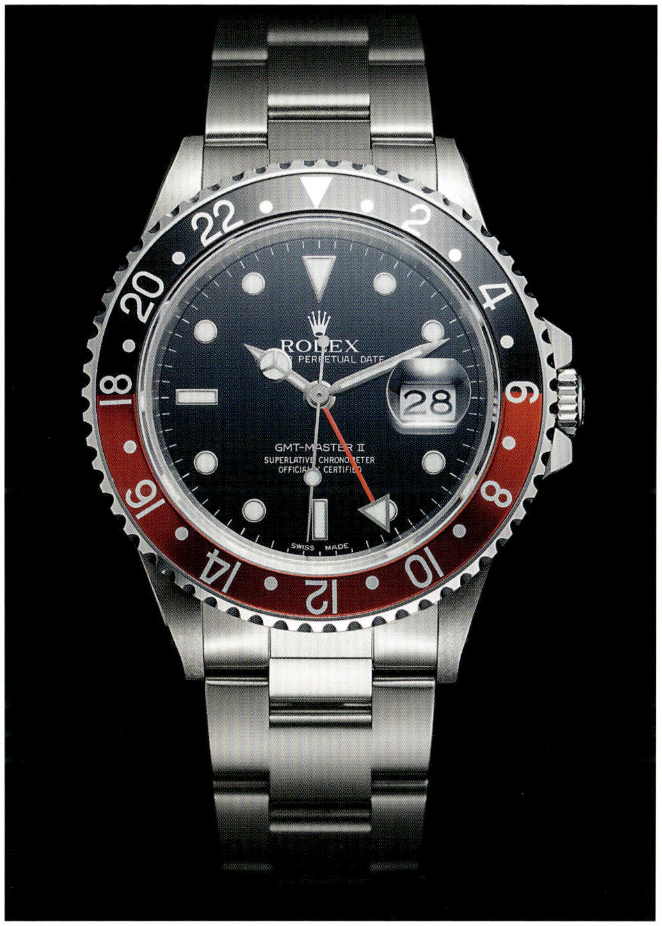

Steel version of the Oystersteel GMT-Master II model, with a two-tone monobloc insert in red and black ceramic.

2013 marked a milestone for Rolex in employing ceramics for its watchmaking when it presented, in an Oystersteel version of the GMT-Master II, a two-tone blue-and-black monobloc ceramic insert. One year later, the brand produced a two-tone ceramic insert in red and blue for an 18-carat white-gold edition of the GMT-Master II. The two-toned insert in brown and black was produced by Rolex in 2018 for two editions of the GMT-Master II, the first in 18-carat Everose gold and the second in a Rolesor Everose (a pairing of Oystersteel and Everose gold). A recent edition of the GMT-Master II is equipped with a two-toned ceramic disc in green and black, with a winding crown on the left side of the case and a date window at 9 o'clock, an unprecedented configuration in a Rolex watch.

The most recent version of the GMT-Master II boasts a Superlative Chronometer certification. The Oyster case has a 40-millimetre diameter and is guaranteed waterproof to a

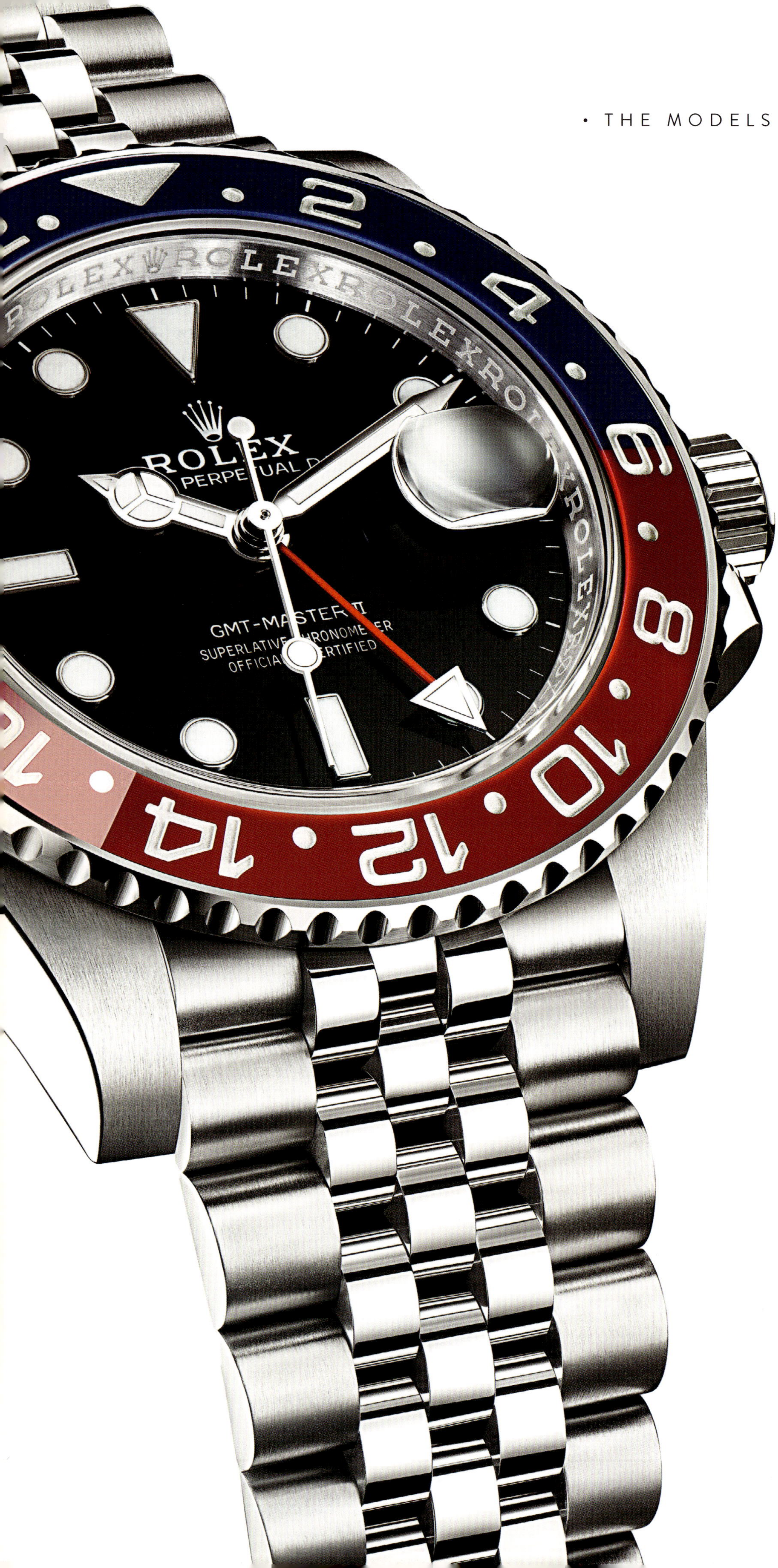

In 2014, the brand introduced the red-and-blue ceramic Cerachrom insert on a white-gold GMT-Master II.

depth of 100 metres. The watch is equipped with the calibre 3285, a self-winding mechanical movement entirely developed and manufactured by Rolex, guaranteeing high performance in terms of precision, autonomy, shock resistance, convenience and reliability. The watch also benefits from a power reserve of about seventy hours, thanks to its barrel architecture and the superior efficiency of its escapement.

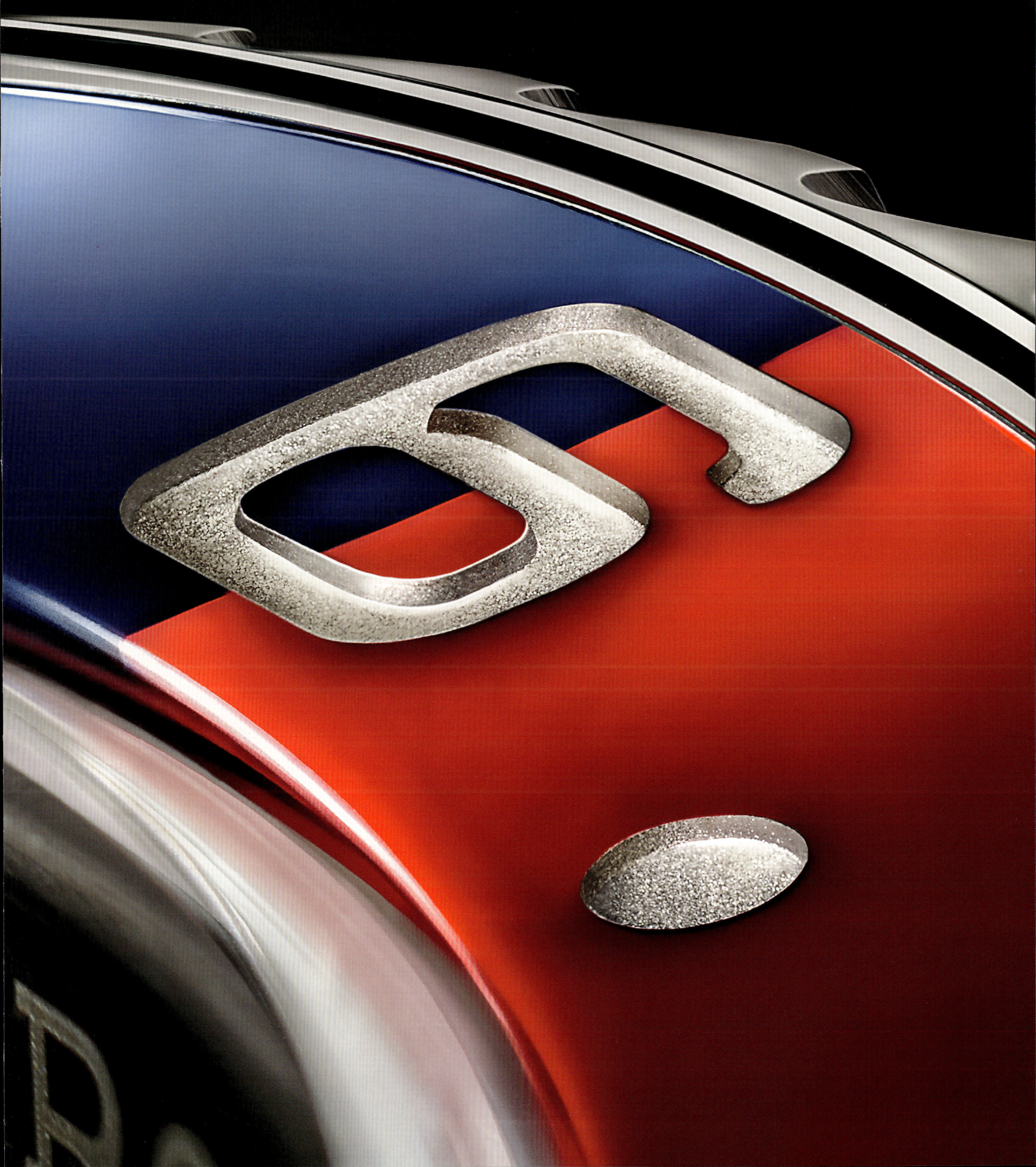

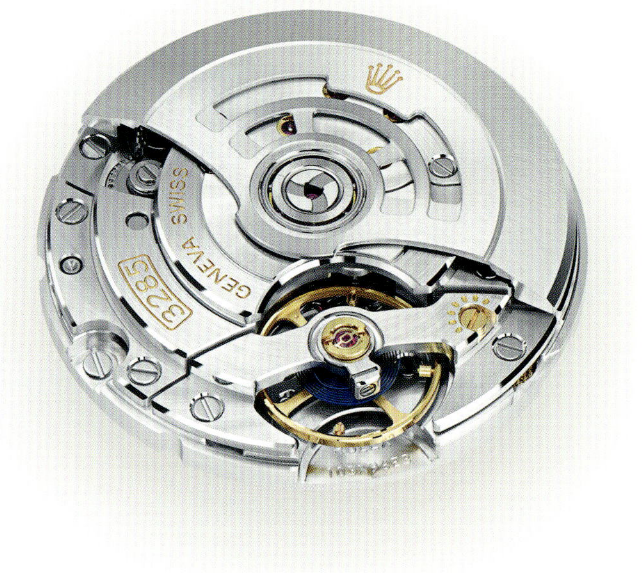

The bidirectional rotatable bezel of the GMT-Master II.

DIAL: The dial is offered in black or blue lacquer with hands and indices in white gold and a lacquered 24-hour hand, all highly visible thanks to its Chromalight display (with a long-lasting blue luminescent glow).

CASE: The GMT-Master II's rotatable bidirectional bezel is equipped with a two-coloured 24-hour graduated monobloc Cerachrom insert. This insert is patented and realised in an especially hard scratch-resistant ceramic, whose colours remain unaltered by ultraviolet rays. Due to their chemical composition, these high-tech ceramics are inert and cannot corrode. The recessed graduation and numerals are coated with gold or platinum via a PVD (Physical Vapor Deposition) procedure. Apart from its functional and technical aspects, the bezel on the GMT-Master II fully contributes to its distinct visual identity.

BRACELET: The modern version of the GMT-Master II is paired with a three-piece-link Oyster bracelet or a five-piece-link Jubilee bracelet, uniquely offered with Oystersteel versions. A concealed attachment system ensures seamless visual continuity between the bracelet and the case.

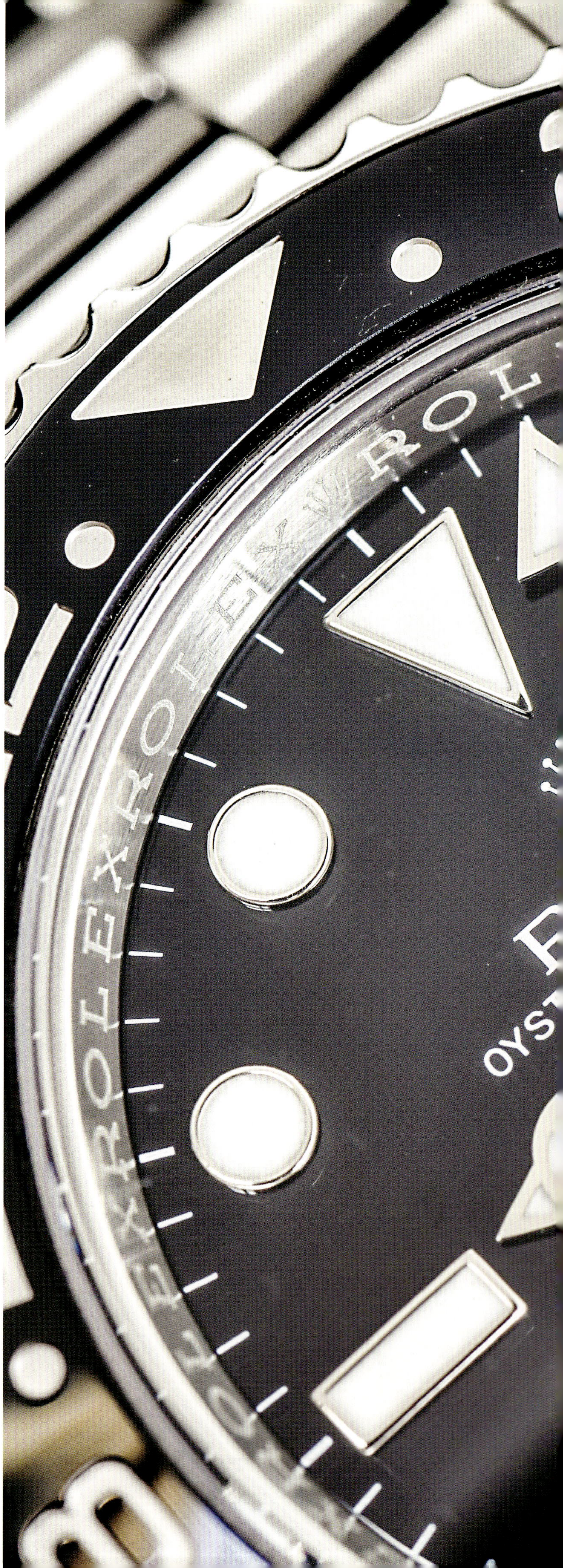

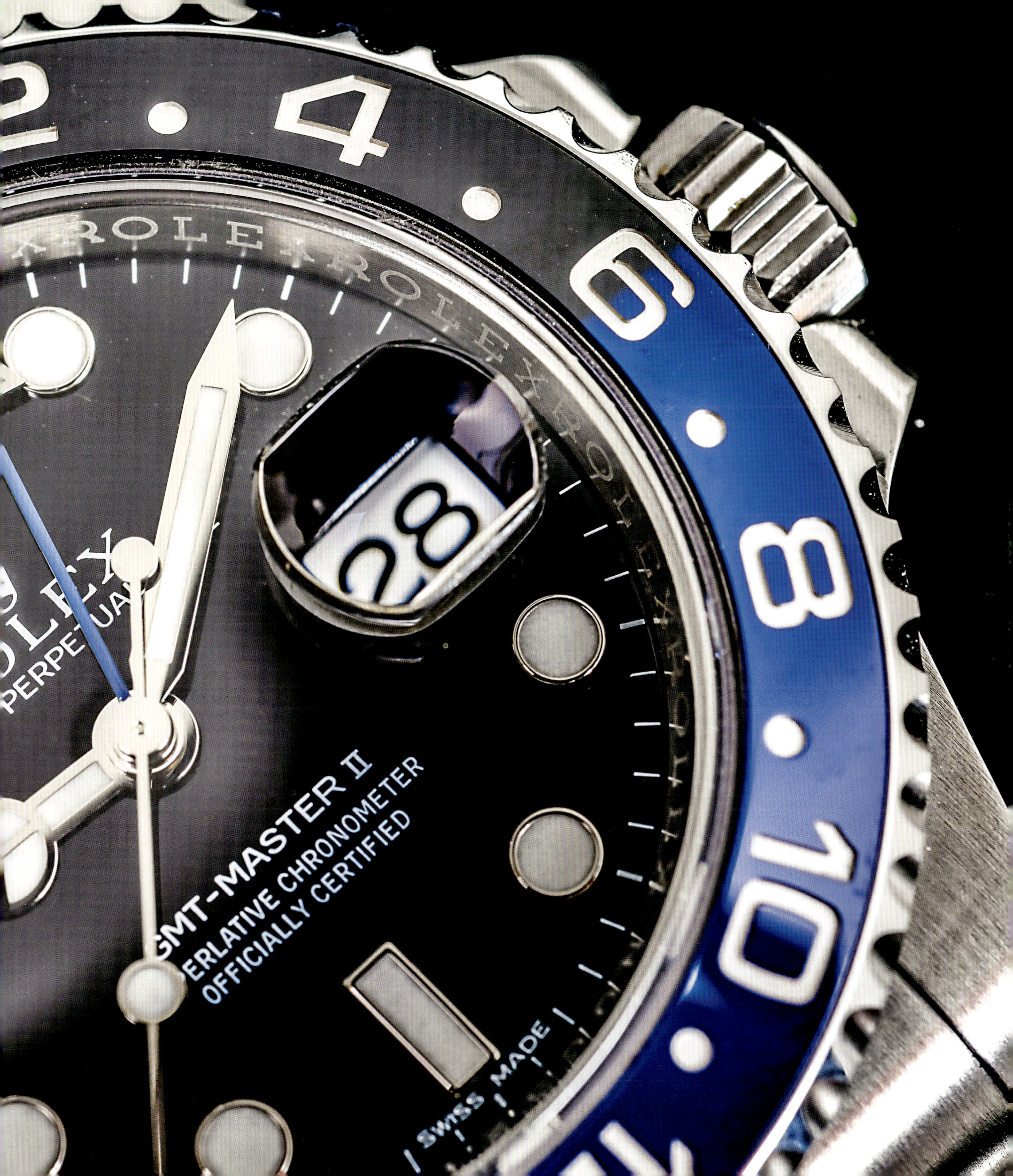
GMT-MASTER II
OFFICIALLY CERTIFIED
SWISS MADE

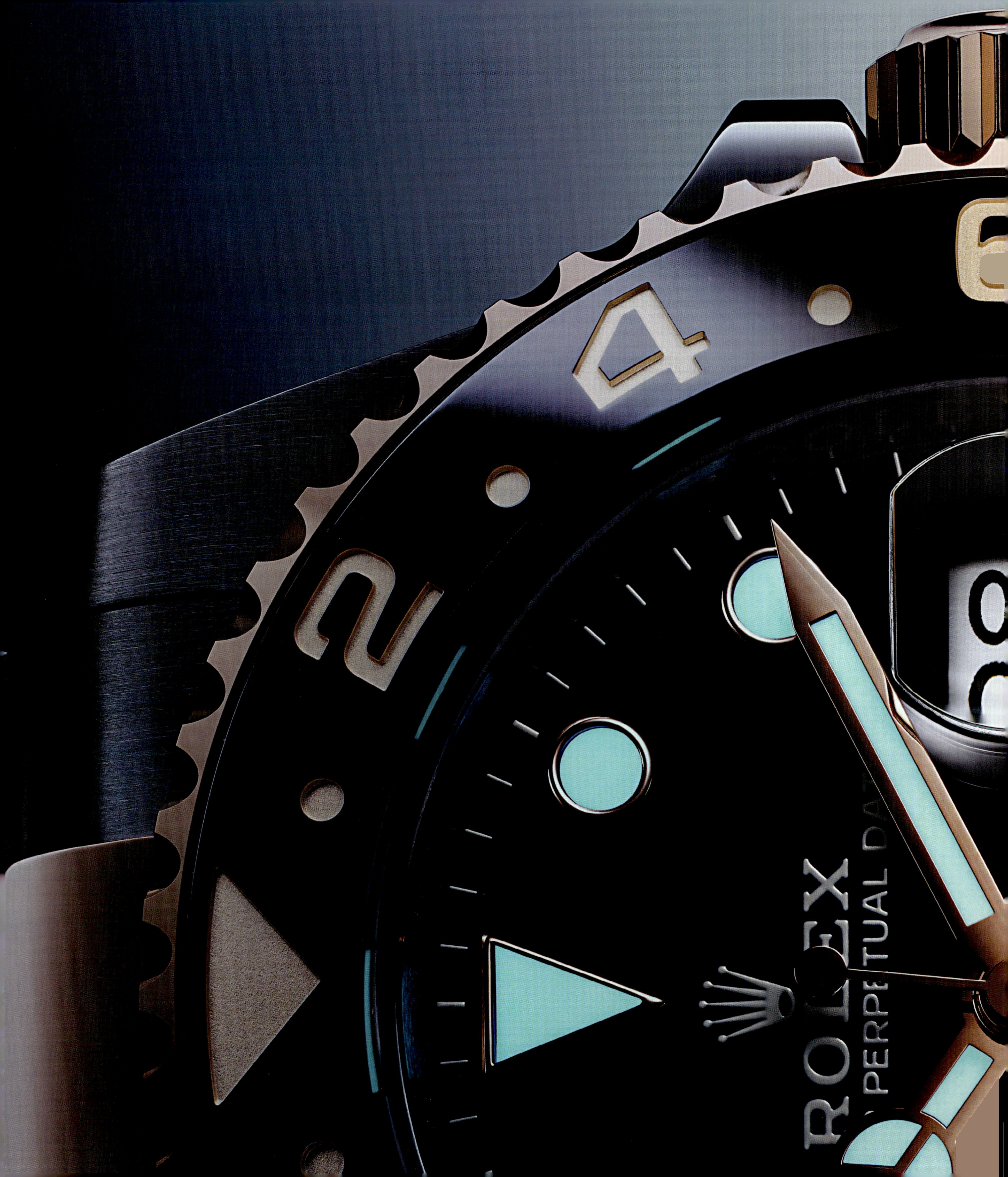
ROLEX
PERPETUAL

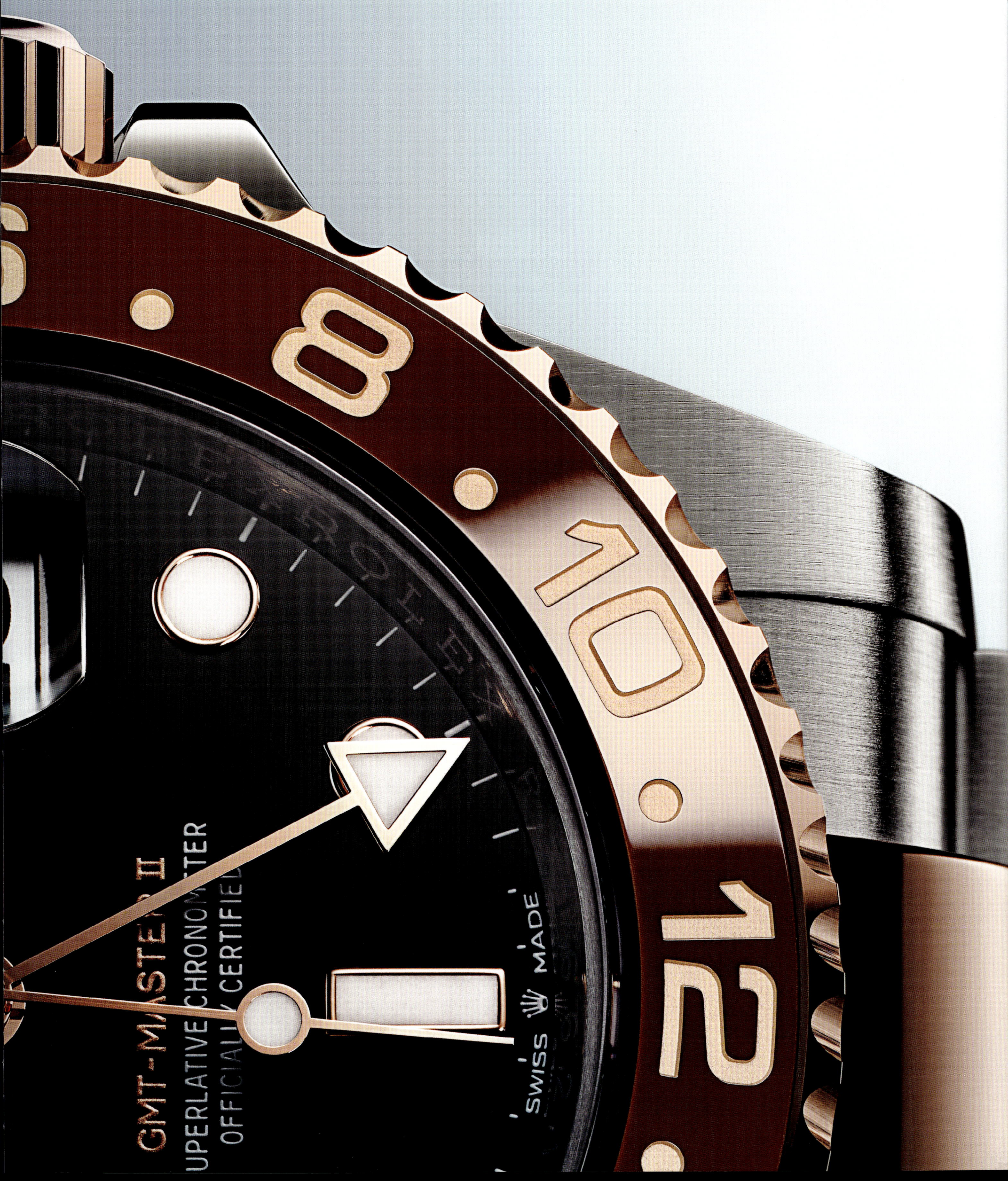
SWISS MADE

Yacht-Master

The Oyster Perpetual Yacht-Master embodies the close ties connecting Rolex and the world of sailing since the 1950s.

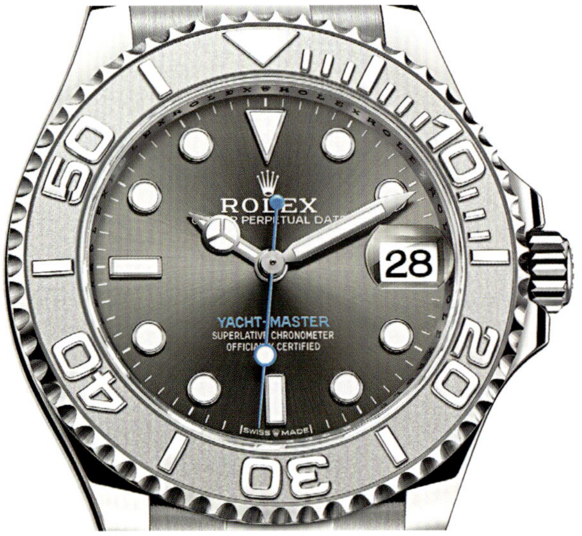

Thanks to their water resistance and chronometric precision, over time Oyster watches have become valued tools for sailing. In 1967, Sir Francis Chichester was the first yachtsman to complete a solo voyage around the globe. He received a hero's welcome upon his return and was knighted by the Queen of England. In a letter from 1968, he wrote: 'During my voyage around the world in *Gipsy Moth IV,* my Rolex watch was knocked off my wrist several times without being damaged. I cannot imagine a hardier timepiece. When using [it] for sextant work and working the foredeck, it was frequently banged, also doused by waves coming aboard; but it never seemed to mind all this.'

Based on these experiences, Rolex developed a model that was suited to travel with sailors.

Presented in 1992, the Yacht-Master was intentionally designed for sailors and skippers. The watch was instantly recognisable for its bidirectional rotatable bezel with raised 60-minute graduation, produced entirely in precious metals (gold or platinum) or fitted with a high-tech Cerachrom ceramic insert. The polished graduation and numerals stand out against their matt background. The characteristic functional bezel allows the wearer to calculate an interval of time – how long it takes to cover the distance between two buoys, for example – fully contributing to the timepiece's unique identity. Thanks to its knurled edge, providing excellent grip, the bezel can be controlled easily.

The Yacht-Master guarantees optimal legibility in every circumstance, including the dark, thanks to its unique Chromalight display: the large indices and broad hands are respectively filled and coated with a long-lasting luminescent material (in the colour blue), which glows up to two times longer than classic phosphorescent materials.

The Yacht-Master's Oyster case, either 37, 40 or 42 millimetres in diameter, is guaranteed waterproof to a depth of 100 metres. The Triplock winding crown, with a triple waterproofing system and protected by an integral crown guard, screws down firmly against the case. The crystal, with a Cyclops lens above 3 o'clock for easier reading of the date, is in scratch-resistant sapphire and benefits from an antireflective coating. Yacht-Master models are equipped with the calibre 2236 (Yacht-Master 37) or calibre 3235 (Yacht-Masters 40 and 42) self-winding mechanical movements entirely developed and produced by Rolex.

The oscillator in calibres 2236 and 3235 has

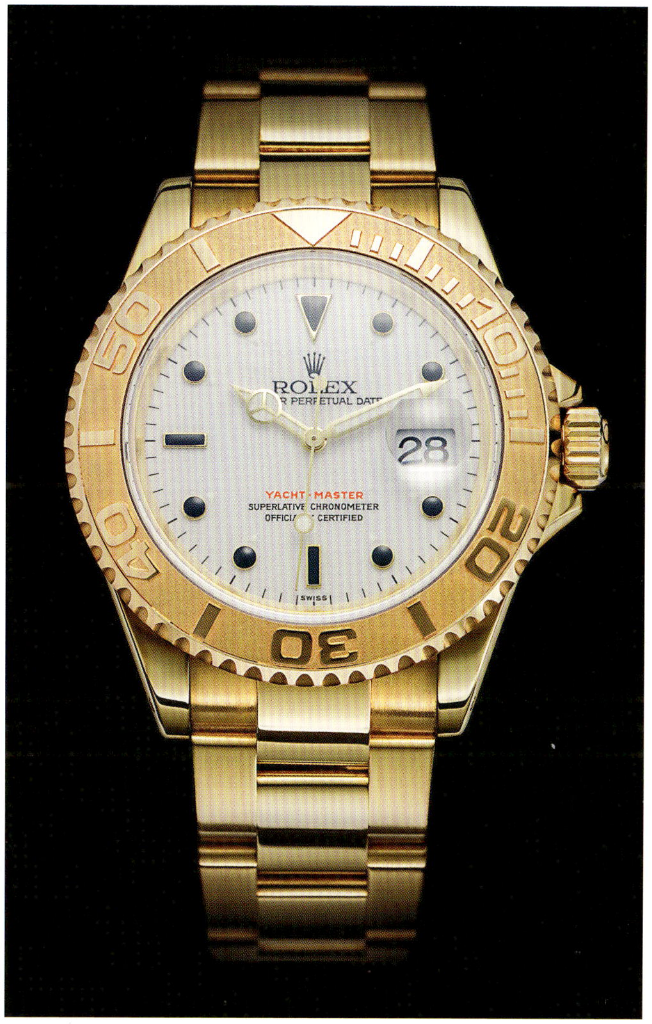

The Yacht-Master model with a bidirectional rotatable bezel with raised 60-minute graduation, made entirely of gold.

a balance wheel with variable inertia, calibrated with extreme precision by Microstella nuts. It is securely held in place by a height-adjustable traversing bridge for extremely stable positioning and better shock resistance. The oscillator is additionally fitted on high-performance Rolex-developed and -patented Paraflex shock absorbers. Calibre 2236 is equipped with a Syloxi silicon hairspring, also patented and produced by the brand, while calibre

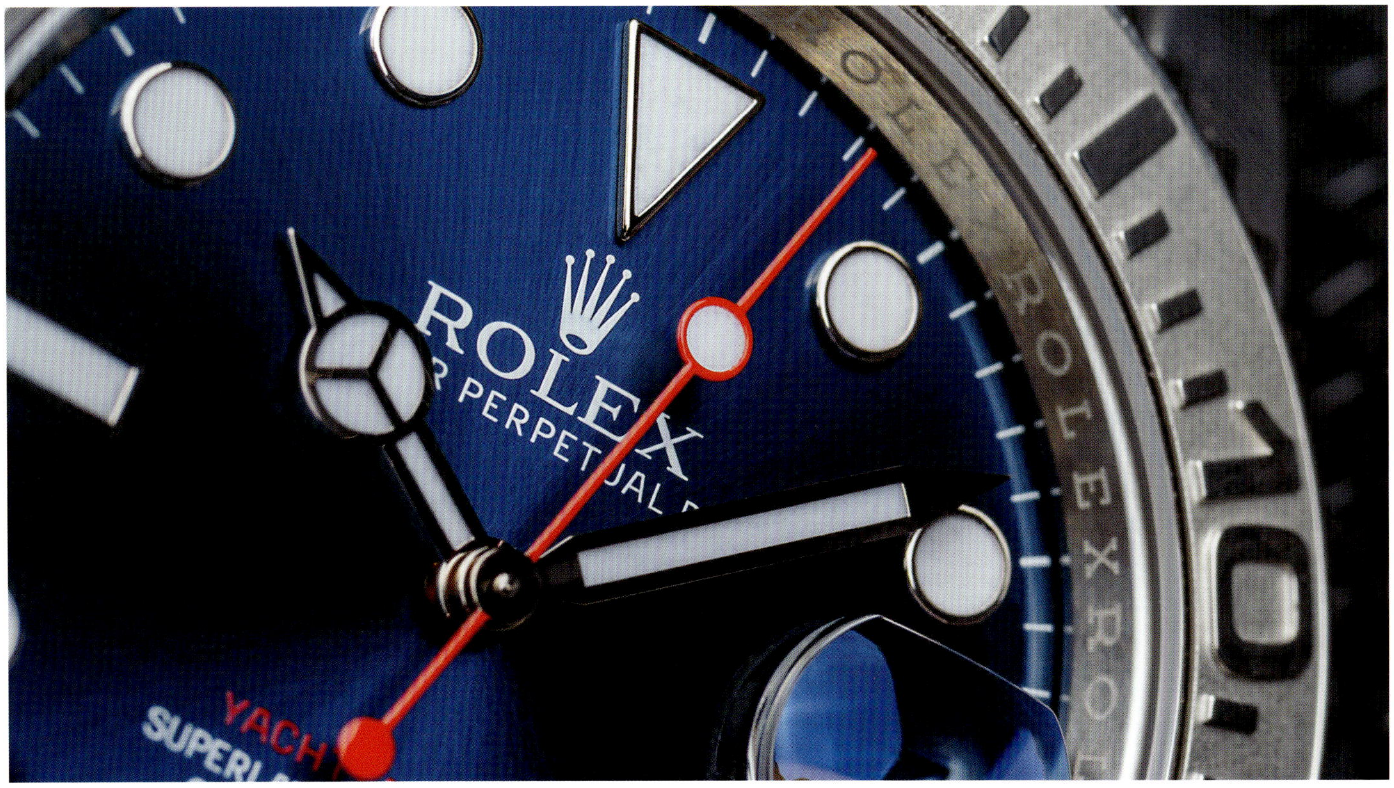

The 60-minute graduated bidirectional rotatable bezel of the Yacht-Master is crafted from precious materials. The raised graduations and numerals with a glossy finish stand out against the matt background with a sandblasted finish. This functional bezel, which allows the wearer to read a time interval, such as the sailing time between two buoys, is also a key component of the Yacht-Master's unique identity.

3235 contains an optimised version of the blue Parachrom hairspring, manufactured by Rolex from a paramagnetic alloy. In addition to being insensitive to magnetic fields, these two hairsprings boast a high degree of stability with swings in temperature. The Syloxi hairspring's geometry ensures that the movement runs regularly in any position, while the blue Parachrom hairspring is fitted with a Rolex overcoil, guaranteeing the same level of regularity.

Calibre 2236 has a paramagnetic escapement wheel made of nickel-phosphorus, and calibre 3235 contains the patented Chronergy escapement, made of the same alloy. This escapement, simultaneously providing for high energy efficiency and extreme dependability, is also insensitive to magnetic fields. Both calibres feature the self-winding module with Perpetual rotor, which ensures continuous winding of the mainspring, delivering constant energy through movements of the wrist. Calibre 2236 guarantees an energy reserve of about fifty-five hours; the 3235 benefits from an energy reserve of about sixty hours, thanks to the barrel's architecture and the escapement's superior efficiency. Like all Rolex watches, the Yacht-Master boasts a Superlative Chronometer certification.

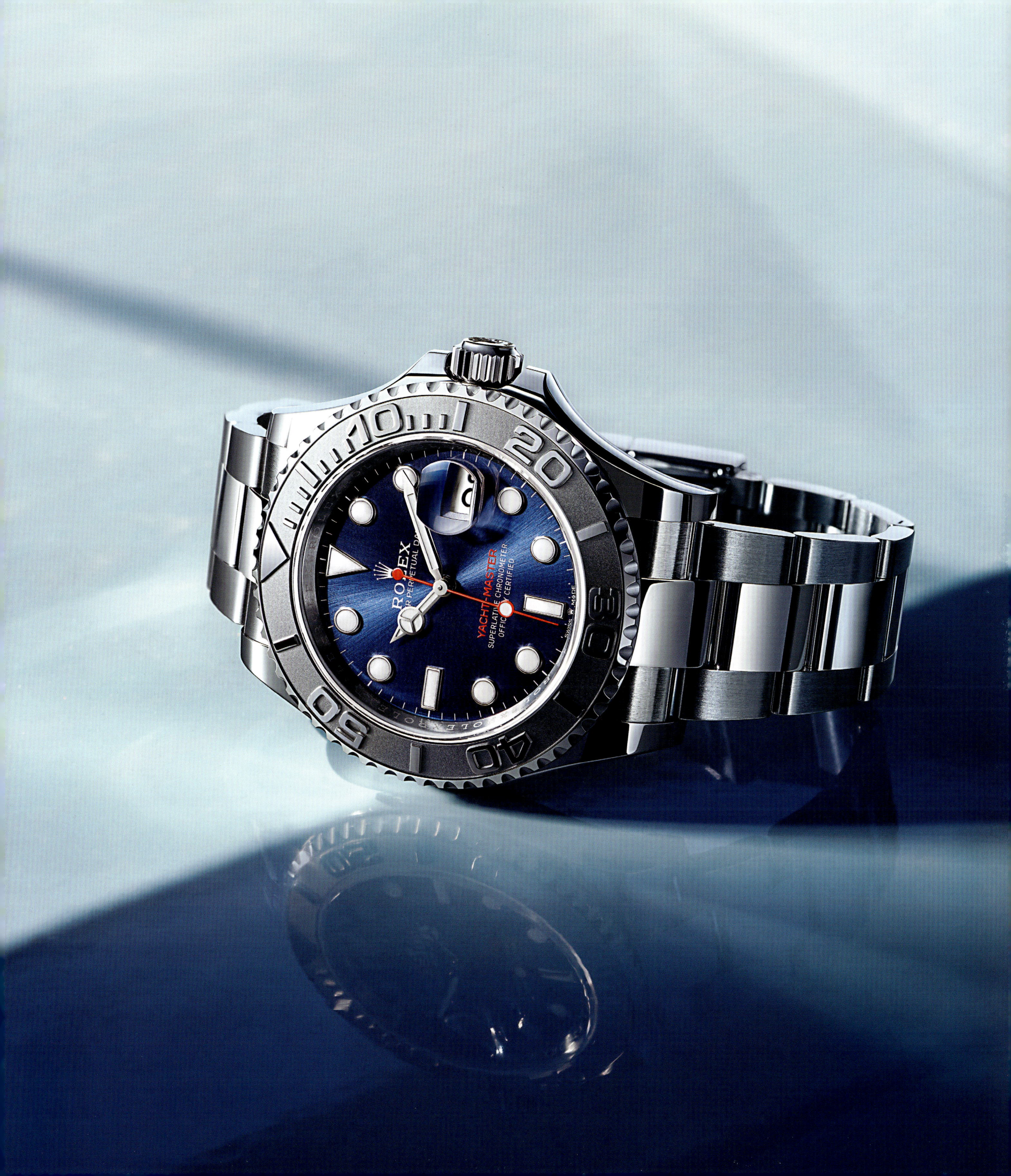
YACHT-MASTER
SUPERLATIVE CHRONOMETER
10
20
30
40
50

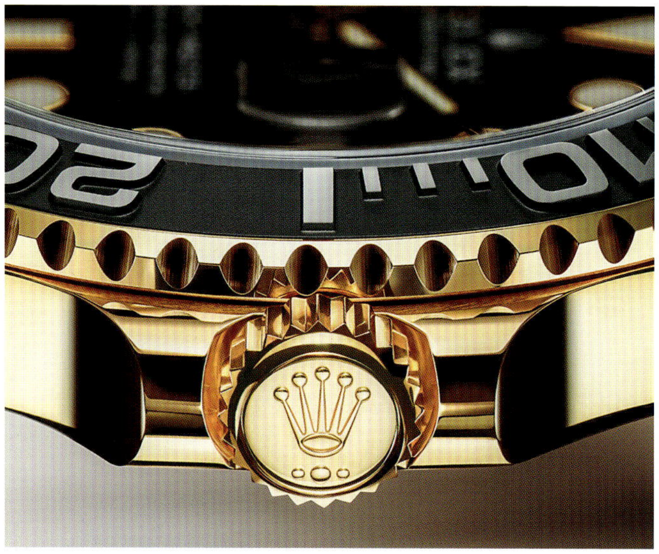

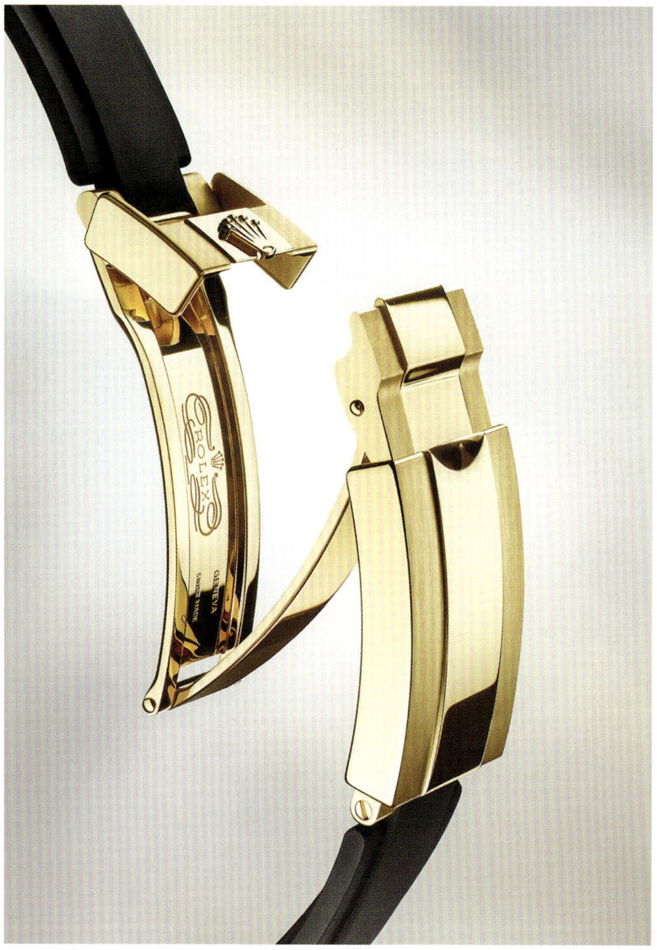

Above: Detail of the winding crown.

Below: Yacht-Master Oysterflex bracelet, with an Oysterlock safety clasp that prevents accidental opening.

DIAL: Like all Rolex Professional watches, the Yacht-Master 40 guarantees an optimum legibility in every circumstance, especially in the dark, thanks to its Chromalight display. Its broad hands and large indices in simple shapes – triangles, circles and rectangles – are coated and filled in a luminescent material that provides a long-lasting glow.

CASE: This watch, destined for sailors and skippers, brings together character and performance. It is the only watch of the Professional category that is available in three different diameters: 37, 40 and 42 millimetres.

BRACELET: Yacht-Masters 37 and 40 in Rolesium (a combination of Oystersteel and platinum) or in Rolesor Everose (a combined Oystersteel and Everose gold) are fitted on a three-piece-link Oyster bracelet. The bracelet comes with the Easylink comfort extension link, which allows for easy adjustment of the bracelet's length by about five millimetres. The Yacht-Masters 37 and 40 in Everose gold and Yacht-Master 42 in yellow or white gold are fitted on an Oysterflex bracelet with the Rolex Glidelock extension system; the system has six notches about 2.5 millimetres apart and allows for the length to be easily adjusted by about 15 millimetres. The Oyster and Oysterflex bracelets on the Yacht-Master models are fastened using an Oysterlock safety clasp which prevents involuntary opening.

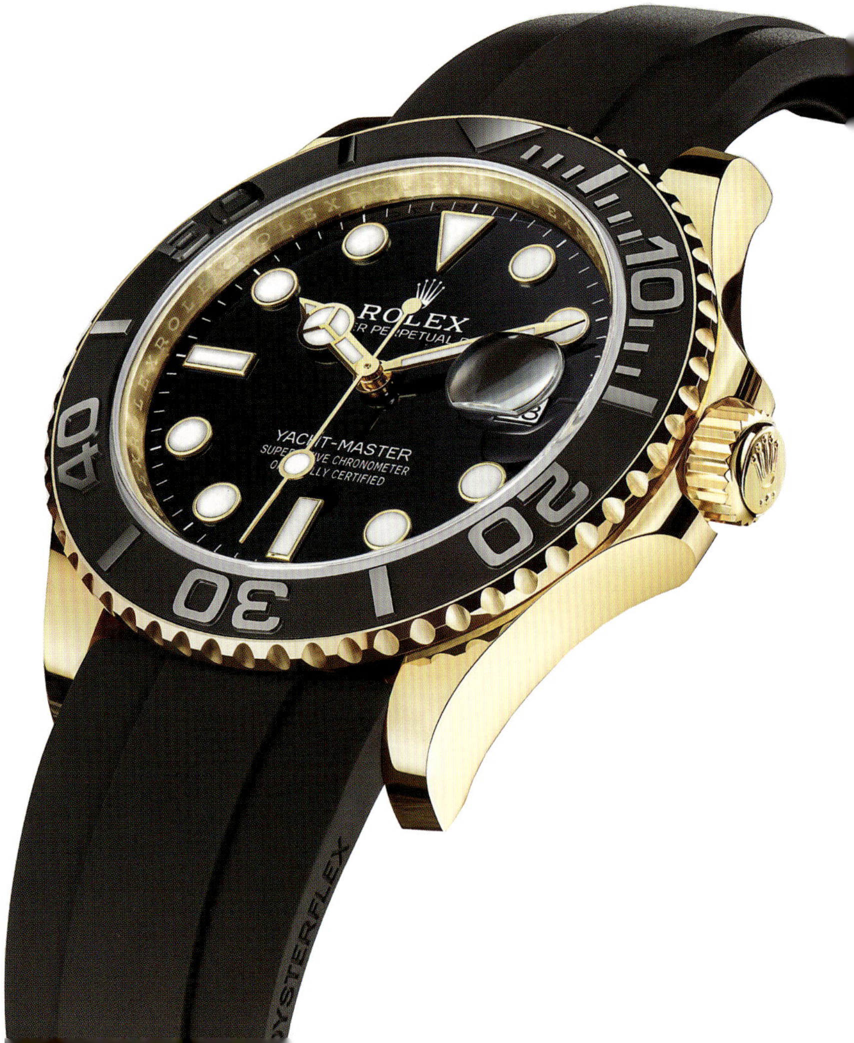

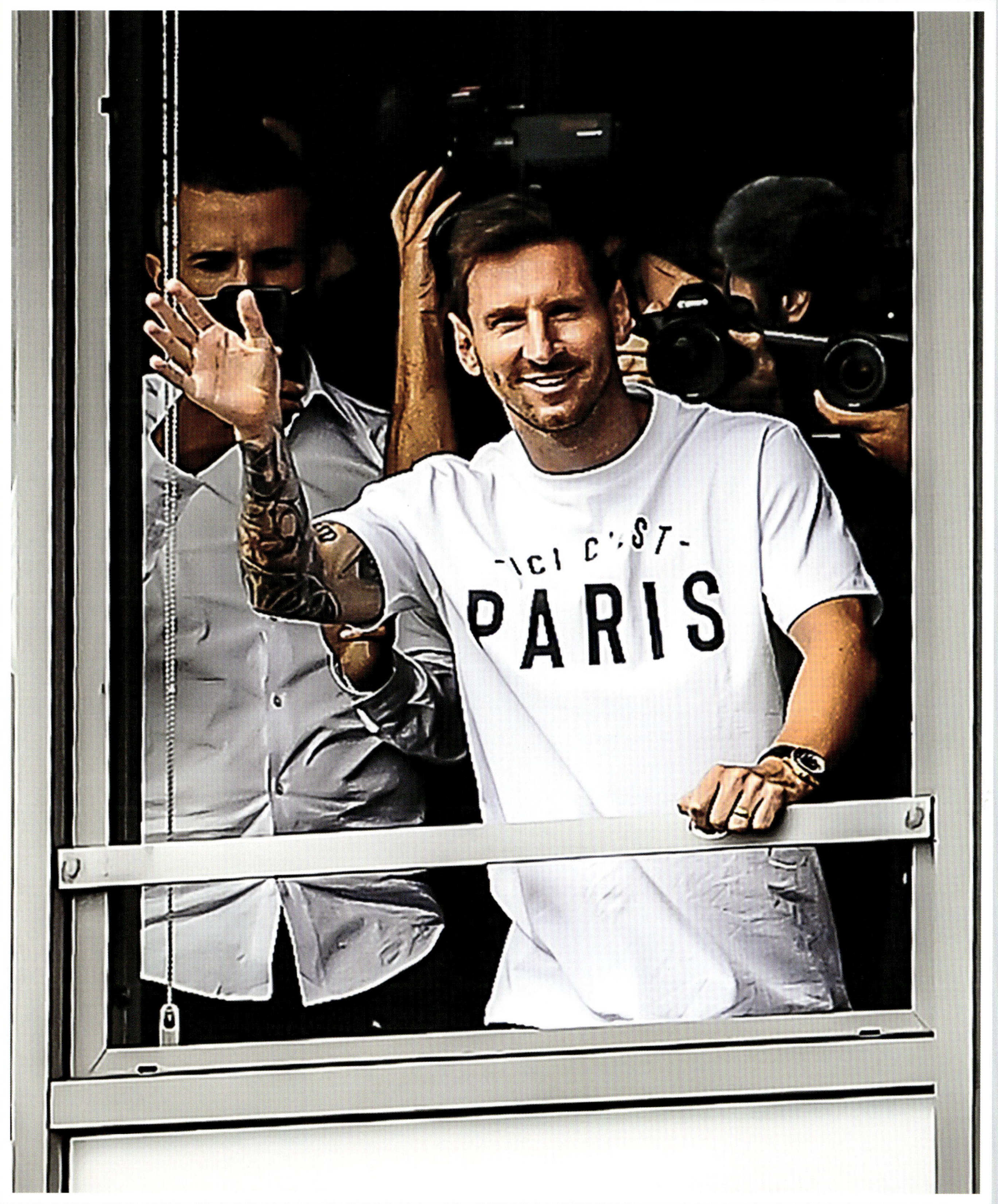

Soccer player Lionel Messi wearing the Yacht-Master model.

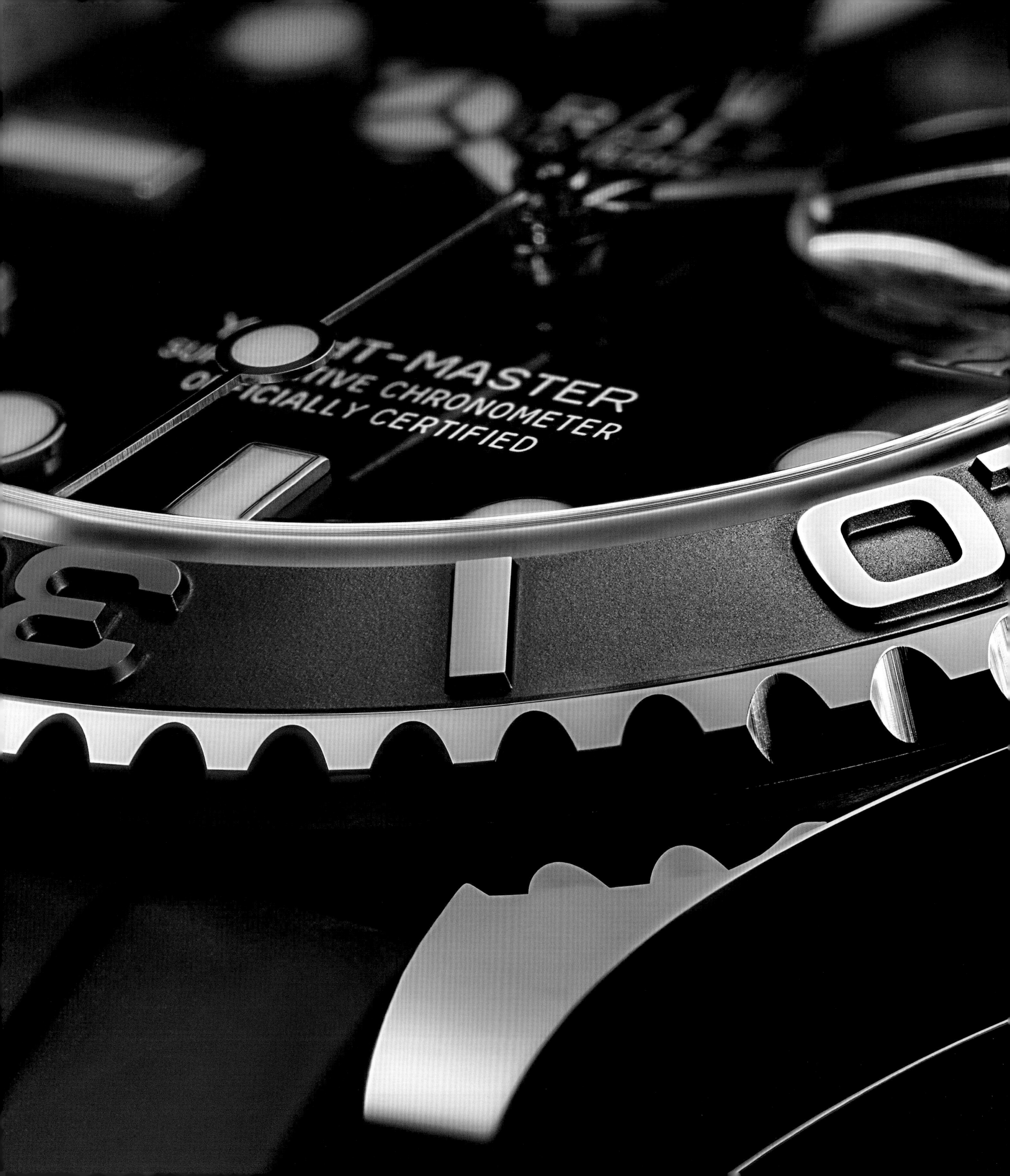
CHRONOMETER
CERTIFIED

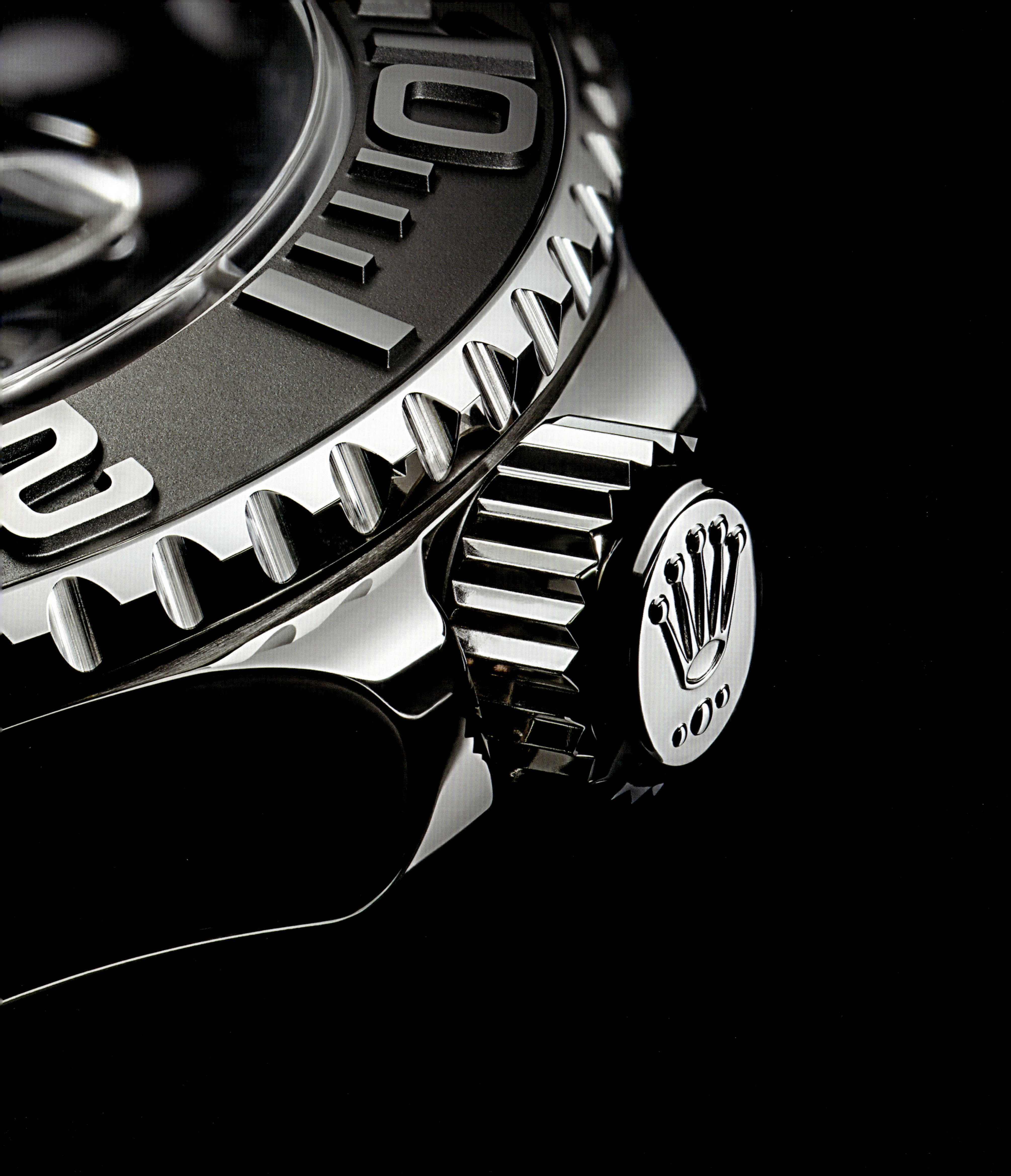
10

Yacht-Master II

Unveiled in 2007, this high-precision chronograph is equipped with a programmable countdown function with mechanical memory and can be synchronised on the fly.

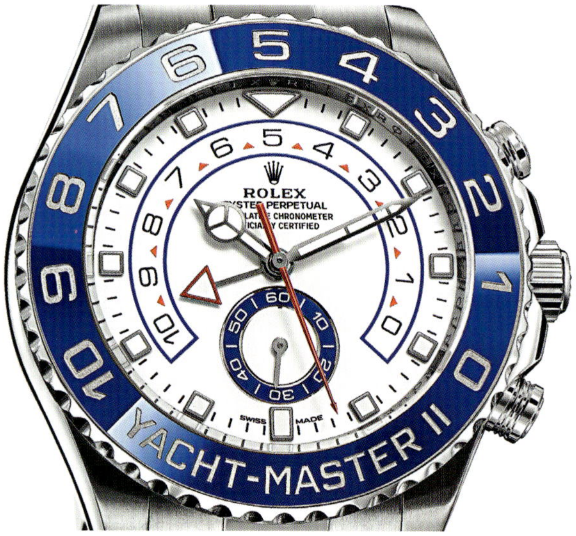

In a regatta, as in any speed competition, the matters of precision and synchronisation are fundamental. The Oyster Perpetual Yacht-Master II was the first watch in the world to offer a programmable countdown and mechanical memory with on-the-fly synchronisation. The function is highly complex from a technical standpoint but extremely simple to use, providing the precision necessary at a crucial phase prior to the start in a regatta. The Yacht-Master II is fitted with a countdown function that can be programmed from 1 to 10 minutes, allowing the wearer to synchronise with the specific sequence of departure based on the type of regatta. These events are generally divided into three phases, indicated through the use of various visual and sound signals. Here is one example of a starting sequence using the corresponding functions of the Yacht-Master II. Before the regatta, the skipper programs the countdown time of their Yacht-Master II according to the regatta rules (for example, 10 minutes). Turning the watch bezel via the Ring Command, a system of interaction between the rotatable bezel, crown and movement, one can program the movement, then lock it. Once the duration has been programmed, the mechanism allows the watch to memorise it, so that when the countdown is reset, it is automatically reset to the programmed duration. The countdown can be reprogrammed as often as one wishes for any length of time between 1 and 10 minutes.

At 10 minutes to start time, the fire of a gun and a hoisted flag signal the start of the regatta's official countdown. The skipper starts the countdown on their Yacht-Master II, pressing the pusher at 2 o'clock. The seconds hand of the chronograph – the central seconds hand – begins to move instantly, indicating the seconds elapsed, while the countdown hand, with a triangular tip, tracks the minutes left on the graduated 10-to-0-minute scale on the dial.

At 5 minutes to start time, a preparatory signal is made with a second firing of the gun and a new hoisted flag. The skipper can now synchronise the countdown on their Yacht-Master II with the official count. Pressing the pusher at 4 o'clock enacts an instant reset of the ('flyback') seconds hand, which immediately begins its course again once the button is released. At the same time, the countdown minutes hand synchronises with the closest minute mark, compensating for the time behind or ahead when the chronograph had been started at the first signal.

At 1 minute to start time, the second preparatory signal is sounded with a trumpet, and the 5-minute flag is lowered. The skipper glances to confirm that the Yacht-Master II's countdown is well synchronised with the official time. The watch provides a precise count, to the second, of the time they have before crossing the starting line.

Finally, at the moment the race begins, the start signal is given – a final gunshot and a lowering of the 10-minute flag. Having been able to optimise their manoeuvres to the last second using the countdown on their Yacht-Master II, the skipper finds themselves in the correct position to cross the line immediately after the

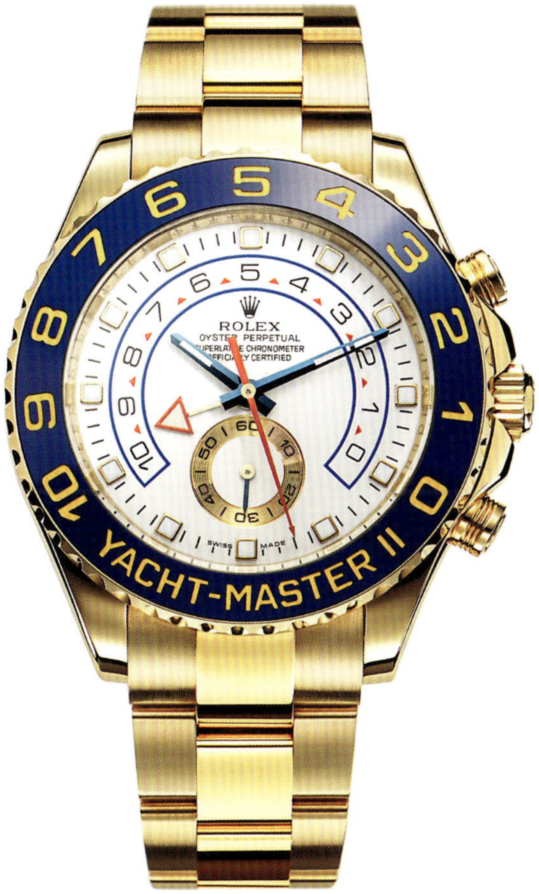

The Yacht-Master II model, gold version.

signal, executing the perfect departure.

It is precisely this flexibility that is the Yacht-Master II's primary advantage. Moreover, programming is memorised mechanically, which means one can start a new countdown without having to reprogram the watch for the next leg, for example, or when the next regatta's rules impose the same timing. Having triggered the countdown, the owner can synchronise the watch with the official count (using the instantaneous reset of the 'flyback' seconds hand and adjustment to the nearest minute). The functions of this countdown timer are designed to be easily operated. The preliminary programming up to 10 minutes is activated and then locked via the rotating bezel thanks to Ring Command, a system of interaction between the bezel, the winding crown and the movement, developed and patented by Rolex. The bidirectional rotatable bezel of the Yacht-Master II is fitted with a monobloc Cerachrom insert in blue ceramic. This disc, also patented, is made of an especially hard anti-scratch high-tech ceramic, whose colour remains unchanged by the effect of ultraviolet rays. It is also inert and cannot corrode due to its chemical composition. The numbers and inscriptions are molded into the ceramic and coloured with gold or platinum using a PVD (Physical Vapor Deposition) technique.

The Yacht-Master II guarantees excellent legibility in all circumstances, even in the dark, due to its unique Chromalight display: its indices and hands are respectively filled or coated with a long-lasting (blue) luminescent substance which glows up to two times longer than traditional phosphorescent substances. The Yacht-Master II's Oyster case, with a diameter of 44 millimetres, is guaranteed waterproof to a depth of 100 metres; the Triplock winding crown, with a triple waterproofing system, screws firmly to the case. The crown is protected by a special guard integrated into the middle case. The crystal is in scratch-resistant sapphire and benefits from an anti-reflective coating.

The Yacht-Master II is equipped with calibre 4161, a self-winding mechanical chronograph movement entirely developed and manufactured by Rolex. The architecture and manufacturing of this movement, as well as its innovative features, provide it with excellent performance in terms of precision and reliability. Calibre 4161 was designed on the basis of the chronograph movement 4130, which powers the Oyster Perpetual Cosmograph Daytona. It features a traversing column wheel and vertical clutch, ensuring an instantaneous and extremely precise start of the countdown function. This calibre's oscillator is equipped with a large balance wheel with variable inertia, regulated with extreme precision via gold Microstella nuts. It is securely held in place by a height-adjustable traversing bridge for stable placement and increased shock resistance. Calibre 4161

The Yacht-Master II model with Chromalight display.

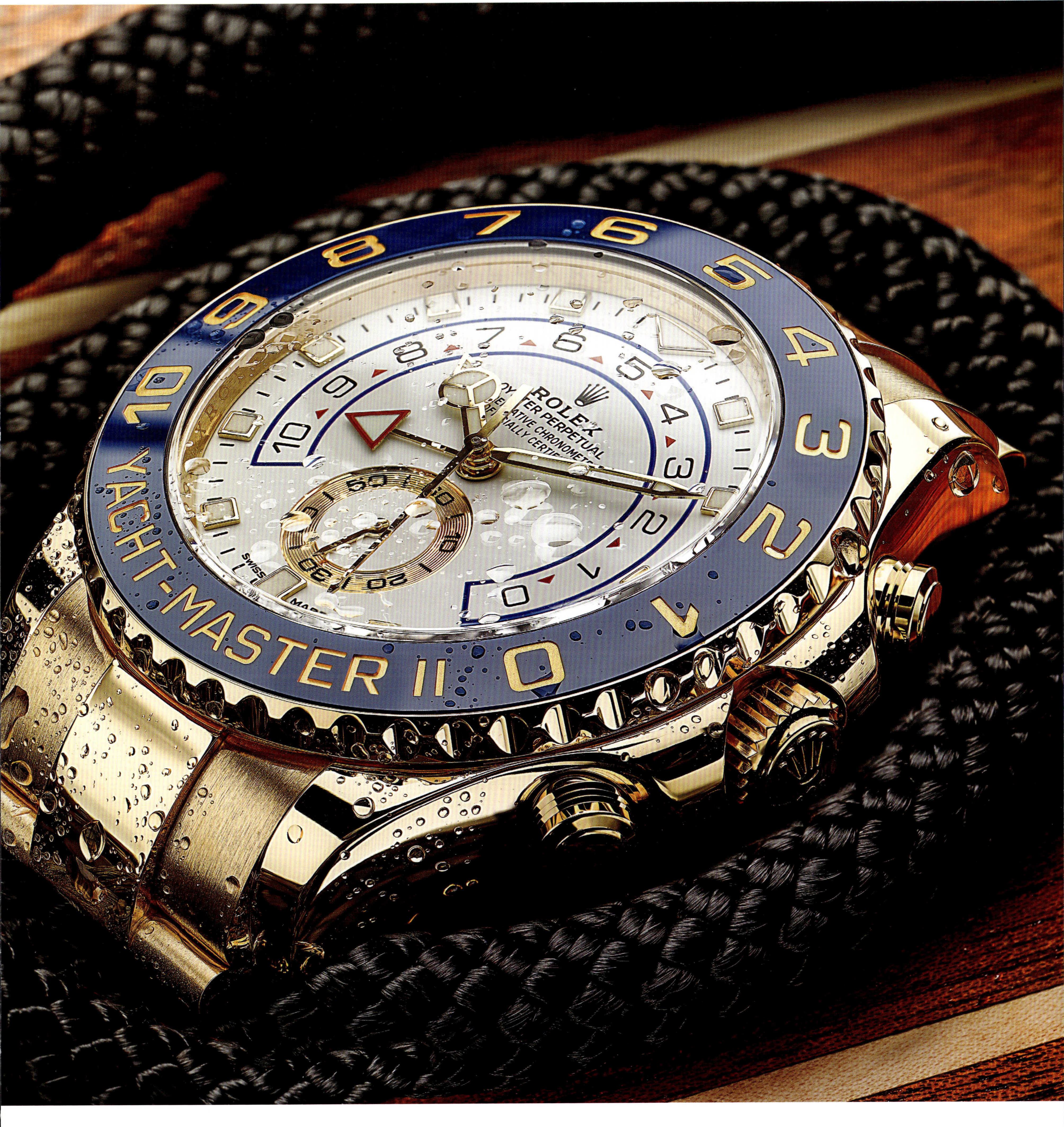
YACHT-MASTER II
ROLEX

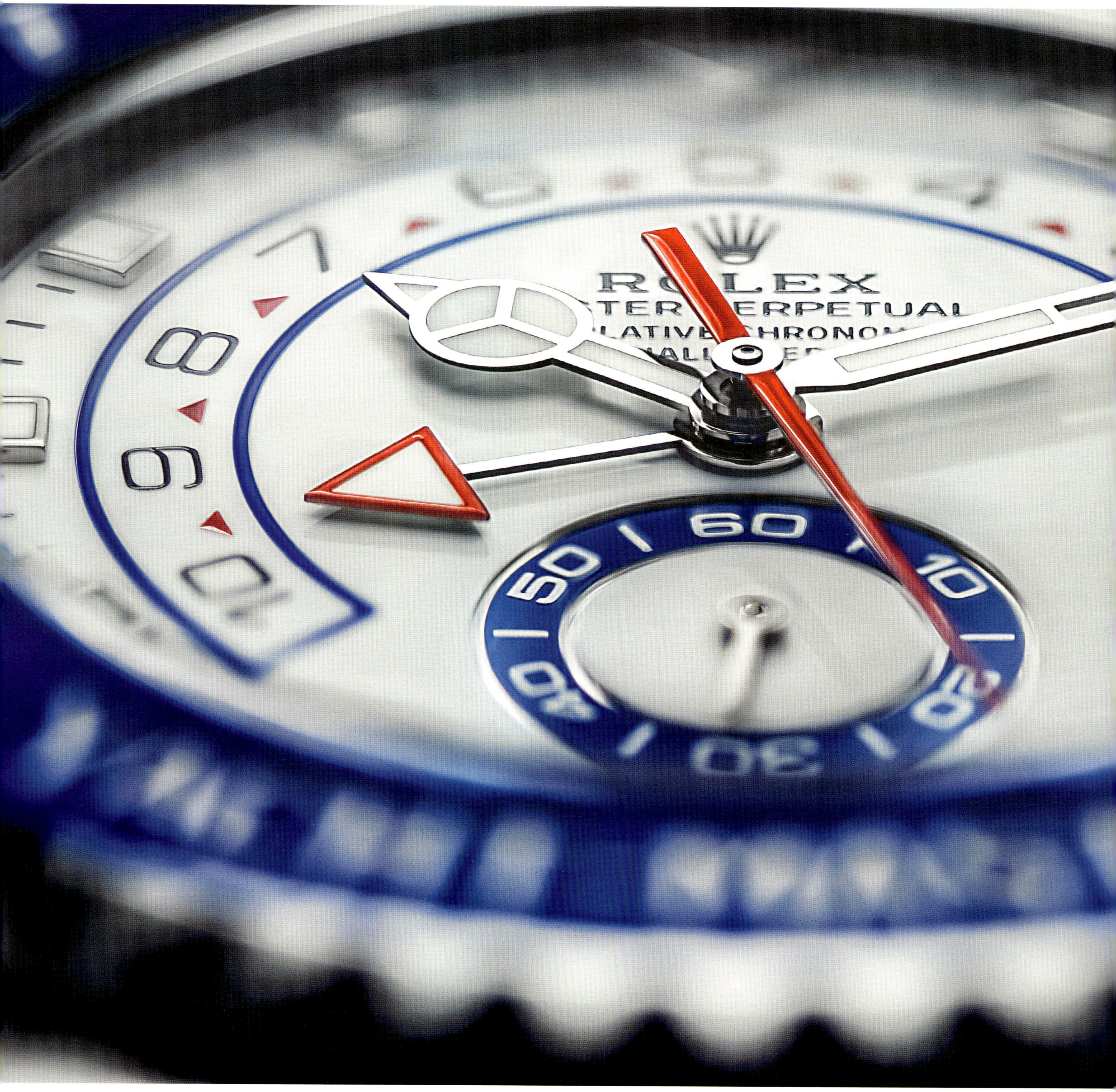
50
60
10

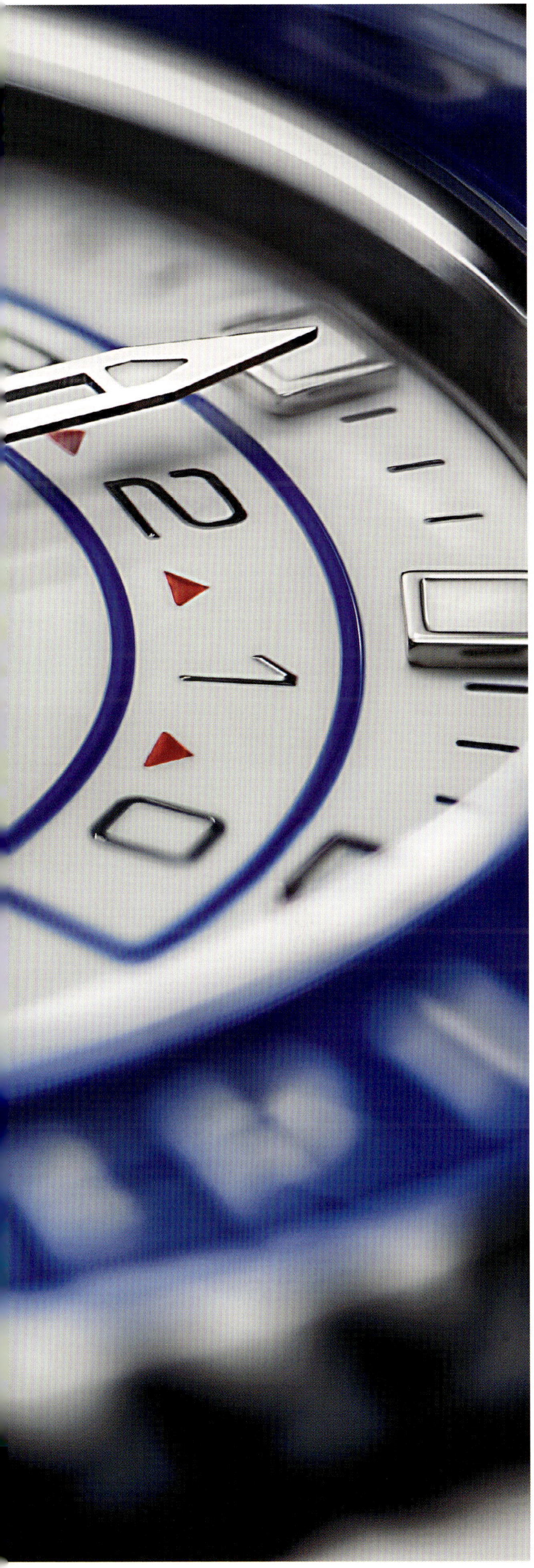

contains a blue Parachrom hairspring, made by Rolex in a paramagnetic alloy, which, while also being insensitive to magnetic fields, boasts high stability with fluctuations in temperature. It is equipped with a Rolex overcoil, which ensures regular running of the movement at every position. Calibre 4161 features an automatic winding module with Perpetual rotor, which ensures continuous winding of the mainspring and thus supplies constant energy using the movements of the wrist. It boasts a power reserve of approximately seventy-two hours. The Yacht-Master II, like every Rolex watch, boasts a Superlative Chronometer certification.

DIAL: Yacht-Master II's design is characterised by its white-and-blue dial with blue bezel. This model features a prominent circular arc with numbers arranged clockwise from 10 to 0, along which a small hand with a triangular tip travels. The bezel features the same sequence of numbers, complete with the inscription 'Yacht-Master II'. Its decreasing numerals make it immediately clear that timing is not accomplished by adding the times but, rather, like a countdown. In addition to its central seconds, minutes and hours hands, it has small seconds at 6 o'clock.

CASE: The Yacht-Master II, fitted with a 44-millimetre-diameter Oyster case, features the Ring Command system, an interaction mechanism between the case and the movement, which allows the wearer to adjust the duration of the countdown by means of its rotatable bezel and winding crown.

BRACELET: The Yacht-Master II is paired with a three-piece-link Oyster bracelet. This bracelet features an Oysterlock safety clasp, which prevents any inadvertent opening. It also features the Easylink comfort extension link, developed by the brand, which allows the wearer to easily increase its length by about 5 millimetres for additional comfort in any circumstance.

Deepsea

Introduced in 2008, the Deepsea satisfies professional divers' highest expectations concerning underwater pressure resistance, accuracy and reliability.

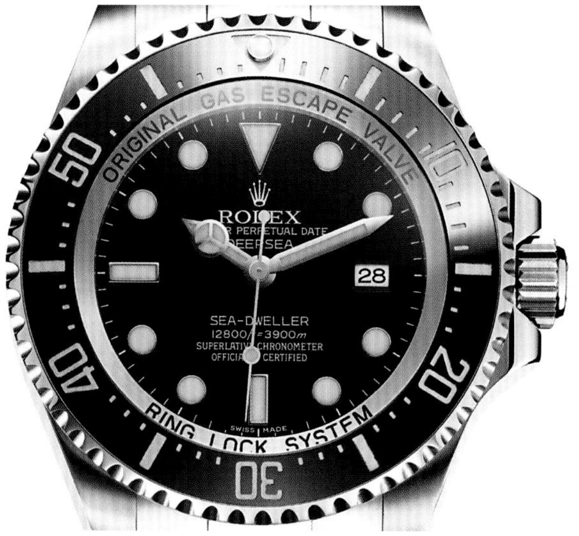

In 2010, about a century after the conquest of the North Pole, eight extreme divers took on the challenge of a polar expedition to conquer the submerged areas of the Arctic. Their mission was to discover, and bring others to discover, what lay hidden beneath the sea ice. For forty-five days, the team skied to the edge of the globe in the Canadian high Arctic, making repeated dives to witness the richness of an ecosystem falling victim to global warming. The expedition members' equipment included cameras for video and photography and five Oyster watches. Emmanuelle Périé, the only woman on the team, declared at the end of the expedition: 'The only diving instruments which performed all the time were our Rolex watches – the Oyster Perpetual Rolex Deepsea model.'

The Deepsea recalls the history of Rolex and the challenges associated with exploring the depths of the oceans. It can resist water pressure at great depths, thanks to two characteristics (see next two paragraphs). It more specifically references the Deepsea Special, a watch that accompanied Jacques Piccard and Don Walsh in 1960 when they dove into the Mariana Trench.

The Deepsea features the Ringlock system, a case architecture developed and patented by Rolex which allows it to withstand the colossal pressure present at a depth of 3,900 metres, equivalent to the weight of around three tons on the watch crystal. The Ringlock system is based on the combination of three elements: a slightly domed sapphire crystal 5.5 millimetres thick, with an anti-reflective coating; a high-performance compression ring in nitrogen-alloyed steel placed inside the middle case (between the glass and the case back), which is able to withstand the water pressure on its own; and an RLX-titanium case back held against the ring by another screw-down ring in Oystersteel. These three components are superimposed inside the middle case machined from a solid block of Oystersteel, an alloy especially resistant to corrosion. The Triplock winding crown, protected by an integral crown guard and screwed into the case, contributes to the water resistance of the watch.

The Rolex Deepsea's Oyster case, measuring 44 millimetres in diameter, is also equipped with the helium escape valve patented in 1967, which acts as a safety valve in watches designed by the brand for deep-sea diving. During decompression phases in a hyperbaric chamber used in saturation diving, it allows excess pressure accumulated in the case to escape without compromising its impermeability.

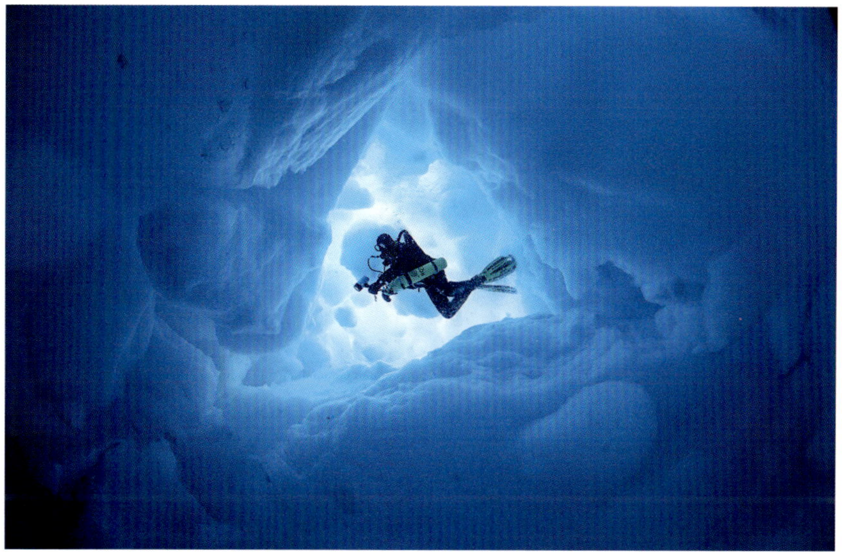

Deepsea Under The Pole by Rolex, a pioneering expedition undertaken in 2010 to explore the submerged part of the Arctic.

The unidirectional rotating bezel of the Rolex Deepsea is fitted with a 60-minute graduated monobloc Cerachrom insert in black ceramic, which allows divers to monitor their dive times. This patented insert is made of an especially hard anti-scratch ceramic whose colour remains unaltered by the ultraviolet rays. Moreover, due to its chemical composition, its high-tech ceramic is inert and cannot corrode. The graduations and recessed numerals are coated with platinum using a PVD (Physical Vapor Deposition) process. Thanks to its knurled edge, offering an excellent grip, the bezel can be handled easily even while wearing gloves.

Subject of much attention has been the legibility of this professional diving watch. On its dial, the Chromalight indices and hands are filled or coated with a luminescent material emitting a long-lasting blue glow, lasting up to two times longer than traditional phosphorescent materials. On the bezel, the 'zero' marker of the graduation, in the form of a triangle, is visible in the dark, thanks to the same luminescent material.

Scuba diving requires absolute reliability and safety, which is why every Deepsea watch is subjected to the very stringent waterproofing checks required as a standard for diving watches. For this purpose, a special hyperbaric tank was developed in collaboration with COMEX (Compagnie Maritime d'Expertises), a French company renowned internationally for its specialisation in underwater engineering and hyperbaric technologies. It is used to test and guarantee the watertightness of each Rolex Deepsea to a depth of 3,900 metres, with an additional safety margin of twenty-five percent.

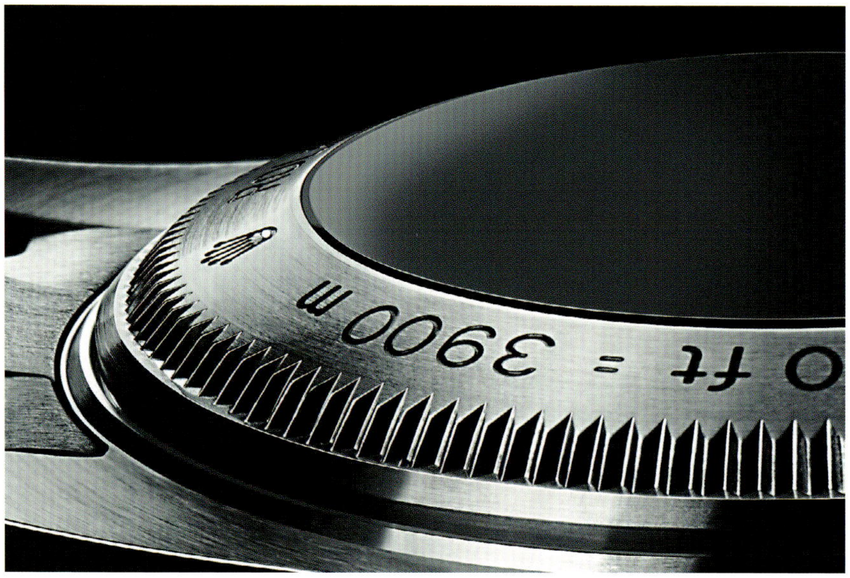

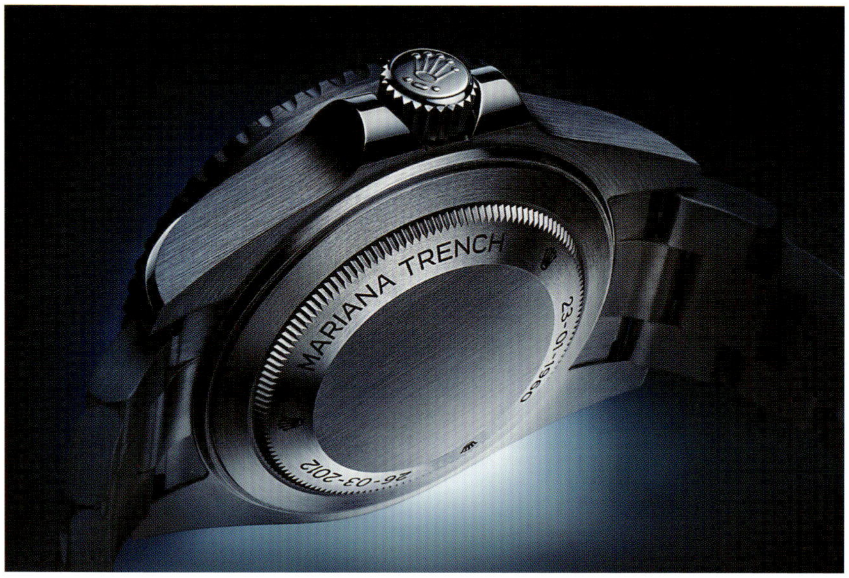

The Deepsea Challenge model.

The Deepsea inspired an experimental watch model called the Deepsea Challenge, which on 26 March 2012 was fixed to the manipulator arm of the submarine Deepsea Challenger accompanying explorer and filmmaker James Cameron in a solo dive at 10,908 metres below the surface of the Pacific Ocean. This watch, guaranteed waterproof to an extreme depth of 12,000 metres, incorporates all the brand's technical innovations in terms of waterproofing. During the test phases, it was successfully subjected to the pressure present at 15,000 metres below the surface. At this depth, the central ring of the Ringlock system is subject to a pressure equivalent to a weight of twenty tons.

In 2014, Rolex presented a commemorative version of the Deepsea. Its 'D-blue' dial, with delicate shades ranging from midnight blue to the most intense black, recalls spaces of the underwater world where even the last ray of sun from the surface disappears, leaving only the darkness of the abyss. The dial is a recollection of the Deepsea Challenge expedition,

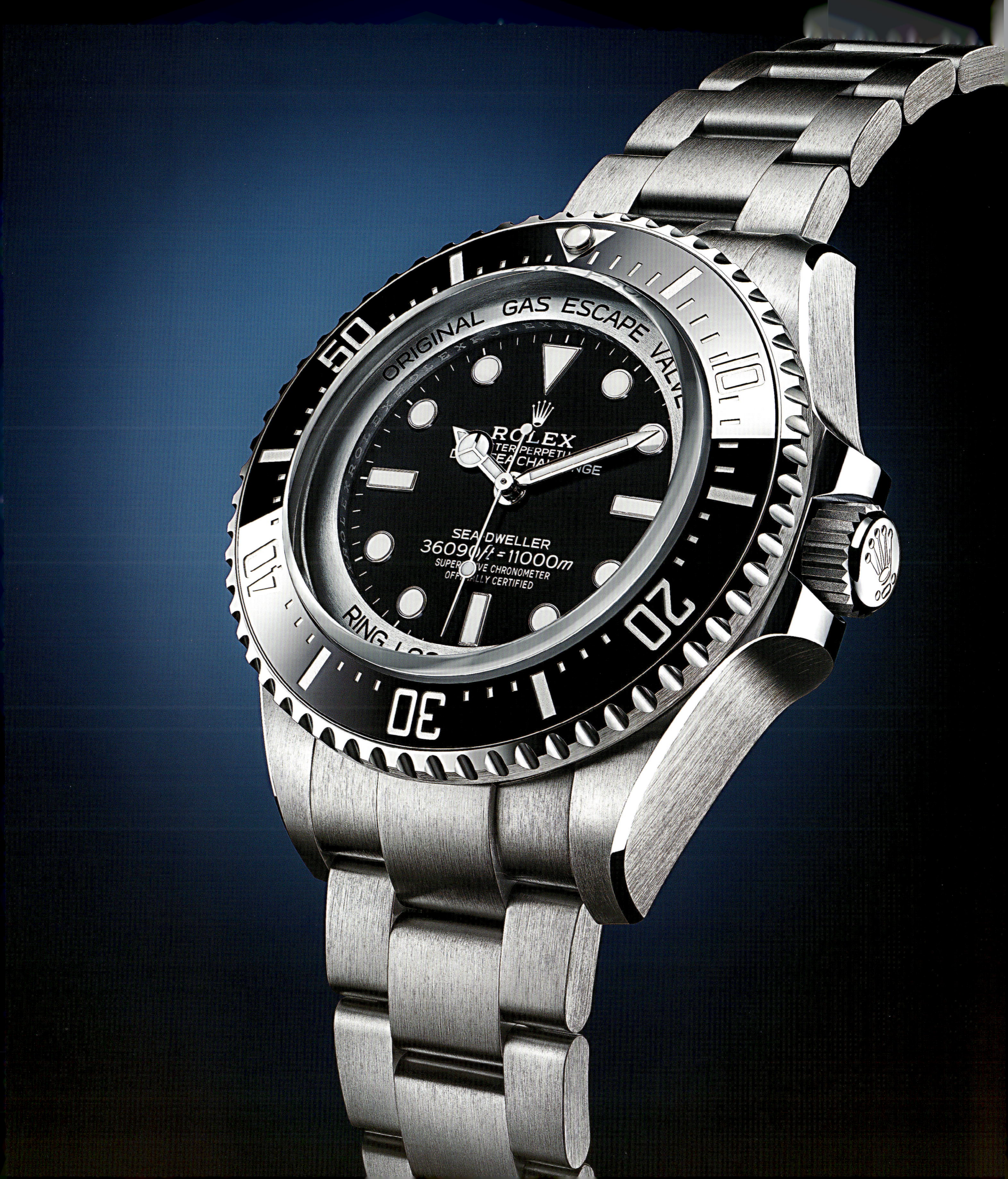
ORIGINAL GAS ESCAPE VALVE
ROLEX
SEA-DWELLER
36090ft = 11000m

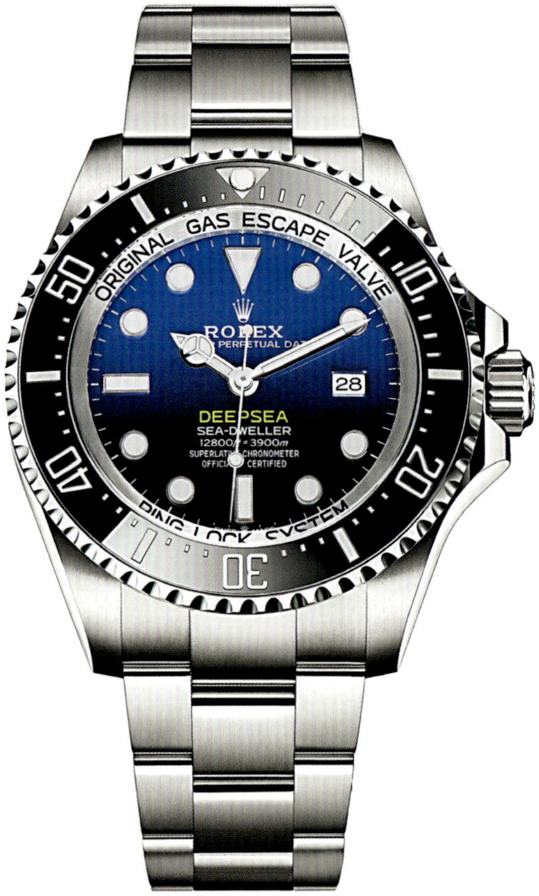

The Oyster Perpetual Deepsea in Oystersteel with a black ceramic Cerachrom bezel insert and Oyster bracelet. It features a unique D-blue dial and large luminescent hour markers.

DIAL: The Deepsea's blue Chromalight hour markers and hands boast long-lasting luminescence, legible even in extremely low lighting conditions.

CASE: Among the innovations of its entire outfit, the Deepsea's 44mm case in 904L stainless steel is a significant highlight.

BRACELET: The Glidelock fine-adjustment system of the Deepsea's Oyster bracelet has ten notches approximately two millimetres apart, allowing the bracelet's length to be adjusted by approximately twenty millimetres. This way, the watch can be worn over a diving suit up to three millimetres thick.

sponsored by Rolex and the National Geographic Society, when James Cameron reached the deepest part of the ocean. To pay homage to this partnership, 'Deepsea' has been added to the dial in a colour inspired by the green of James Cameron's submarine.

The Rolex Deepsea, like every Rolex watch, boasts the Superlative Chronometer certification and offers a level of precision in the order of –2/+2 seconds per day, the deviation tolerated by the brand for a finished watch. The Rolex Deepsea is equipped with calibre 3235, a self-winding mechanical movement entirely developed and manufactured by Rolex. The quintessence of technology, this movement carries numerous patents and guarantees high performance in terms of precision, autonomy, shock resistance, convenience and reliability. Calibre 3235 features a self-winding module with a Perpetual rotor ensuring continuous winding of the mainspring, thus supplying constant energy using movement of the wearer's wrist. It benefits from a power reserve of approximately seventy hours, thanks to its barrel architecture and the superior efficiency of its escapement.

The Rolex Deepsea is paired with a three-piece-link Oyster bracelet fitted with an Oysterlock safety clasp, designed and patented by Rolex to prevent any inadvertent opening. It also features the Rolex Glidelock fine-adjustment system; this ingenious toothed mechanism, integrated into the clasp (also designed and patented by the brand) allows for fine adjustment of the bracelet's length without the use of tools.

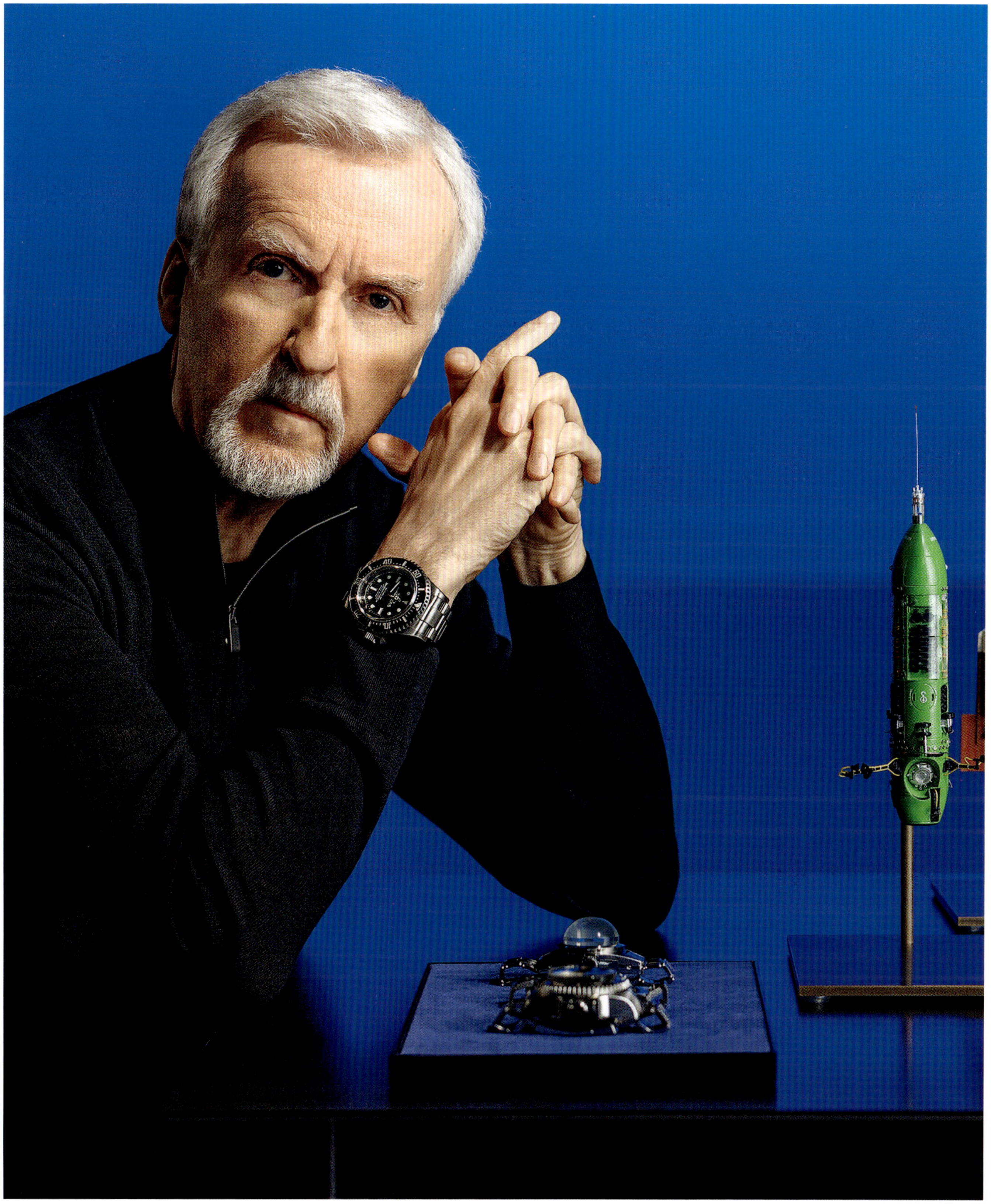

Following his historic feat, director and explorer James Cameron commented that 'The Rolex Deepsea Challenge was the reliable companion throughout [my] dive.'

Sky-Dweller

Presented in 2012, this model was the first Rolex timepiece with an annual calendar.

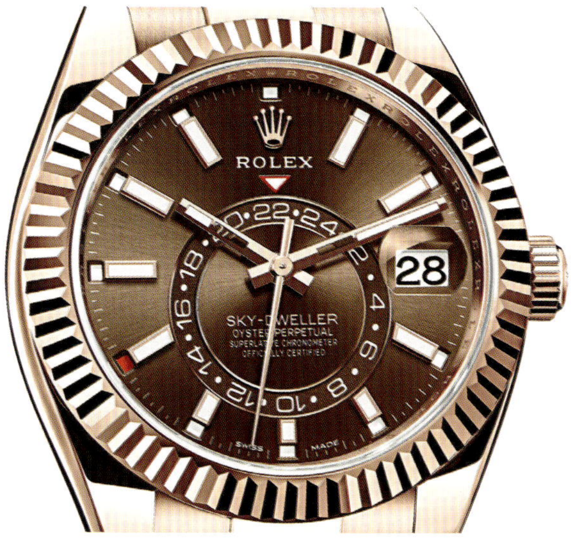

Rolex does not traditionally produce complications. The Oyster Perpetual Sky-Dweller is, however, one of the most complicated watches by the 'crown company': the watch had, in fact, an annual calendar. An annual-calendar complication is able to indicate the exact date, automatically recognising months with 30 and 31 days, the only exception being February, when a correction will need to be made based on whether the month is 28 or 29 days long.

The Rolex Sky-Dweller uses the Saros annual calendar, a mechanism that moves in coordination with the date, displaying the correct month as well as the day. Its name derives from the phenomenon (called Saros) that takes place over the course of 18.03 years, after which the Sun, Earth and Moon align.

Protected by fourteen patents, the Sky-Dweller is able to provide travellers frequently moving from one time zone to another with all the information needed to easily orient themselves in time. It allows for the time to be read simultaneously in two time zones, in addition to providing an annual calendar. The reference time can be read via the 24-hour display in an off-centre disc while the traditional central hands indicate the local time. The tip of a red triangle just below the 'Rolex' lettering indicates the second time on the off-centre 24-hour dial, which differentiates between the day and nighttime hours. The annual calendar with month display is shown inside the twelve windows along the edge of the dial; these correspond with the hour markers, beginning with January at 1 o'clock, February at 2, and so on, continuing to December. The Saros mechanism displays the current month in colour via the appropriate aperture.

The Sky-Dweller uses the Ring Command system, a mechanism of interaction between the rotatable bezel, the winding crown and the movement, which allows the wearer to select and regulate functions simply, quickly and securely, one by one.

The fluted bezel is also used in other watch designs, including the Day-Date and various Datejust models. However, while in the aforementioned watches the bezel's task is a purely aesthetic one, in the Sky-Dweller it performs a very important function: when rotated, it allows one to access the model's various complications and to regulate them. Its bezel acts as a selector once unscrewed, operating together with the winding crown. There are three positions that the bezel can assume, turning it clockwise, in order to adjust the same number of functions: setting the calendar, local time (hands) or reference time (the off-centre disc).

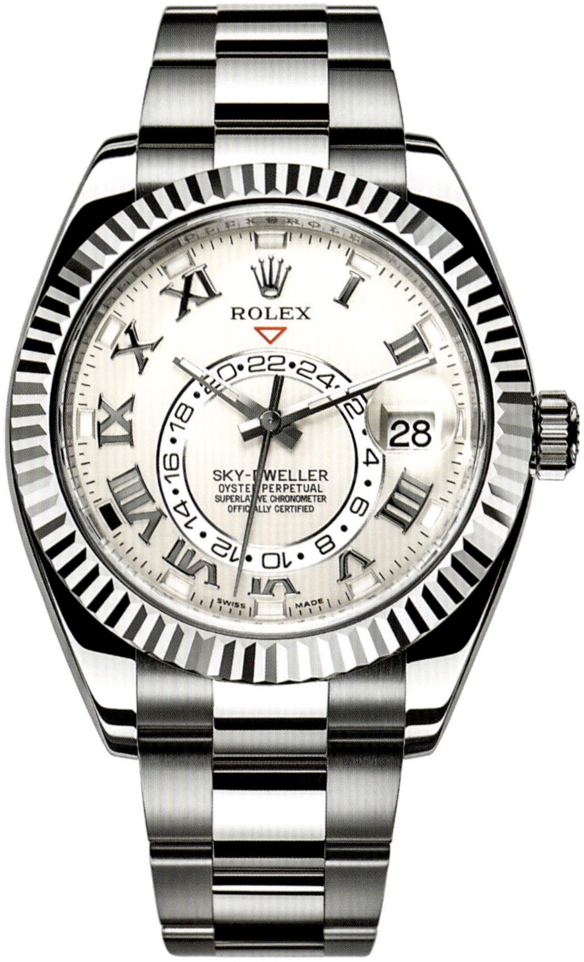

Version of the Sky-Dweller model with a steel bracelet.

Several versions of the Sky-Dweller were introduced following its launch in 2012: bracelet versions are available in steel, yellow gold, pink gold, steel with yellow gold and steel with white gold, and versions with a strap are offered with a yellow- or pink-gold case. Each model, with a 42-millimetre case, is also paired with dials of different shades.

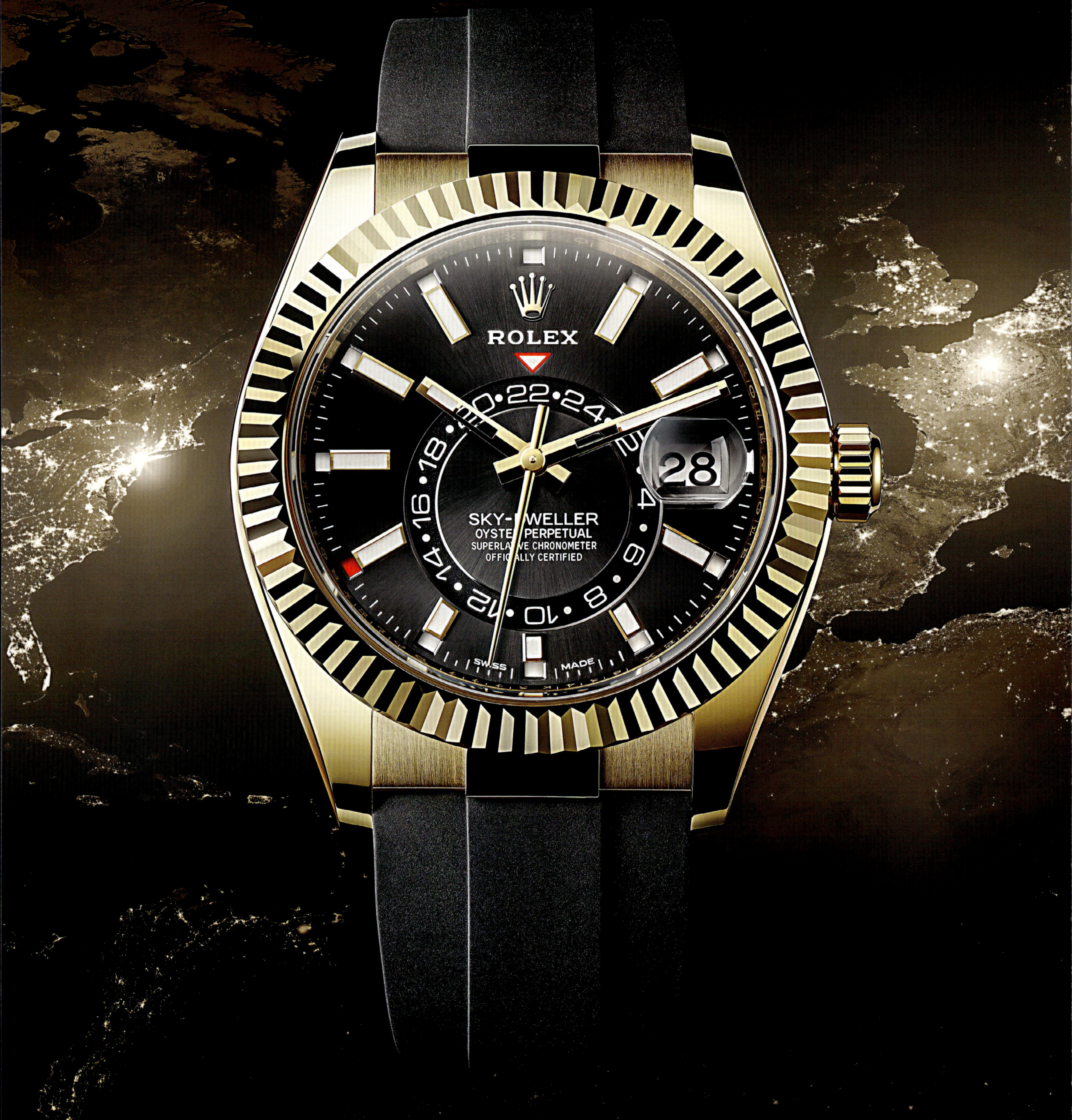
ROLEX
SWISS
MADE

In 2020, Rolex presented a yellow-gold version that was paired with an Oysterflex bracelet. Developed and patented by Rolex, this innovative bracelet is made up of two flexible metal blades – each for one of the band's two halves – made of a titanium-and-nickel alloy. The blades are overmolded with high-performance black elastomer, a material that is long-lasting and resistant to environmental effects. For maximum comfort on the wrist, the Oysterflex bracelet features 'cushions' on its inside surface. In the most recent version of the Sky-Dweller, it is fitted with a Rolex-designed and -patented folding Oysterclasp in yellow gold. It also features a Rolex Glidelock extension system, also developed and patented by the brand. This device contains a toothed panel beneath the clasp cover; it can be adjusted by means of a sliding link fitted with a tooth that can be fixed into a chosen notch. The Rolex Glidelock on the Oysterflex bracelet has six notches about 2.5 millimetres apart and allows for its length to be easily adjusted by about 15 millimetres, without the aid of any tool.

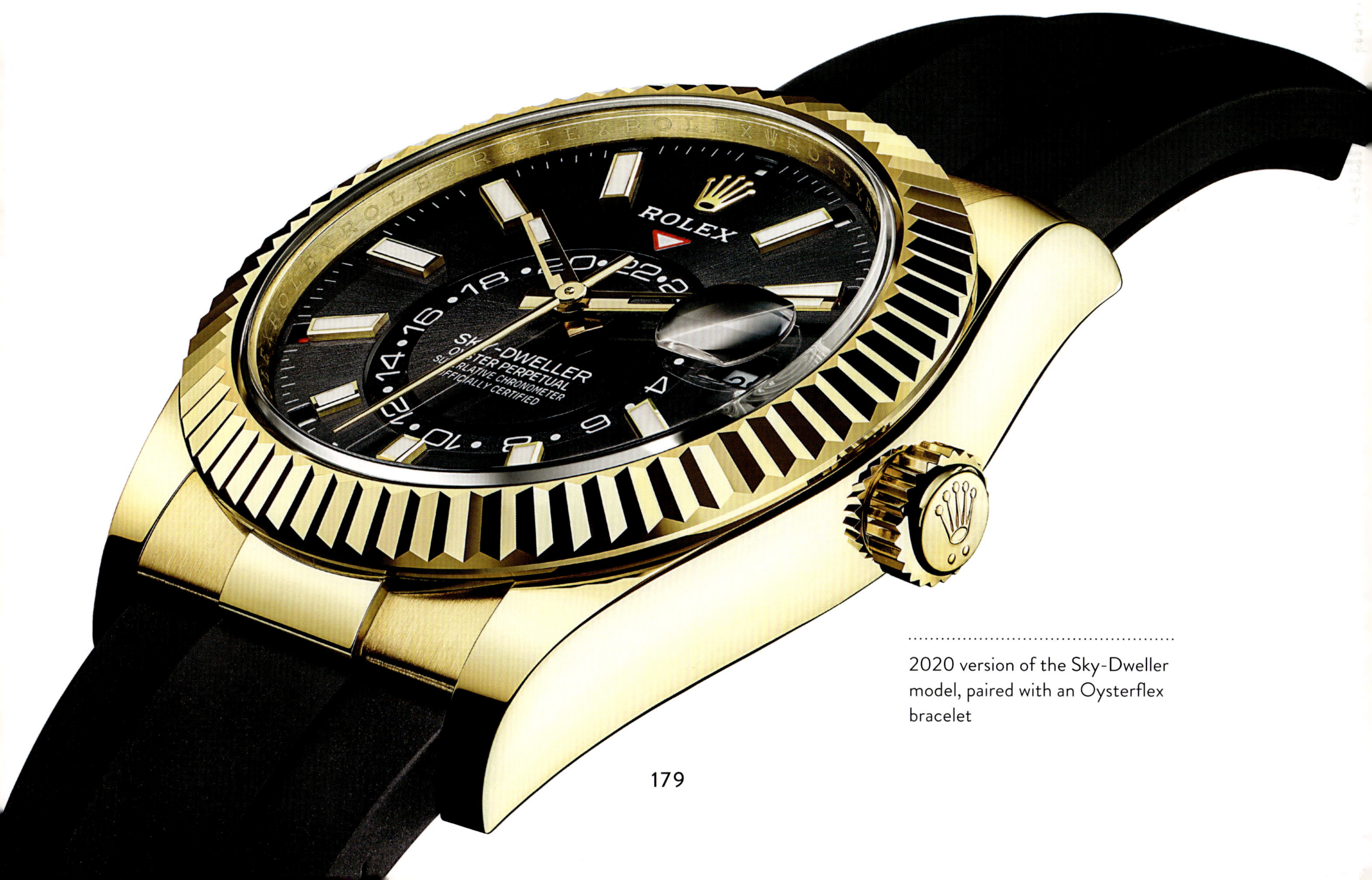

2020 version of the Sky-Dweller model, paired with an Oysterflex bracelet

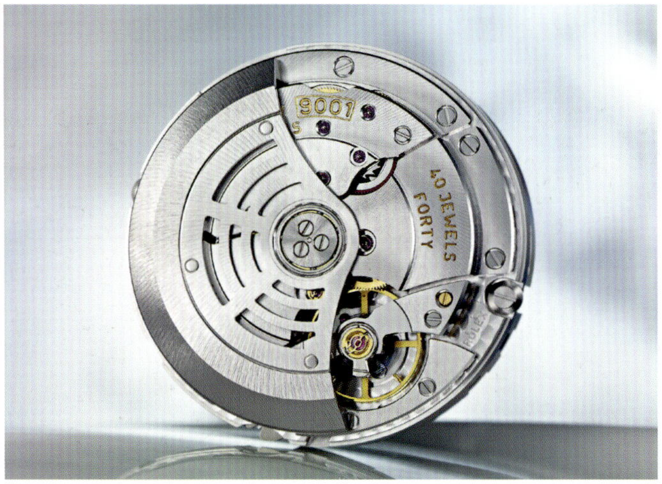

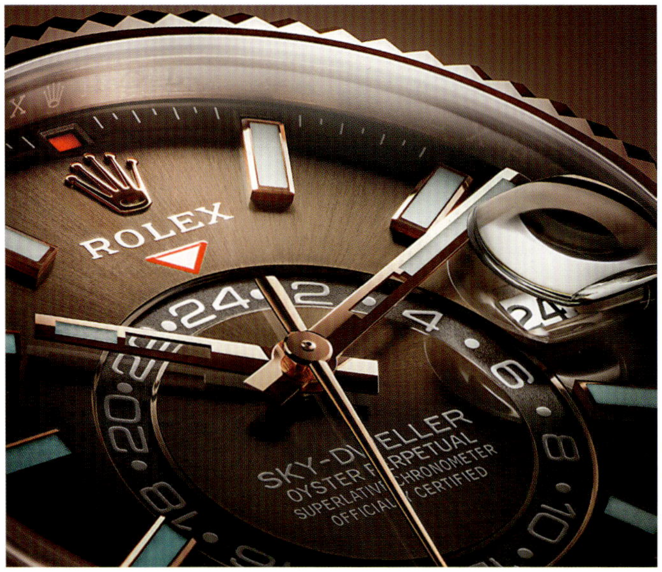

DIAL: The reference time, in a 24-hour display, can be read on an off-centre disc. The local time is meanwhile presented via the traditional central hands. Months of the year are displayed inside twelve apertures along the edge of the dial, with the current month highlighted in red.

CASE: The Oyster case is one of its most recent versions, guaranteed waterproof to a depth of 100 metres. The middle case is carved from a solid block of yellow gold.

BRACELET: The yellow-gold edition of the Sky-Dweller is coupled with an Oysterflex bracelet, creating an unprecedented union of the robustness and reliability of a metal bracelet with the flexibility, comfort and aesthetic of an elastomer strap.

The timepiece, the first in the classic watch category to adopt the innovative high-performance elastomer bracelet, presents a dial in vivid black with a sunray finish, with yellow-gold hands and indices. Reference 326238 has a Chromalight display, with hands and indices coated or filled with luminescent material for a long-lasting glow, emitting a blue light when the watch is in the dark.

The Sky-Dweller is animated by calibre 9001, a self-winding mechanical movement manufactured entirely by Rolex. Boasting several patents, it is one of the most complex movements ever designed by the brand. Its design and manufacture, with the innovations it incorporates, combine to give it extraordinary performance in terms of precision and reliability. The Twinlock winding crown, equipped with a double waterproofing system, screws securely to the case. The watch crystal, with a Cyclops lens above 3 o'clock to facilitate reading of the date, is made of a scratch-resistant sapphire. Calibre 9001 is animated by a self-winding module with Perpetual rotor and boasts a power reserve of 70 hours. Like all watches from Rolex, the Oyster Perpetual Sky-Dweller boasts a Superlative Chronometer certification, signified by a green seal that accompanies each watch alongside an international five-year guarantee.

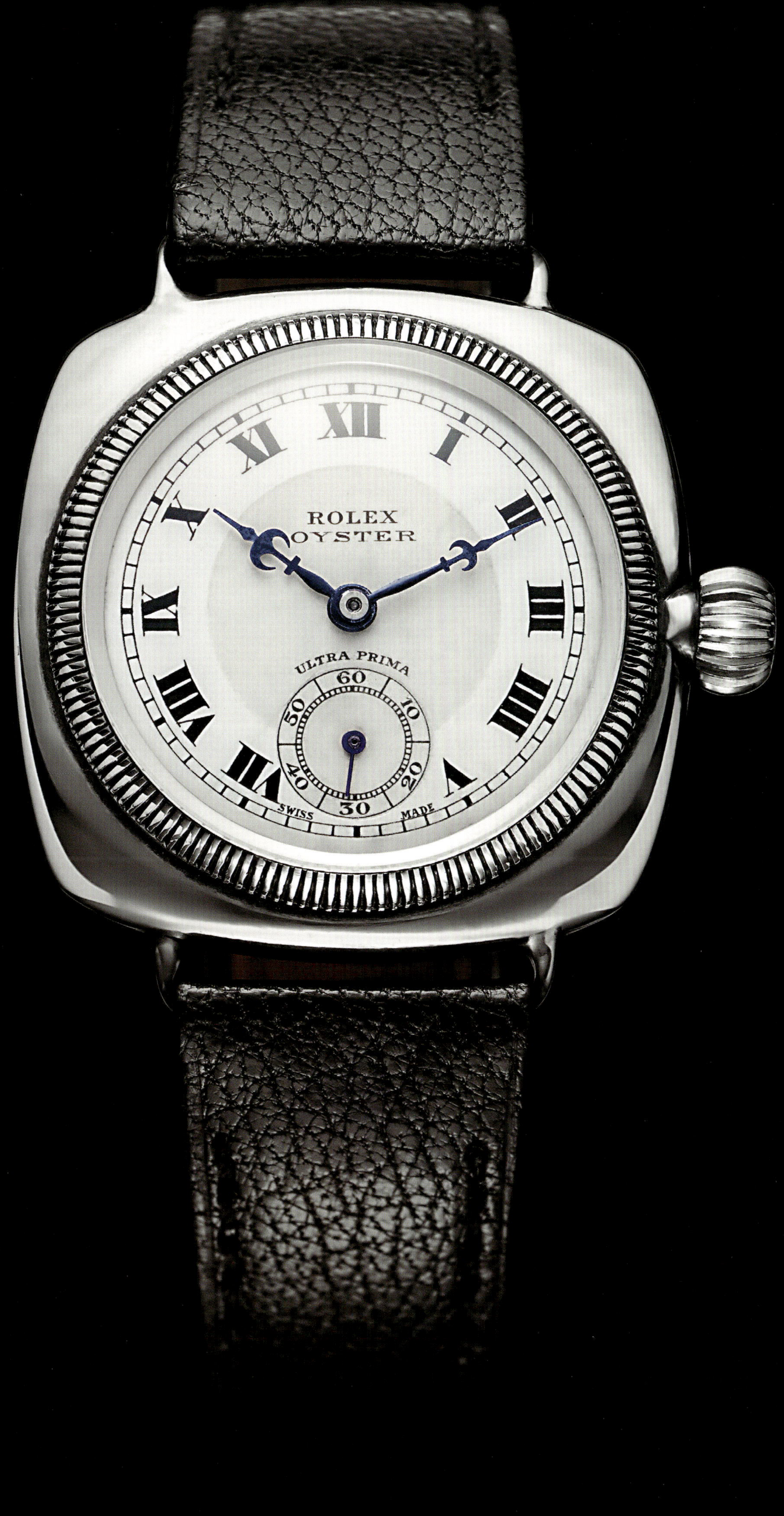
ROLEX
OYSTER
ULTRA PRIMA
SWISS
MADE

The Concept of the Oyster Perpetual

Watches in the Oyster Perpetual collection are characterised by a hermetically sealed case. The most recent version of the Oyster case is waterproof to a depth of one hundred metres – which becomes three hundred metres in the Submariner and Submariner Date, 1,200 metres in the Sea-Dweller, 3,900 metres in the Rolex Deepsea, and an actual 11,000 metres in the Deepsea Challenge. The case can be crafted from a solid block of Oystersteel, RLX titanium, gold or 950 platinum. The subtly grooved case back is hermetically screwed using a dedicated instrument. The Twinlock winding crown, equipped with a double waterproofing system, and the Triplock crown with a triple waterproofing system, specifically included in the brand's diving-watch models, are also securely screwed against the case. All Oyster Perpetual models are equipped with a self-winding mechanical movement with a Perpetual rotor. Today, the 'Oyster Perpetual' name appears on all watch dials except for those with diamond paving. All watches in the collection are submitted for official COSC certification (from the Contrôle Officiel Suisse des Chronomètres). The movements that successfully pass all COSC tests earn official chronometer certification. Rolex, however, evaluates the precision of each finished timepiece in accordance with a second, more rigorous criterion of tolerance than that of the official certification: according to the *maison*, the watches' precision must not deviate more than -2/+2 seconds per day, while COSC tolerates, in the movement on its own, a deviation of -4/+6 seconds per day. Rolex watches therefore boast a Superlative Chronometer certification, redefined by the brand in 2015. The exclusive title attests that each watch exiting its workshops has successfully passed a series of tests conducted on a fully automated chain using high-tech equipment. These test the watch as a whole following the installation of its movement inside its case, guaranteeing its performance and the level of its precision, water resistance, automatic winding and autonomy. Superlative Chronometer status is represented by a green seal that accompanies every Rolex watch.

Watches in the Oyster Perpetual collection are also distinguished for their rigorous planning and meticulous selection of materials. For their cases, dials and bracelets, Rolex only employs precious or highly robust materials: 18-carat gold, cast in the company's own foundry, then shaped and milled in its own workshops; 950 platinum; Oystersteel, with extremely high anticorrosive properties; RLX titanium (a lightweight grade-5 titanium alloy that is also resistant to warping and corrosion); or high-tech ceramic for Cerachrom bezels and inserts. The same reliability in these Oyster Perpetual models can be found in their bracelets and the clasps Rolex has developed and manufactured, ensuring wearers the highest level of comfort and security.

Over the years, Rolex has carried on the vision of its founder Hans Wilsdorf: that a timepiece must, above all, privilege function and reliability, under all circumstances.

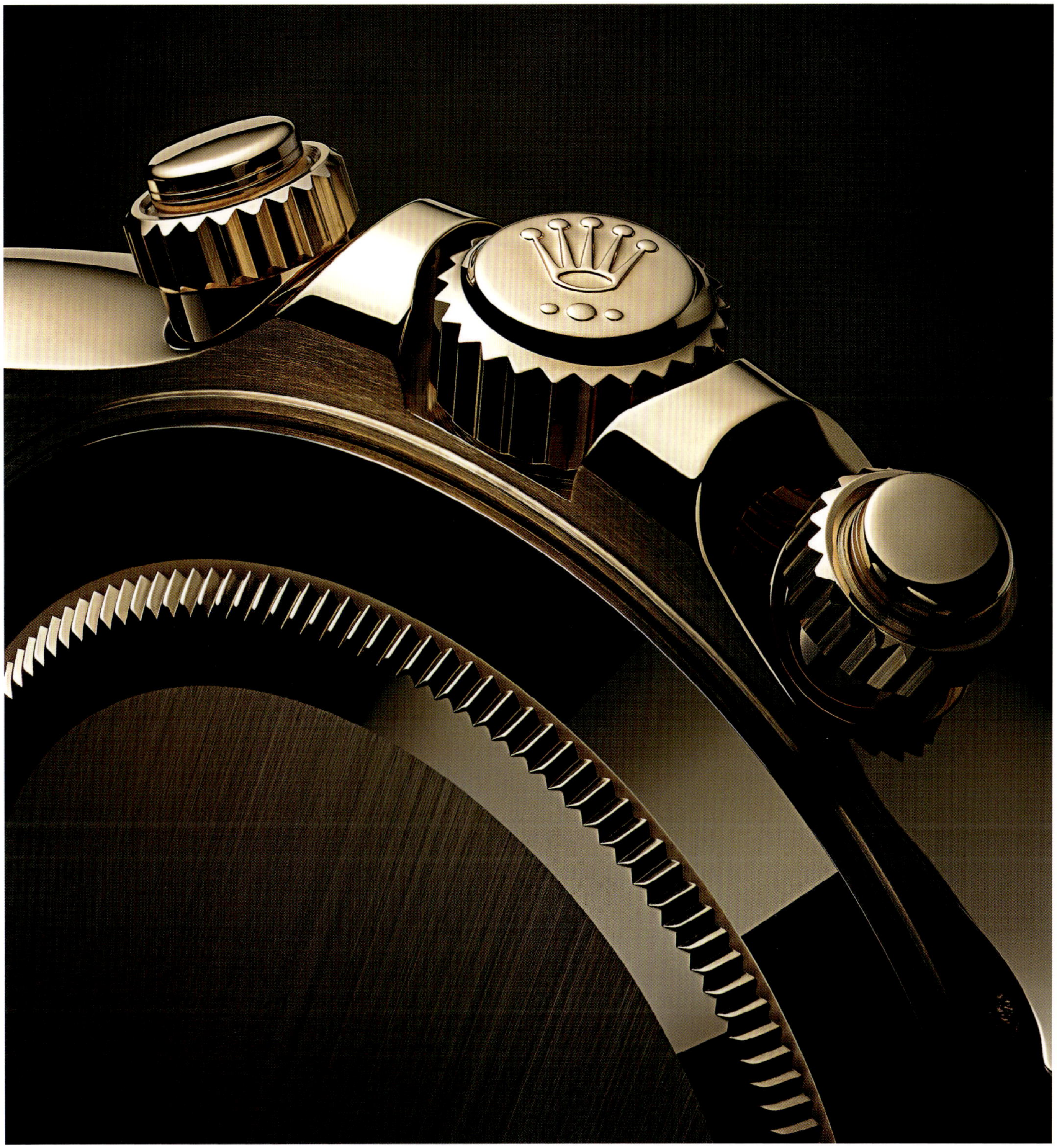

The Cosmograph Daytona is fitted with the calibre 4130, a self-winding mechanical chronograph movement entirely developed and manufactured by Rolex, which offers a power reserve of approximately 72 hours. This movement incorporates a chronograph function with fewer components, further enhancing its reliability. The chronograph is engaged by a mechanism with a column wheel and vertical clutch. This allows for an instantaneous and highly accurate start to timing.

The Oyster's fluted bezel originally had a specific function: it was used for screwing it down to the case, helping to ensure the timepiece's water resistance. Consequently, the bezel's fluting was an exact reproduction of the flutes on its case back, which, for the same reason, was screwed to the case using special Rolex tools. Over time, the fluting has turned into an aesthetic feature, a hallmark of Rolex only available in luxury editions made of gold or platinum.

The Rolex Oyster Case, which was revolutionary for its perfect watertightness, was tested some years following its launch by the celebrated polar explorer Henry George 'Gino' Watkins, who brought various Oyster Perpetual specimens with him on his long expedition along the coast of Greenland between 1930 and 1931. On his return, Watkins shared his admiration for these timepieces with Rolex, as they had been submerged in water many times and continued to function perfectly.

In 1933, pilots on the Houston expedition wore Rolex watches while conducting the first flight over Everest, over ten thousand metres above sea level. One of them, Lieutenant-Colonel Stewart Blacker, later stated 'I can hardly imagine that any watches have ever been subjected before to such extremes under practical conditions. In spite of all this, the watches continued to work with the utmost accuracy.[…] Not the slightest fault could be found with them.'

The highly talented young mountaineer Annelies Lohner, hailing from a small village called Grindelwald at the foot of the famed Jungfrau in the Bernese Alps, demonstrated remarkable strength of character when she offered to organise the first post-war Swiss expedition to the Himalayas. A lover of adventure and a pioneer in climbing the highest peaks, she managed to convince the Swiss Foundation for Alpine Research that an expedition to the Gangotri Group of mountains, part of the Garhwal Himalayas in northern India, was the right course of action. The expedition, of which she was part, faced its initial stages in the space of five months, from May to September of 1947, successively climbing Kedarnath, Satopanth, Kalindi (from the northeast ridge), Balbala and Nanda Ghunti, not neglecting a reconnaissance in the Chaukhamba massif. Each of these mountains' peaks exceeds six thousand, if not seven thousand, metres in altitude. To support them in their endeavour, Rolex entrusted Oyster Perpetuals to all five members of the expedition and these were kept on their wrists for the duration. The watches accompanied them in each of their movements throughout the day, facing the most extreme conditions and never failing. Upon their return, each mountaineer emphasised their characteristics of water resistance, precision and convenience, this last aspect denoting the movement's self-winding system via the Perpetual rotor. Expedition leader André Roch wrote on 7 July 1947, from Gangotri base camp: 'The Rolex watches that we are each wearing keep surprisingly accurate time. They are very useful, and we are delighted with them. The fact that we do not have to wind them is especially appreciated.'

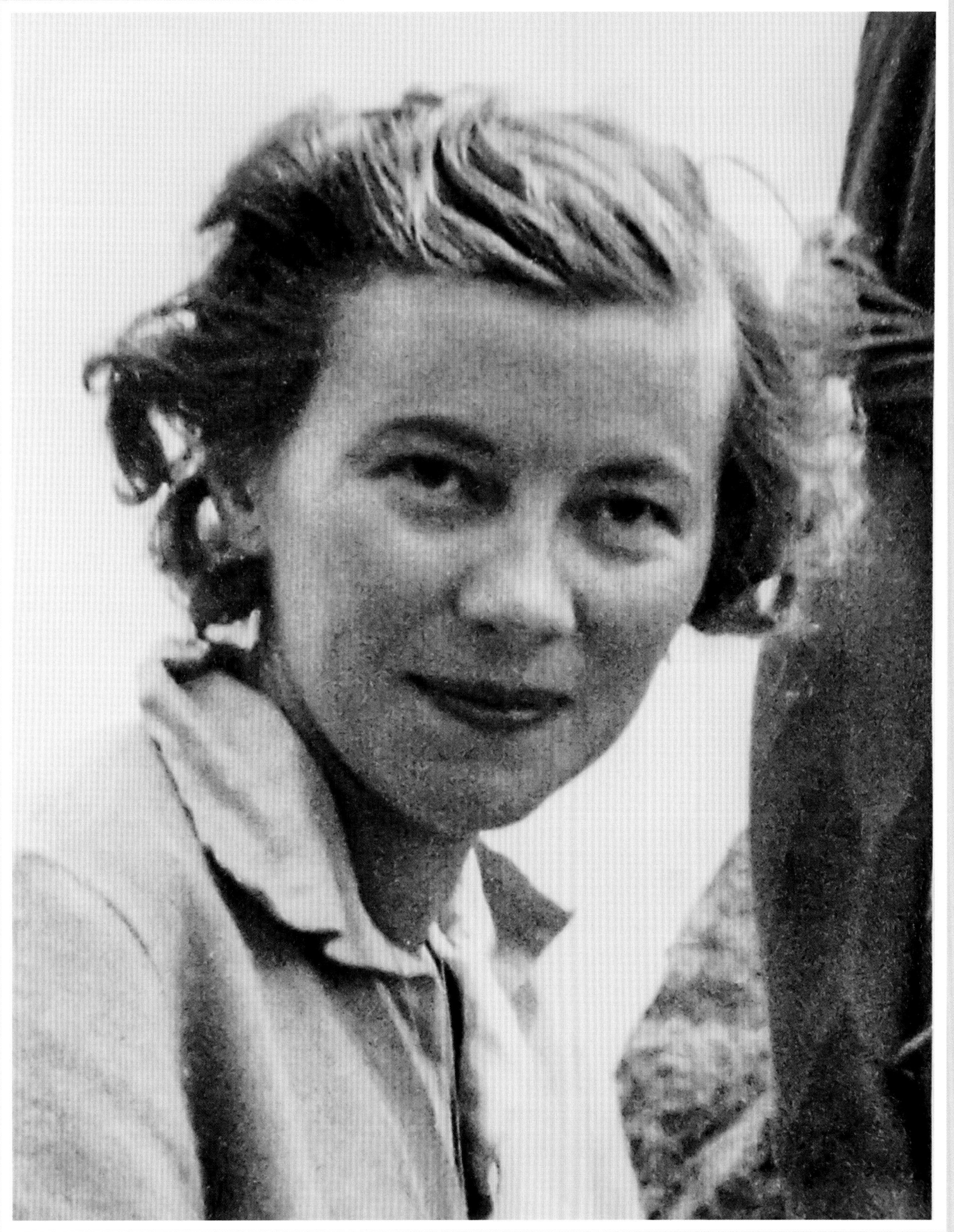

The highly talented young climber Annelies Lohner used an Oyster Perpetual on her historic climbs.

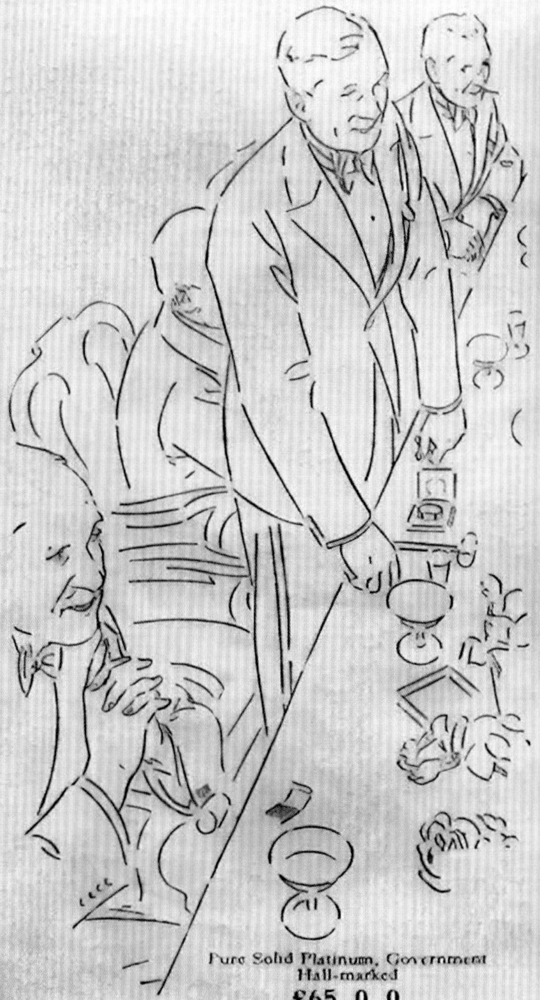

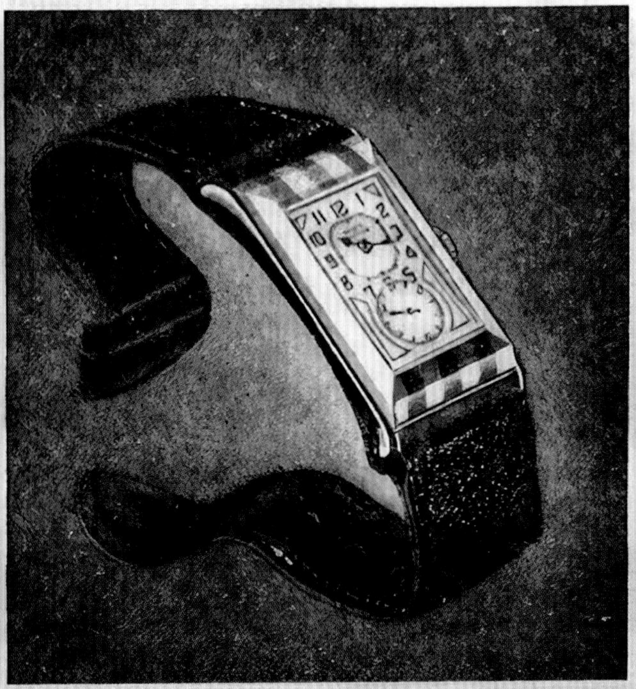

Actual Movement of Rolex 'Prince'

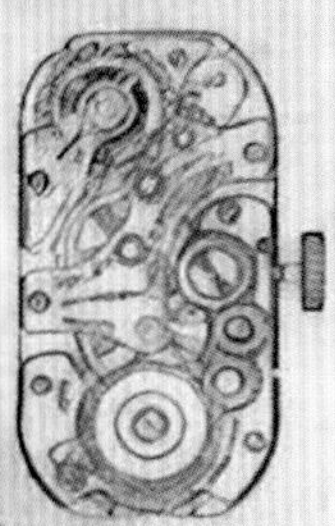

A WORTHY GIFT—*for services rendered*

- - - - - - The Rolex 'Prince' is no ordinary wrist watch—it is a *supergrade* production of outstanding merit and distinction. The 'Prince' is the only wrist watch in the world known to horological science as "Observatory Grade" and its perfect accuracy is verified by Swiss Government Scientists.

It is a watch for men of distinction . . . . a presentation watch in recognition of exceptional service rendered to art, science, or industry . . . . the highest mark of esteem to men of special talents or achievements.

Beautifully curved to the wrist—separate "seconds" dial—set in 15 fine rubies—58 hours with one winding—accuracy not affected by change of climate or temperature.

Manufacture of the 'Prince' wrist watch is limited to a comparatively small number—such fine watches cannot be produced quickly or in large quantities.

Pure Solid Platinum, Government Hall-marked
£65.0.0

18 carat, White and Yellow Inlaid
£21.0.0

9 carat Plain Yellow Gold
£14.14.0

Non-tarnishable Sterling Silver
£8.8.0

These prices include a Swiss Government Observation Certificate of Super-Excellence, a spare strap, and a spare glass.

TRADE MARK

25 WORLD'S RECORDS FOR ACCURACY.

The Rolex 'PRINCE'

The watch for Men of Distinction

25 WORLD'S RECORDS FOR ACCURACY AT KEW (ENGLAND), GENEVA AND NEUCHATEL OBSERVATORIES.

See name on Dial and Movement.

Manufactured by the—

ROLEX WATCH CO., LTD. (H. Wilsdorf, *Managing Director*), GENEVA & LONDON

Stocked by good Jewellers throughout the British Empire.

Sole Agents for India: J. Boseck & Co., Calcutta; Lund and Blochley, Bombay; P. Orr & Sons, Ltd., Madras and Rangoon.

Garland Ad.

COLLECTABLE PIECES

Some timepieces that were produced in small numbers and over limited periods are now assiduously sought by collectors.

In the early 1900s, when men were beginning to wear wristwatches rather than placing them in their pockets, it was immediately clear that the new timepieces would have a much 'harder' life. Pocket watches, protected inside one's clothing, were spared exposure to rain and humidity, as well as dust and shocks. These were elements that inevitably became part of the active lifestyle that was growing more popular in those years.

The introduction of a waterproof watch case and self-winding movement from Rolex, taking place over the 1920s and the beginning of the 1930s, certainly changed the history of watchmaking forever, setting the standard for timepieces of the modern era. If many of the models still being produced by Rolex today date from these early prototypes and are widely known on the market, others belonging to the 'crown' *maison*'s history were produced in small quantities, in limited or personalised editions. Today, these are highly valued on the collector's market.

Among these examples of rare technical features and design, for example, is the first Oyster from 1926, with an octagonal design in homage to the Art Deco style that was in vogue in Europe. Just a few years later, in 1931, the first Oyster Perpetual had a much more streamlined slight tonneau shape; it soon assumed the form which would then remain nearly unchanged until today.

Also following the Art Deco style, the Prince model was a watch in rectangular format with a dual dial; it was also dubbed the 'Doctor's Watch' for its separation of the hours and seconds dials, allowing the doctor to more easily read a heartbeat and pulse. Unlike the Oyster, it had no waterproof case, but its movement allowed for a higher degree of precision. The first Rolex Prince models were the Prince Classic and Prince Brancard: the

Vintage Rolex Prince ad declaring: 'The watch for men of distinction.'

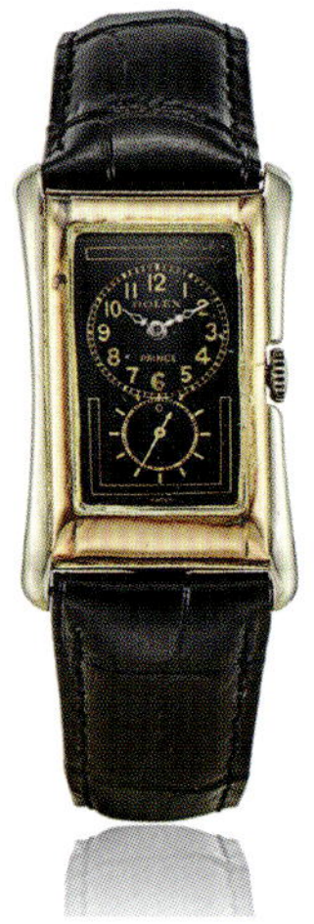

The Prince model.

chief difference was in the sides of their cases, straight in the Classic and flared in the Brancard. Several variations of these two models were produced by Rolex, in terms of both the dials and the metal used for their cases, among them the Railway Prince, with a case developed on several fronts and the Prince with a 'jump hour', displaying a minute hand in its upper dial while the hour was shown on an internal rotating disc via a window on its dial.

It is interesting to note that an advertisement from 1930, apart from highlighting its guaranteed Observatory Grade accuracy, touted the Prince as the 'Watch for Men of Distinction', a premium timepiece suited to those distinguished for skill in the arts, sciences or industry; it highlighted that a watch of such workmanship was produced only in small quantities, an item not suitable for large-scale production. An advertisement from 1932 also proclaimed that the Prince was not an object of aesthetic value alone – a 'gem' like many others on the market – but, despite its simple, sophisticated elegance, it was an instrument of extreme precision, as indicated by the certificates it had garnered. The advertising did not neglect to emphasise that, like the Prince, the women's version called the Princess and the Oyster were intended for an elite and sold by watch vendors as elite, positioning the products within an exclusive segment of the market.

In 1935, Rolex offered a special Jubilee edition, limited to five hundred examples, to commemorate the jubilee of Britain's King George VI. For the launch of this special model, the Jubilee watch was advertised as an object that only five hundred families could pass on to future generations; the lucky children and grandchildren of forward-thinking customers could make a connection to history through this watch. To facilitate the acquisition of this watch with 'perfect precision', some complimentary services were also conceived and guaranteed for three years, which included insurance against theft or loss, repairs and engraving on the back of the case.

Rolex discontinued the Prince models in the late 1940s and, more than half a century later, revived them as part of the Cellini line in 2005.

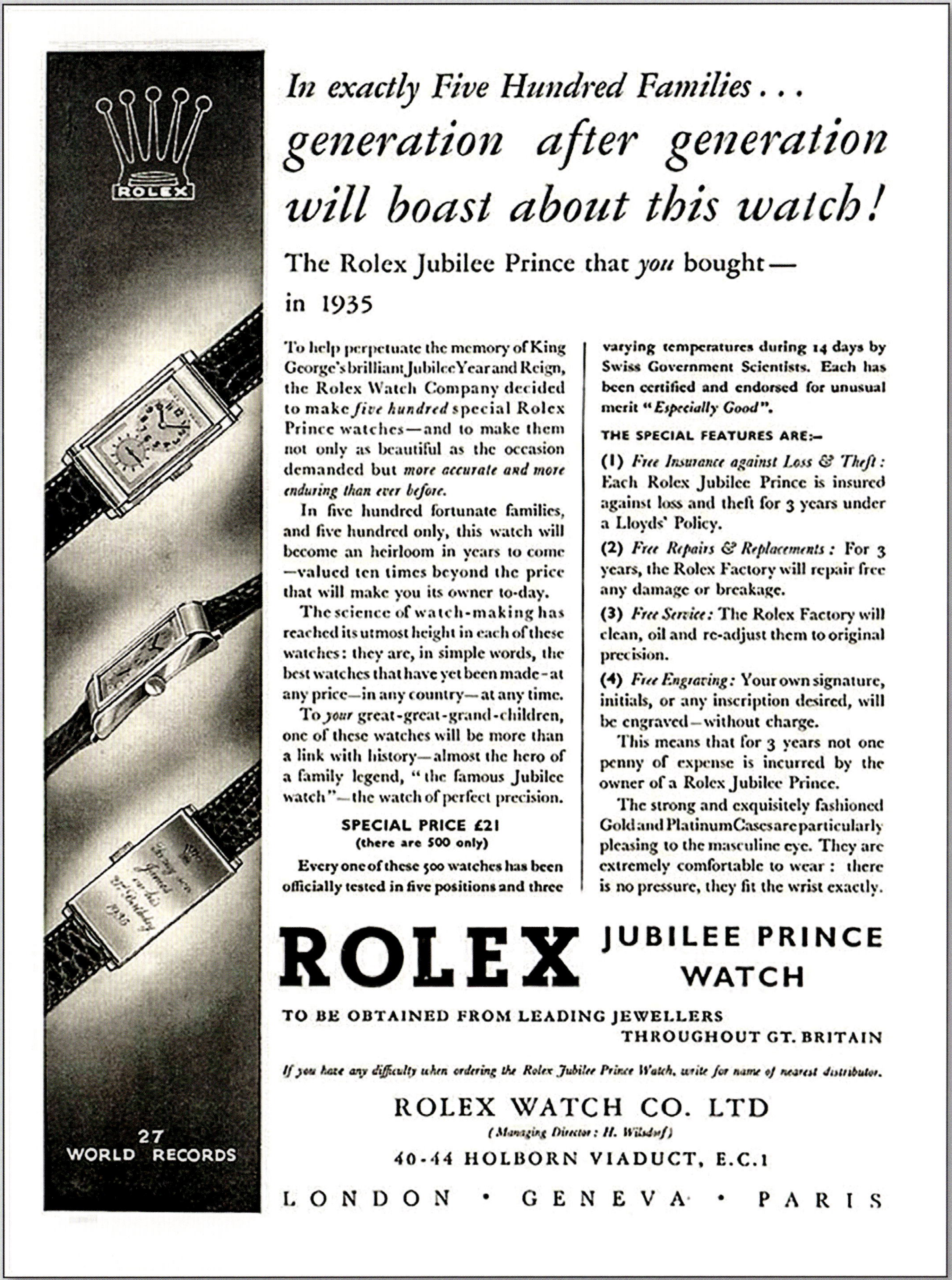

1935. Advertisement for the Jubilee Prince model.

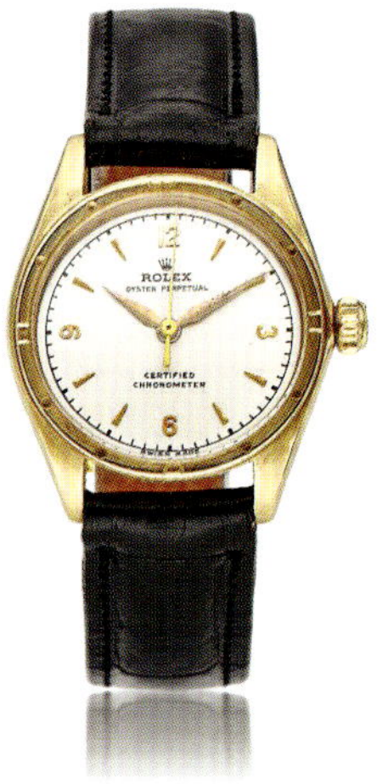

The 'Bubbleback'

In 1931, Rolex introduced and patented the first self-winding mechanism in the world with its Perpetual rotor. This device was the progenitor of all modern automatic watches. The union of an Oyster case with Perpetual winding brought a long and fortunate series of models to be produced and sold. The forefather of these models was the reference 1858: it was called the 'Bubbleback' in English, to emphasise its curved shape, since the rotor was fixed to its movement and the case back needed to be more spacious to accommodate it. The long line of Oyster Perpetual 'Bubbleback' products includes examples with various bezels with different decorative processing and hands in many shapes. Bubbleback dials also evolved quickly, both in terms of colour and graphics. These first models were announced to the public via advertisements that highlighted their water-resistance and reliability. At the same time, Rolex began to provide Oyster watches for explorers, pioneers and athletes. This was to test their timepieces' technical features and improve them, while also furthering relationships between the brand and exceptional people who could serve as endorsements.

The first figure in this vein was Mercedes Gleitze, who in swimming the English Channel in 1927 put the Oyster's water resistance to the most convincing of tests. Its 'aerial test' was meanwhile entrusted to the pilot Charles Douglas Barnard, who held numerous records, among them a nine-thousand-mile flight in one hundred hours from England to Cape Town, South Africa in 1930. The first flight over Everest in 1933, the assignment of the Houston expedition, constituted a new extreme in testing the Oyster. In 1935, 'flying' in his Blue Bird racecar, with an Oyster on his wrist, Sir Malcolm Campbell was the first man to reach the legendary land speed of 300 miles an hour.

In 1945, Rolex produced a waterproof self-winding model with a small window on its dial at 3 o'clock indicating the day of the month: the 'Big Bubbleback'. It had the same rounded case back as the original Bubbleback to contain the bulky rotor, but was four millimetres wider (36 millimetres), with the date display: this was the precursor of the Datejust.

From 1949, the Oyster evolved with new shapes and mechanisms, very much nearing the design of today's models, with a slightly wider and slimmer case.

An example of the Oyster Perpetual model nicknamed the 'Bubbleback' for the oval profile of its case.

Chronographs

The earliest Rolex chronographs were birthed toward the middle of the 1930s. These would evolve until the arrival in 1963 of the Cosmograph Daytona. Viewed as professional watches, these chronographs were produced in very small quantities, so much so that today they are considered Rolex icons. Toward the end of the decade, models with two push pieces were also introduced. Beginning in 1937, the first chronographs were sold with pulsometer and telemeter scales on the dial. Chronographs from the 1930s and 1940s had no Oyster case and were therefore not waterproof. Chronographs with the reference numbers 6234 and 6238 from the early 1960s were, however, equipped with an Oyster case: number 6238 in particular is considered a 'pre-Daytona' model. Simple and minimal, at its peak there were versions of this model with rare, personalised dials. Advertising at the time underlined its technical superiority. Also rare and highly sought after are the twelve *rattrapante* ('flyback') chronograph specimens, providing two different seconds readings, produced in 1942. This was the reference 4113, a unique model in the history of Rolex that was never placed on the market. These twelve watches were given as honourary gifts to racecar drivers and their teams.

Rolex 'prisoner of war watches' were also limited; these watches could be acquired by British soldiers captured by the Germans in World War II. Rolex offered to replace watches that were confiscated from soldiers captured and sent to prison camps, not asking for payment until after the war ended. In order to receive them, soldiers needed only to write Rolex explaining the circumstances of their loss and indicate where they were being detained. The watches were then sent to the prison camp, typically through the Red Cross. This form of credit contributed no small boost to morale among Allied prisoners of war, as the operation indicated that the manufacture and its owner did not believe that the Nazis would win the war – had it ended otherwise, the watches would not have been reimbursed. Hans Wilsdorf personally signed letters to servicemen who requested a timepiece: this was the reference 3525, one of the first chronographs produced with an Oyster case, apparently produced from the end of the 1930s or the beginning of the 1940s.

Other rare examples from wartime include models born of a collaboration between Rolex and Panerai. In the mid-1930s, the Italian navy turned to Panerai in Florence for watches that could equip its divers.

This is how the diving watches designed and made by Panerai were created: they were equipped with a special dial treated with Radiomir to meet necessary parameters for legibility, while their case, movements and crown were signed by Rolex. A very high value is assigned to all these watches as collectables.

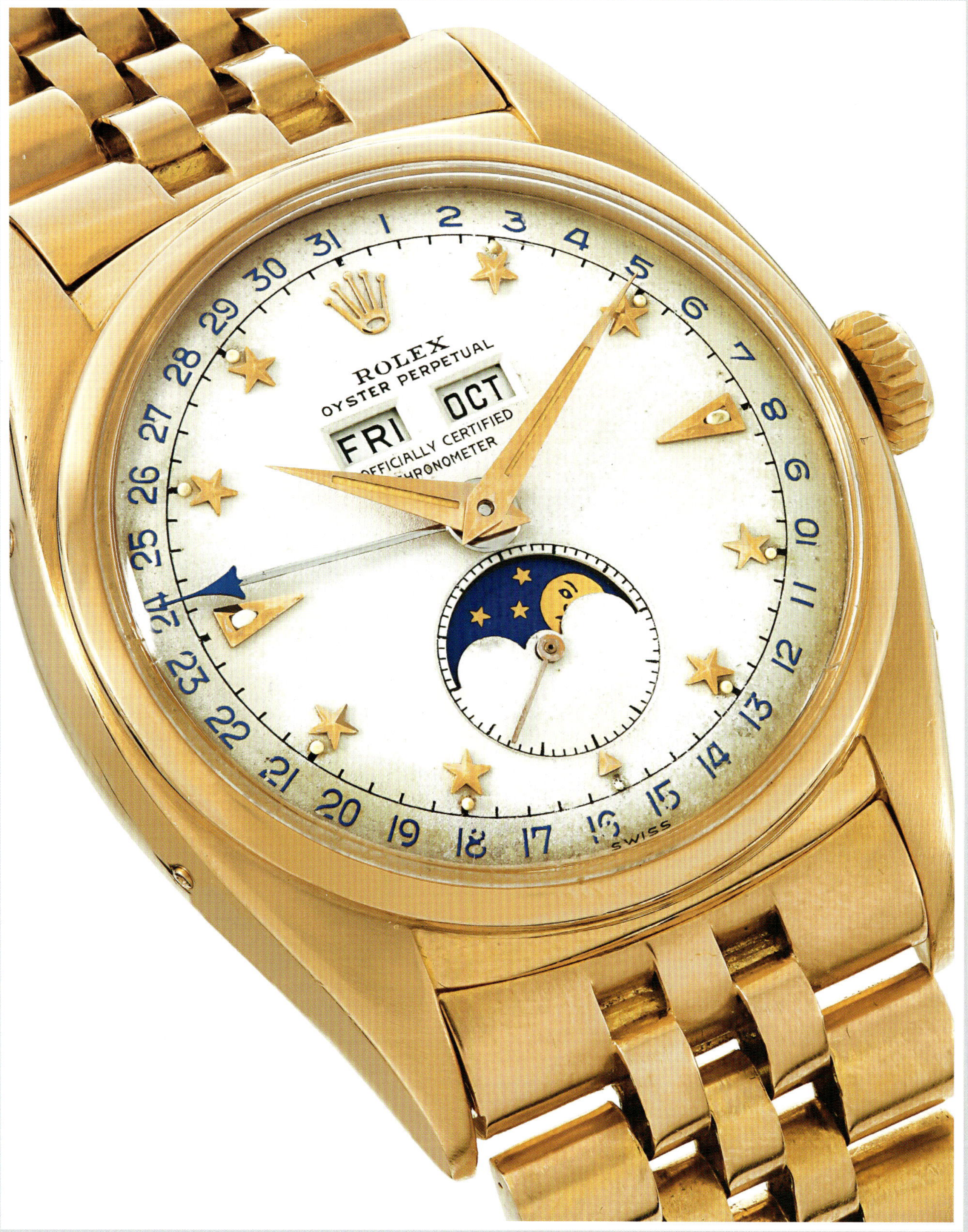

The model nicknamed 'Stelline' due to the shape of the indices. It is one of the few complicated models produced by the *maison*.

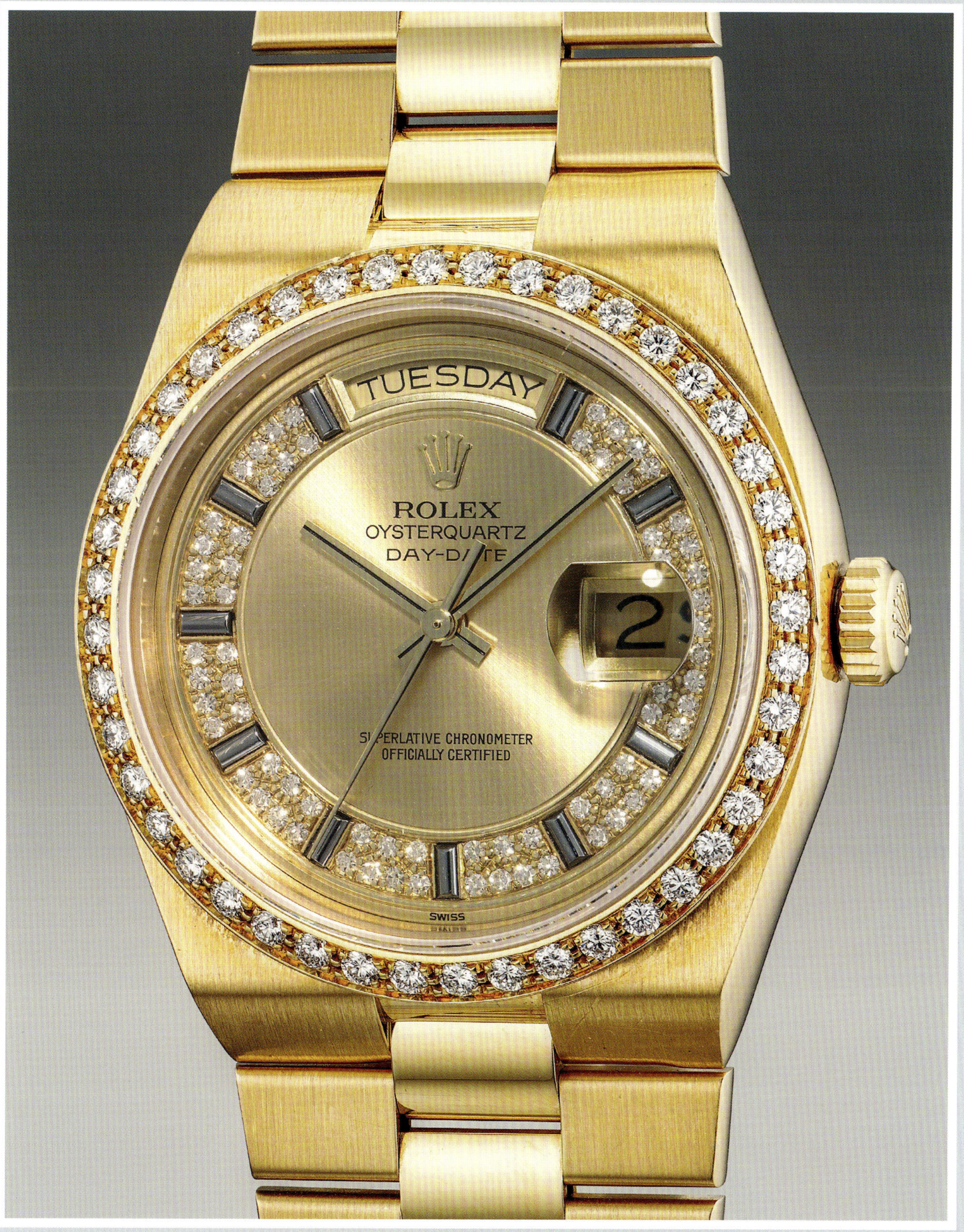

A rare Oysterquartz reference 19048 in yellow gold and diamonds.

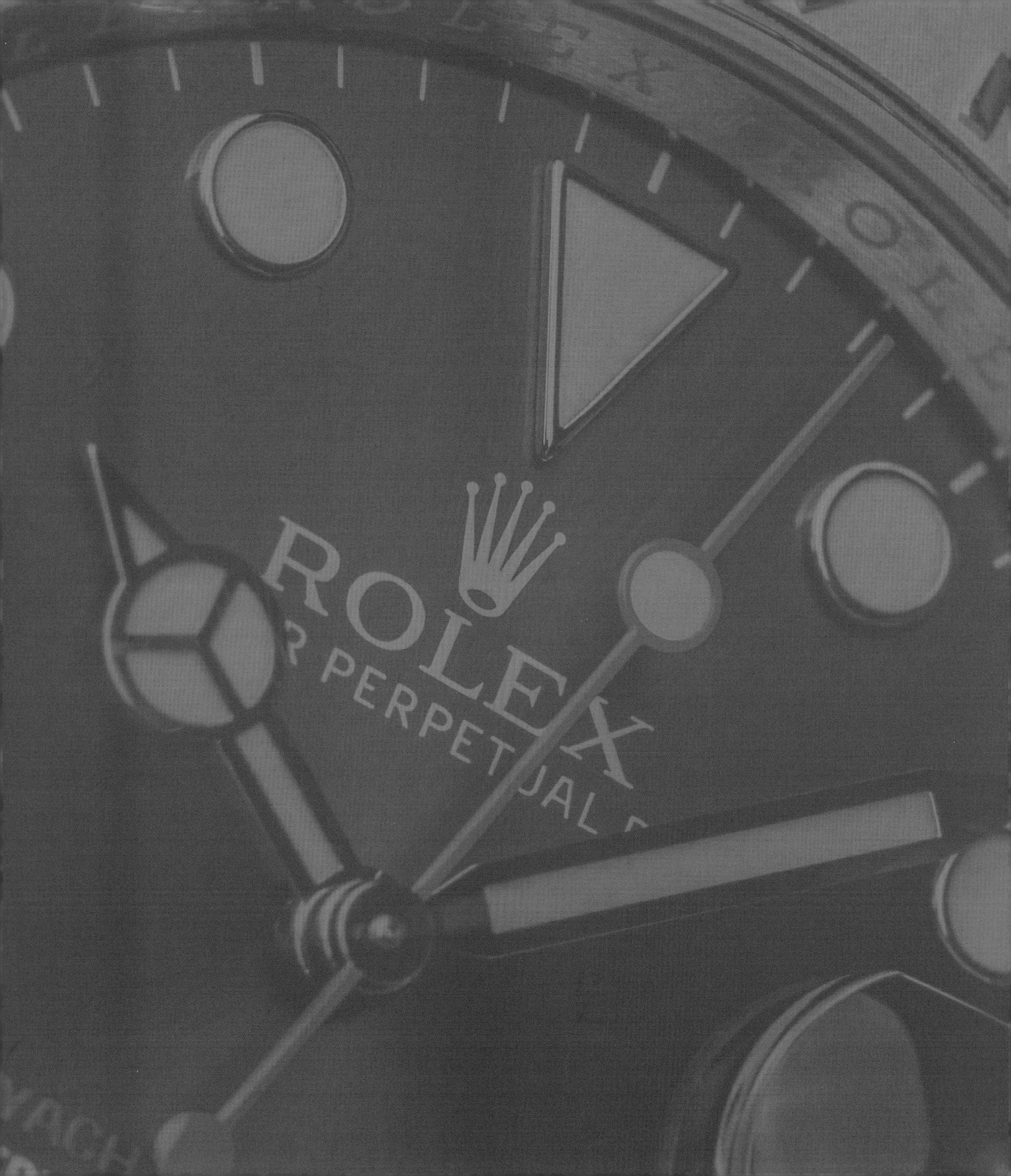
ROLEX

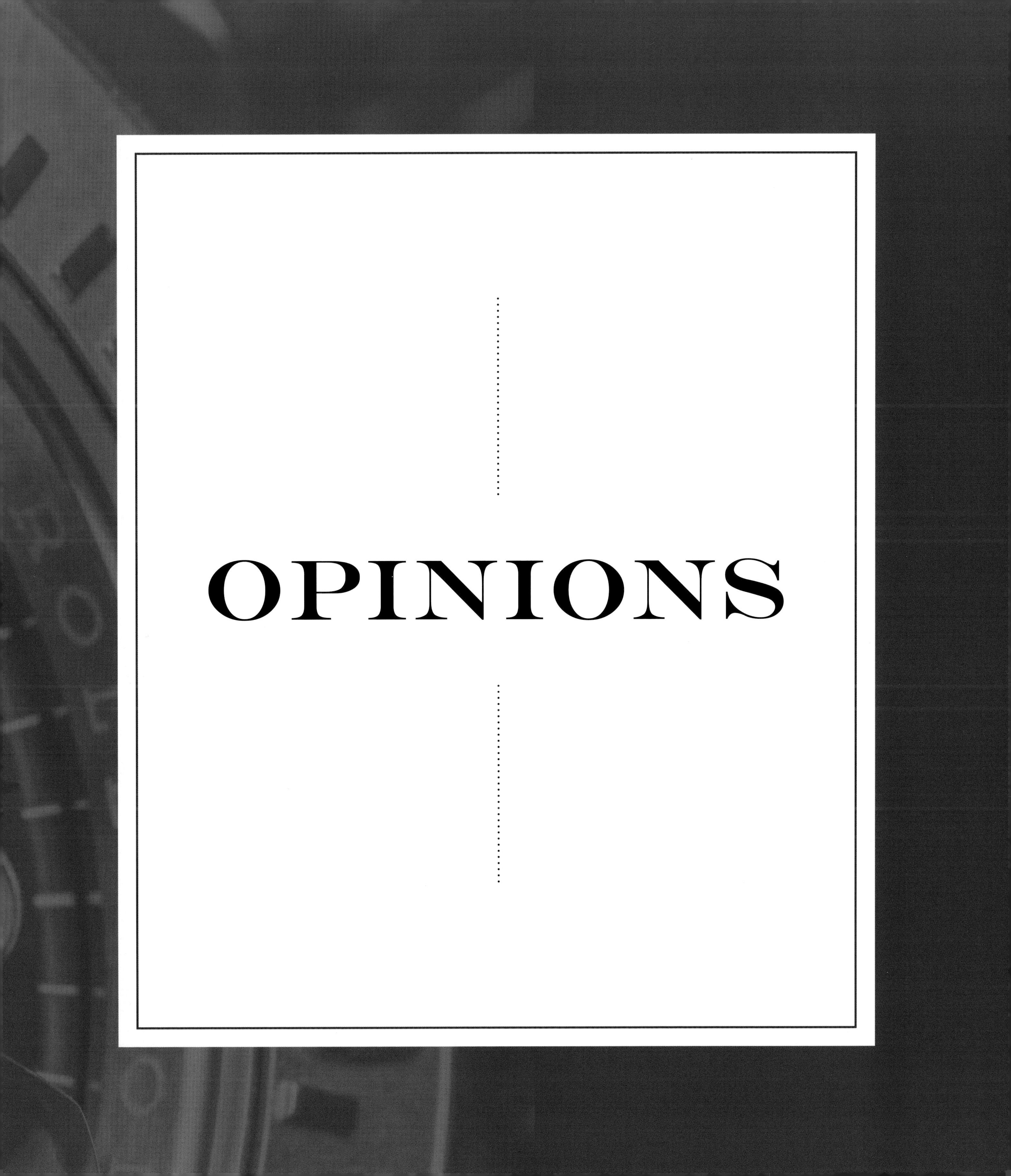

OPINIONS

Osvaldo Patrizzi

THE POWER OF A NAME (OR NICKNAME)

Leading horology expert Osvaldo Patrizzi analyses the importance of names the brand has given to models, cases and bracelets. Meanwhile, the nicknames and denominations of certain references and characteristic features, including the famous 'Patrizzi dial', are owed to collectors.

One of the aspects determining Rolex's modern character from its very beginning is the use of names in lieu of the reference numbers that all watchmakers have used historically. According to Patrizzi, 'the names would assume an extremely important role over the years and would be indicative of the Rolex brand, which got its start in the early twentieth century, later than that of most big brands. And while the other *maisons* were still largely creating pocket watches, the specific goal of Hans Wilsdorf was to produce wristwatches, addressing a new audience.' The first waterproof case was called Oyster, a name that certainly recalls its subaquatic nature, in addition to its preciousness as a technical 'pearl'. Perpetual, meanwhile, was the name given to the mechanical movement allowing the model to self-wind automatically. To follow were model names referring to a concept of exclusivity, exceptionality and precision in the domain of time: Prince, Datejust, Submariner, Explorer, GMT-Master, Yacht-Master. . . . Standing out among all these is the iconic Daytona, associated with the speed of motor racing. Even the bracelets would be named Jubilee or President. One of Rolex models' most interesting features is that they have an unusually high chance of being given a nickname.

It was traditionally collectors who began to name certain models after products or movie characters and by referring to the specific characteristics of an individual model. 'Whether spontaneous or marketing-driven, nicknames have helped to fuel the myth of the brand, giving models a greater emotional proximity to buyers while not necessarily making the watches more accessible.'

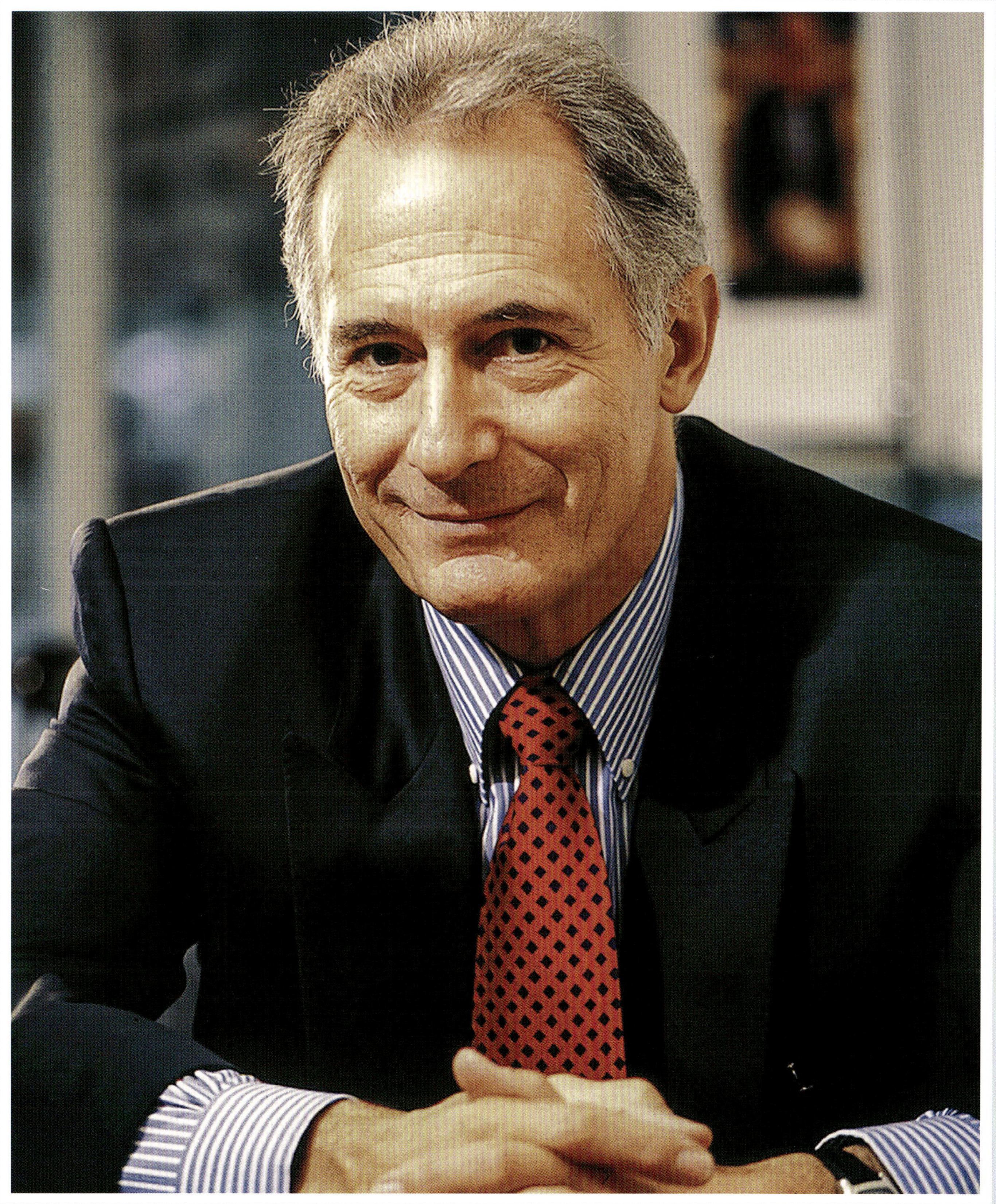

Osvaldo Patrizzi, watch-industry expert and consultant.

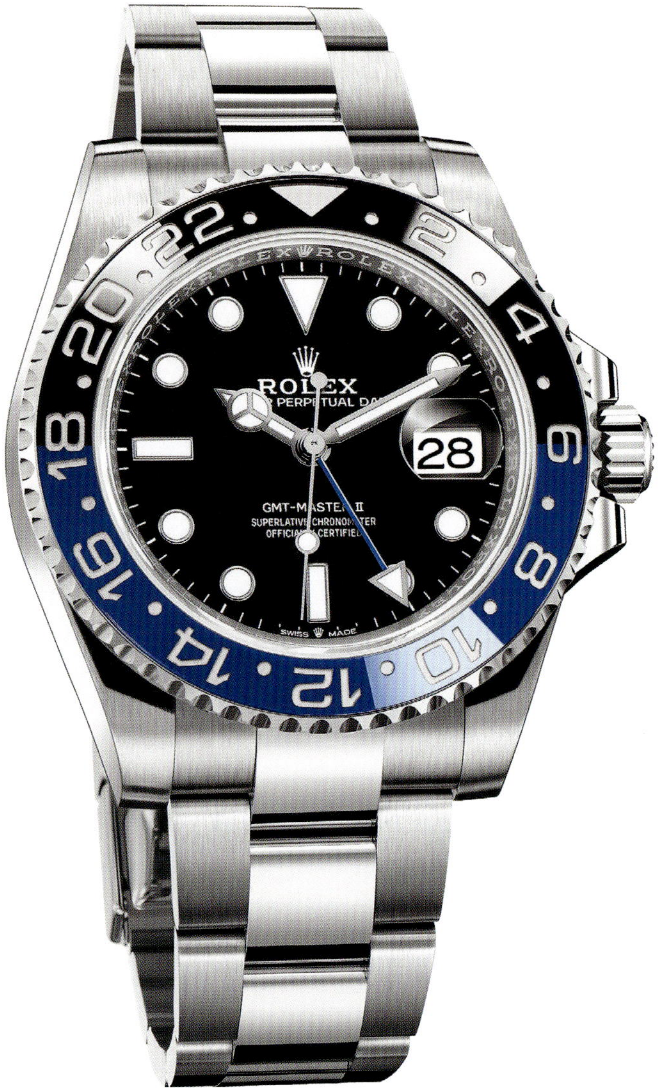

The 'Batman' edition of the GMT-Master model.

Colour Variants

Rolex GMT-Masters with a red and blue bezel are called 'Pepsi'. This name comes from the similarity between the colours of the watch bezel and of the logo belonging to the famed American soft drink. The GMT-Master with a red and black bezel insert is called 'Coke' for its affinity with the logo of another soda giant, Coca-Cola. The Rolex GMT-Master with a black and blue bezel is called 'Batman', in this case evincing its connection with the DC Comics superhero, who sports a distinctive costume in this colour combination. The GMT-Master II reference 16713 is commonly called 'Tiger's Eye' by virtue of its dial's colouring, recalling the gemstone's various brown hues, but it has also earned the nickname of 'Root Beer', as it captures the drink's characteristic shade of brown.

The Submariner featuring a green ceramic dial and bezel is named 'Hulk', after the Marvel superhero. However, the version with a green bezel introduced in 2003, made to celebrate the model's fiftieth anniversary, is known as 'Kermit', named after the Muppet character created by Jim Henson.

Specifics of Shape and Size

Some models derive their nicknames from their distinctive characteristics. In 1933, Rolex presented a model with an Oyster case and Perpetual movement: this nameless model would be called the 'Bubbleback' – or, to Italian collectors, the '*Ovetto*', meaning 'little egg' (in the version with a 32-millimetre case) and '*Ovettone*', or 'large egg' (with a 36-millimetre case) – because the case back is uniquely rounded to create space for the rotor.

The reference 8171 is known as *'Padellone'*, produced over a period of just four years, between 1948 and 1952, with a case measuring 38 millimetres – an unusual size for the time, hence its name, translating to 'large frying pan' in Italian. This was the first complicated watch by Rolex, featuring an annual calendar and an automatic movement. The model with an annual calendar complication showing the date, month and weekday, as well as moon phases, dates back to 1952 – the technical masterpiece went down in history with the nickname of 'Stelline' (Italian for 'little stars'), thanks to the star shape of its hour markers. Meanwhile, the coloured arrow-shaped hand is what gives the Explorer II the name of *'Freccione'* in Italy, meaning 'big arrow'. From 1971 onwards, it was also called 'Steve McQueen' after the famous actor – who does not, however, appear to have worn it. The Sea-Dweller's name, *'Polipetto'* ('small octopus' in Italian), can be easily understood by glancing at its dial, which bears an octopus, the emblem of the Italian Police Divers Corps.

The Daytona and its Variants

One of the most famous models by the crown-bearing *maison* is the Daytona, whose success is also linked to its association with the Hollywood idol Paul Newman. This association between the watch and the actor is in fact owed to Osvaldo Patrizzi: 'I had the idea of naming the Rolex Daytona reference 6239 after Paul Newman after I saw a photograph of the American actor wearing that model. It was a Rolex Cosmograph with an unusual two-tone dial, different from other classic dials, which his wife Joanne Woodward had given him. It was a watch that he wore often, and it even appeared on the Italian poster for the 1969 film *Winning*.' It was at an auction in Basel in April 1990 that the watch in question, dubbed the 'Paul Newman', was made available to the public, who welcomed it with great enthusiasm: in that moment, the iconic Daytona Paul Newman was born.

Another example of this form of 'emotional' communication that Patrizzi implemented was his idea to name references 4767, 6036 and 6236 after the famous Olympic ski champion Jean-Claude Killy. A great Rolex enthusiast, Killy loved to wear this triple-date chronograph model. In this case as well, the model became part of the history of collecting. 'The decision to replace reference numbers with names was certainly a winning move in facilitating their identification. It allows you to go beyond the simple technical aspect and stylistic features; myth, desire and the concept of exceptionality take over.'

The Patrizzi Quadrant

The Daytona Patrizzi is a Rolex Daytona reference 16520 with a black dial whose white counters have turned various shades of brown. 'Over the years, I understood why some dials on the reference 16520 tended to change colour over time. The colour change is due to Rolex's use of a transparent varnish called 'Zappon' to protect the dial. Due to the organic origin of the varnish, the counters were exposed to natural oxidation. This work of nature has created unique pieces of great value. Collectors have honoured me by giving my name to this variation.'

Today these specimens are especially coveted by collectors, and their valuations are significantly higher than those of other models.

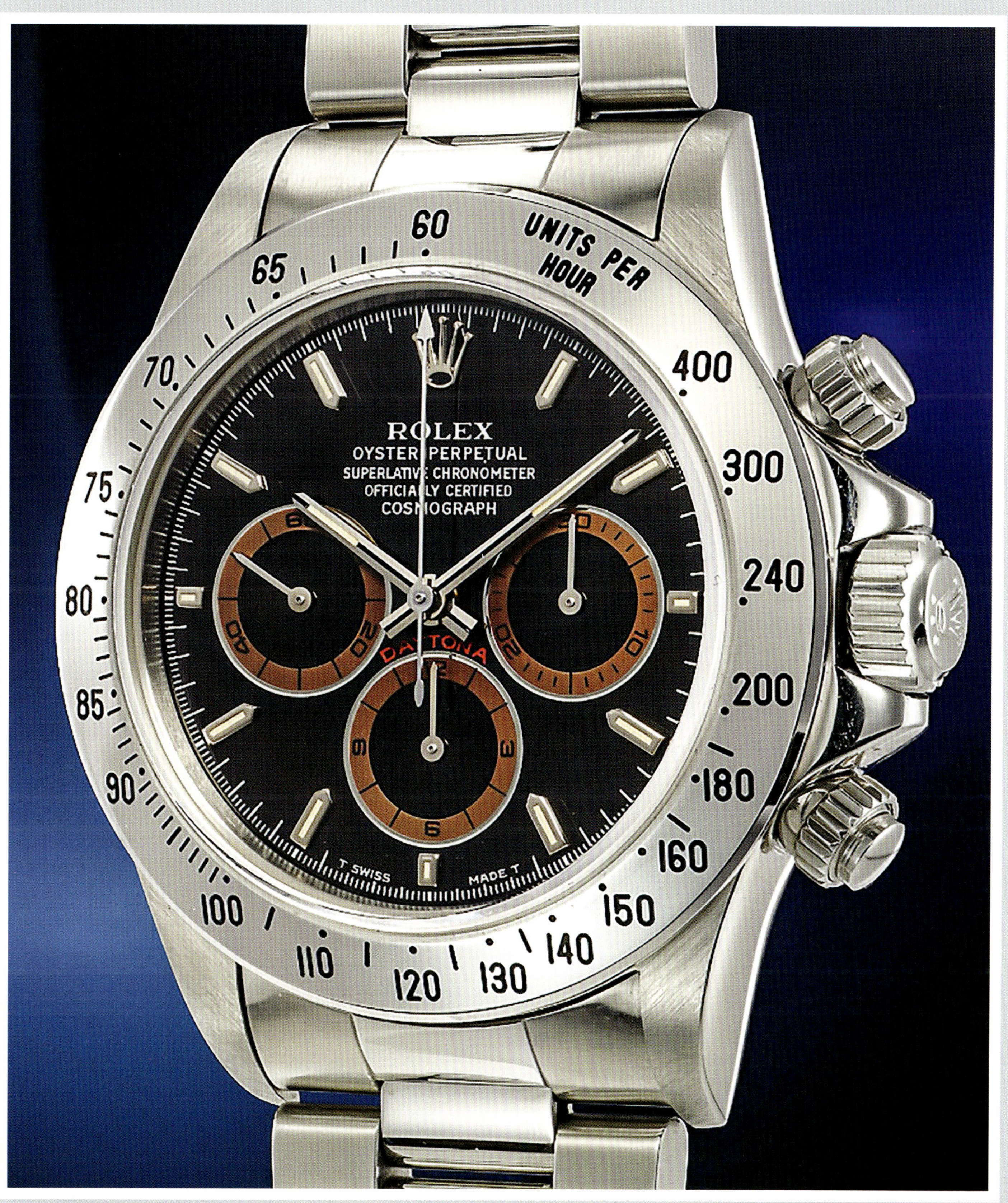

The Rolex Daytona Patrizzi model is a Daytona reference 16520.

Alexandre Ghotbi

SELLING THE HOLY GRAIL

Alexandre Ghotbi, Head of Watches, Continental Europe and Middle East, of the Phillips auction house, discusses the evolution of Rolex's positioning on the auction market, recalling the historic sale of Paul Newman's Daytona.

According to the consulting firm McKinsey & Company, the secondary watch market is set to become the fastest-growing segment of the luxury industry, reaching a value that will settle between 29 and 32 billion dollars by 2025.

We asked Alexandre Ghotbi of Phillips what role Rolex will play in this context.

'We must firstly bear in mind that there are two types of secondary market. The first represents a simplified access route to the watches present in the retail sector, which, however, not everyone can purchase because they are not immediately available. In this area, the prices are more regular, and the desires of customers reflect the desires of the primary market. In the market for historical pieces, on the other hand, prices and selections follow various forms of logic, comparable to those in the art sector: certainly, a watch is comparable to a work of art, of which its aesthetics and its technical genius can be recognised. Rolex plays a central role in this context, as there has been, and continues to be, a significant demand for watches from this brand.'

Over the years, Rolex has introduced innovations that have influenced the market. In the fifties especially, it created a series of models, initially conceived as professional watches, that went on to become icons of watchmaking and a landmark for the sports-watch market.

Moreover, Rolex watches have a distinctive design, always characterised by certain stylistic elements.

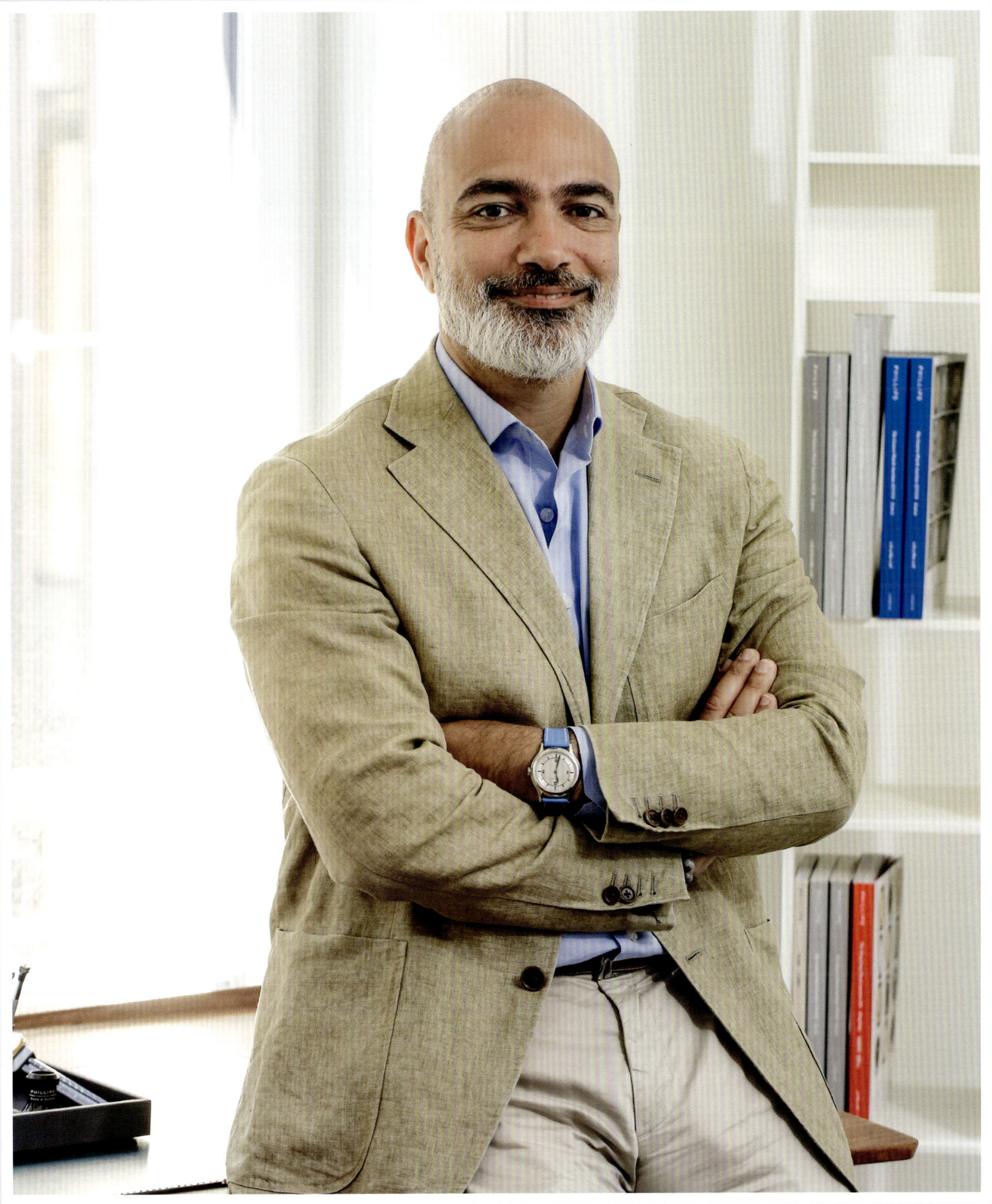

Alexandre Ghotbi, Head of Watches, Continental Europe and Middle East, at Phillips auction house.

'Formal consistency certainly plays an important role in adding value to a company's branding as a whole, and Rolex is recognised as having a consistency that supports the brand's values.'

The reinforcement and pursuit of this coherence has been a feature of Rolex's strategy in its hundred years-plus of activity, during which only a small number of high-level executives have taken their turn, committing themselves to reinforcing, over time, the original decisions that have characterised the company's philosophy.

'The CEOs that have inherited management of the company have been able to develop the intangible values associated with the brand, evoking quality, a quest for beauty, and the desire to succeed in life. The Rolex brand is associated with luxury, and yet it reaches a plurality of people: from the student to the 'man of success', everyone aspires to own and wear a watch bearing the crown emblem.'

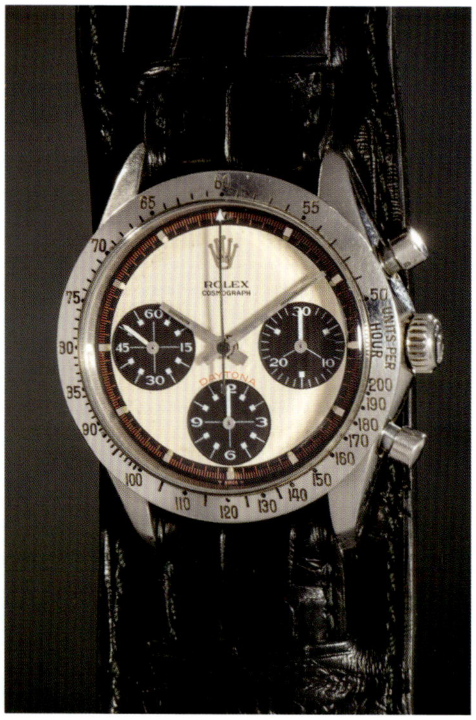

Daytona reference 6263, produced from 1963 – the year the Daytona was first produced – until about 1970. The actor's wife, Joanne Woodward, had chosen, in great taste, a Daytona with the dial that Rolex termed 'exotic', a highly sought-after variant in all vintage watches.

Despite its being a brand that sits at the top of the pyramid of luxury, and therefore exclusive, Rolex creates an effect of inclusivity among those who purchase – a feeling of being welcome to the club of 'those who made it'. The brand does not in fact generate a contrasting reaction, akin to one that occurred in China: when a higher number of women could afford to purchase Louis Vuitton bags, women of higher social classes began to only purchase Hermès to distance themselves from a brand that was now considered too inflated. 'This does not happen with Rolex; the sharing of an ideal, a lifestyle and passion for a product prevails.'

As director of the watch division for both Europe and the Middle East, Ghotbi has a broad view of the market in various parts of the world.

'Our customers are equally distributed between Europe, the United States and Asia, and we don't notice great differences in tastes on these three continents. The Middle Eastern market is certainly the youngest. Just consider that only five years ago, the concept of the secondary market did not exist in Near Eastern countries. Here, customers belong to a group between 28 and 35 years of age, who are oriented toward the purchase of more recent timepieces: watches from the seventies, for example, or more modern ones. Jewelled editions of watches, such as the Rainbow model, are also very popular. It is, however, difficult to say whether this trend is destined to last. Tastes change, and in the future, people could select watches that they do not take into consideration today. On the other hand, the search for quality and rarity is constant.'

Certainly rare, if not unique, was the Paul Newman Cosmograph Daytona that belonged to the actor himself, which the Phillips auction house sold in 2017 for $17.8 million. Cosmograph Daytona watches were introduced by Rolex in 1963 and designed for racecar drivers, hence the name, which refers to the Daytona International Speedway motor-racing circuit in Florida. Rolex offered a wide selection of dials for the Cosmographs, calling them 'exotic' due to their colour variations; these dials later were given the name of 'Paul Newman', since the actor wore one with an exotic dial.

The legend of the Daytona, created for a few and today desired by many, evolved around Paul Newman's popularity. The actor had a passion for speed – to such a degree that his wife Joanne Woodward, genuinely concerned about the risks he faced in personally shooting all of his scenes for the film *Winning* in 1968, gave him a Cosmograph Daytona with the words 'DRIVE CAREFULLY – ME' engraved on the case back. The dedication, in its own right, with the final 'me' revealing the couple's great affection and intimacy, and for the person to whom it was dedicated, was worth tens of millions of dollars.

'Selling Paul Newman's Cosmograph Daytona with Paul Newman dial was like having the primary source of collecting in front of you. It was an object loaded with historical significance – not just generally, but for collecting itself. It was when the value of vintage wristwatches began to appreciate in the 1980s, especially with the quest for unique models that could tell a story or were linked to big names, that the collecting market really grew.'

Another legendary watch auctioned by Phillips was the reference 6062, nicknamed 'Bao Dai' after its owner, the last emperor of Vietnam. In the spring of 1954, a meeting of the world's powers was held in Geneva to negotiate the future of Vietnam following the long-fought First Indochina War. The Vietnamese emperor, better known as Bao Dai, or 'keeper of greatness', was also present. Between conferences, the emperor went to Philippe Beguin, a historic Rolex dealer, where he declared his intention to purchase the most exclusive Rolex ever made. After several politely declined offers, Beguin contacted Rolex headquarters directly, from which an emissary was sent, bringing with him a timepiece that perfectly suited the emperor's tastes. Produced in just a few hundred pieces from 1950 to 1953, and with an annual calendar complication, it is the most complicated model ever made by Rolex and is still considered a milestone of collecting today.

In 2002, when the original owner's family auctioned it, again with Phillips, it sold for $235,000; but fifteen years later it would fetch 5 million francs, far exceeding the initial estimate of more than 1.5 million.

Among the many record-breaking Rolexes is a GMT-Master worn by Marlon Brando on the set of the movie *Apocalypse Now* in 1979. While filming in the Philippines, a member of the production staff advised Brando not to wear the easily recognised Rolex in order to make his character more believable, since he was playing an army colonel lost in the jungle. Brando then decided to have the overly conspicuous rotating bezel removed, leaving the shiny metal beneath. This peculiarity, and its case back engraved with 'M. Brando' by the actor himself using a special electric engraving pen, made this object unique – so unique that it reached the figure of $1.9 million in 2019. Another record-breaking example was the white-gold Daytona dubbed the 'Unicorn', created in 1970 on a special order from a major German retailer, which fetched 5.93 million Swiss francs. In 2019, a specimen of the reference 4113 split-second chronograph fetched 2.4 million francs. It was the first and only split-second watch Rolex ever produced, rare and unobtainable.

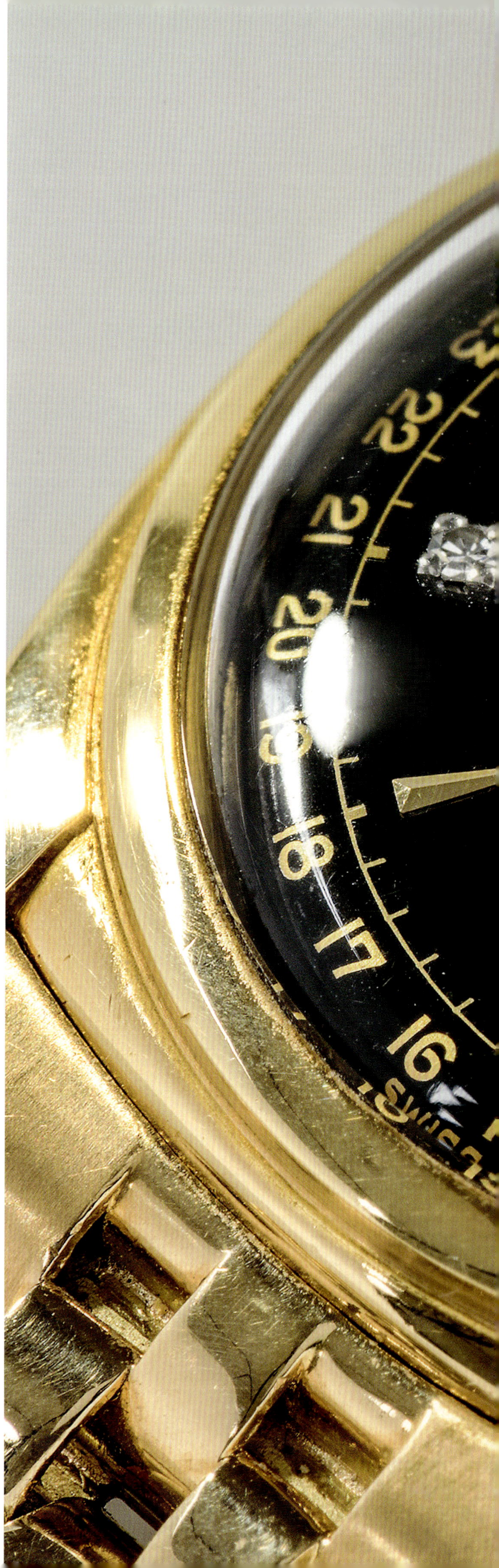

The reference 6062 triple-calendar moon-phase watch made for the Vietnamese emperor, for whom it was named 'Bao Dai'. This specimen is enhanced by a black dial with five diamond indices, present in only three other examples of the 6062. The diamonds' arrangement, however, differs from that of the other three watches, making this timepiece unique in all respects. The black dial with a lacquer finish, paired with the diamonds, lends the watch an unparalleled charm, all of it housed in a 36mm yellow-gold Oyster case.

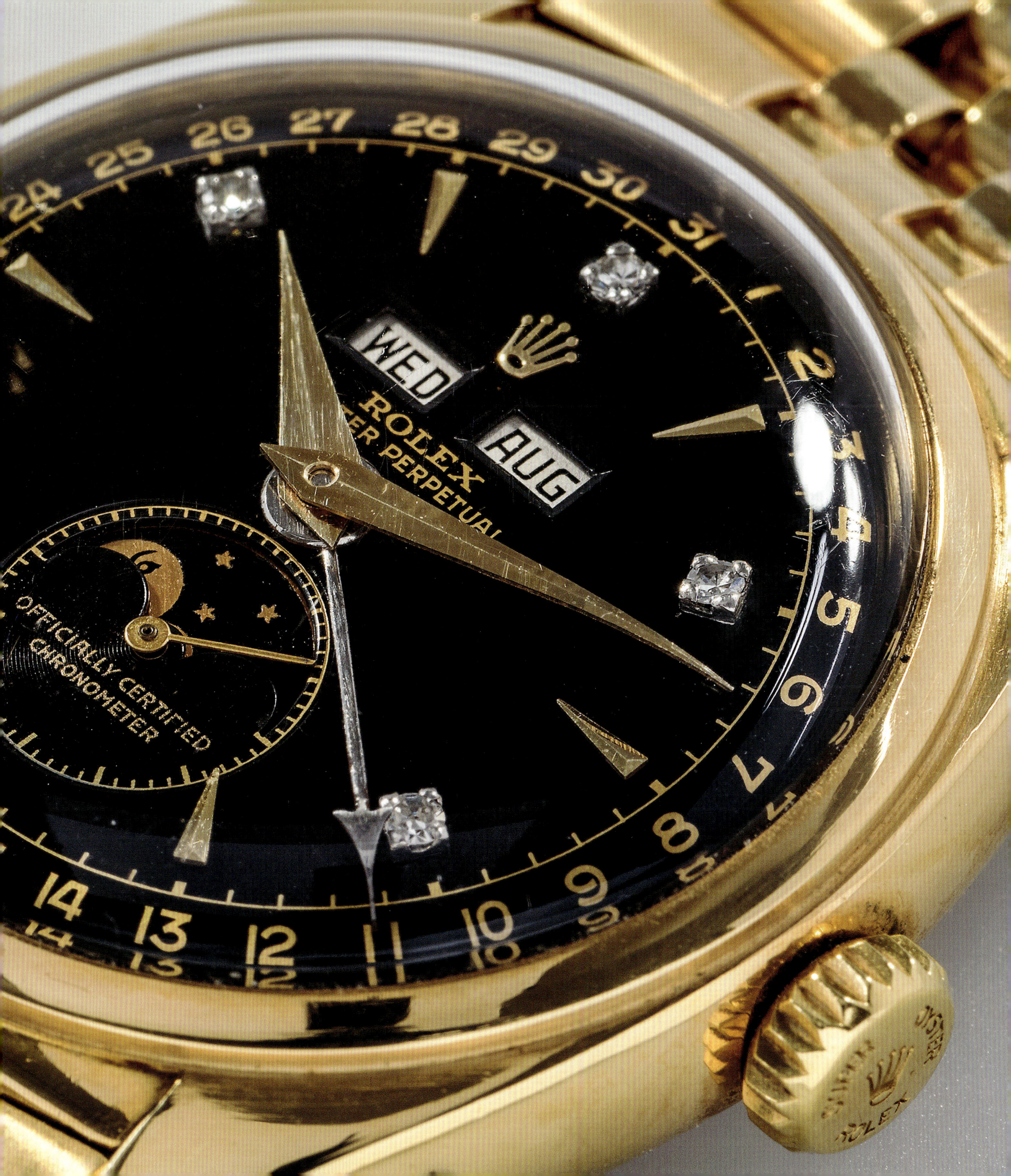
WED
AUG
ROLEX
OFFICIALLY CERTIFIED
CHRONOMETER

Jan Schoormans

A PSYCHOLOGICAL PROFILE OF THE ROLEX LOVER

Human experience is deep, personal, and potentially infinite. From the first cry of the newborn to the last sigh of the old person passing away, it is the person – the unique individual – who creates one's personal world through meaningful objects in one's immediate environment.

Jaan Valsiner, *An Invitation to Cultural Psychology*

The psychological levers that underlie luxury consumption are many and complex. In 1899, the social scientist Thorstein Veblen sharply criticised what he called the 'conspicuous consumption' of upper-middle-class America. The rich were so obsessed with their social status, he wrote, that they would do anything to show it. In more recent times, the philosopher Thierry Paquot has understood luxury consumption as a stimulus for the evolution of humankind, a driving force for progress and economic transformation, liberating them from a vision linked to the accumulation of useless status symbols and instead interpreting luxury as a search for exclusivity through which people express their individuality. Individuality is indeed central to the definition of a style based on personal preference, therefore situating itself outside of fashion and allowing for freedom of choice. From writer Oscar Wilde to sociologist Georg Simmel, fashion has been viewed as a 'tyrant' – yet one of excess – and, due to its ephemeral nature, a fleeting one. It thus makes no sense for an individual to automatically conform to the trends of a period; if it cannot be made personal, it will never be elegant. The search for a personal style is an expression of character, and the choice of a wearable object is an extension of one's own individuality.

According to the theories of Schoormans, a specialist in consumer behaviour from Delft University of Technology in the Netherlands, it is possible to attribute a personality to objects as one would to human beings. Together with Peter van der Bel, who specialises in linguistics, Schoormans applies a

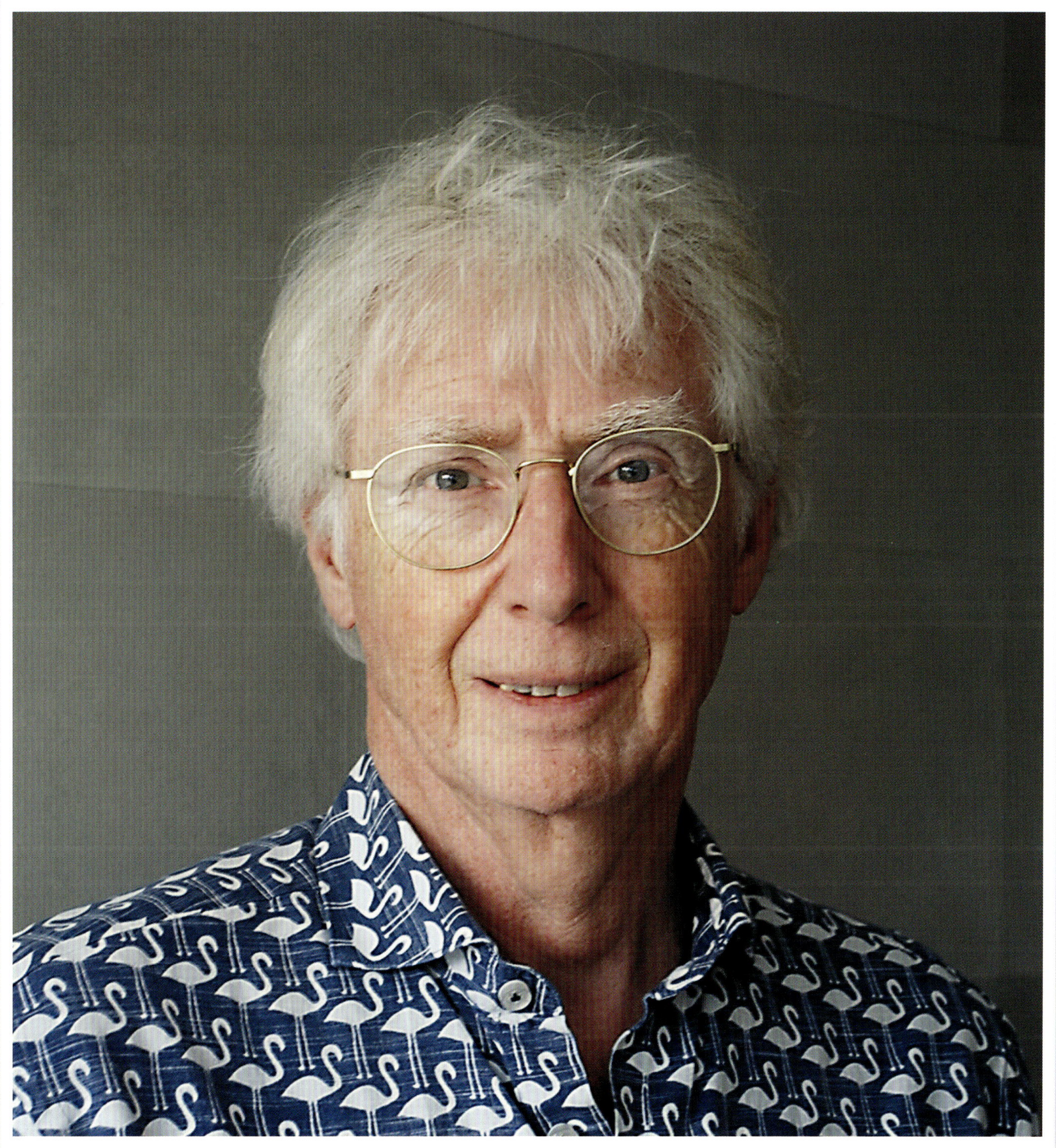

Jan Schoormans is Professor Emeritus of Consumer Behaviour at Delft University of Technology in the Netherlands. His research over the years has focused on the development of the concept and the visual evaluation of a product's personality. With his research, he has shown that there is a correspondence between the personality of a human and the personality of a product. Together with Peter van der Bel, he founded the Centre for Applied Product Personality Research (CaPPr) in 2011 based on these insights. The centre designs and uses visual assessment methods to evaluate human, product and service personality.

model for defining products called OCEAN, delineating the personalities of large brands, Rolex among them. According to this model, their personalities can be divided into five (also called the 'Big Five') traits: Openness to Experience, Conscientiousness, Extroversion, Agreeableness and Neuroticism.

Openness to Experience measures an individual's behaviour in regard to new technological and social phenomena, as well as sensitivity toward aesthetic aspects of life. 'A high score in this trait defines a curious, multifaceted, creative, original and unconventional personality. Conversely, a low score marks people who are conventional, pragmatic, realistic or with few interests. Conscientiousness is the second dimension, referring to behaviour and the way people control and direct their impulses. A high score is typical of organised and disciplined personalities, and a low score is typical of an impulsive personality', Schoormans explains.

The trait of Extroversion evaluates an individual's disposition toward interacting with the external world. 'Extroverts are sociable people who enjoy social relationships. They are talkative and able to assert themselves with others. A low score, on the other hand, is typical for temperate and calm people. This doesn't mean they are antisocial; rather, they prefer privacy in relationships.' The trait of 'Agreeableness' is useful for understanding the value that an individual assigns to harmony and to what degree an individual is capable of placing themselves in another's position. Those obtaining a high score in this respect are compassionate, affable and gentle. A score on the opposite end is received by competitive people. Neuroticism indicates emotional stability. 'People with a high score are nervous and subject to stress, and a low score identifies a non-emotional personality, with high self-esteem and self-confidence.' Research conducted by Schoormans and van der Bel is founded on the application of the OCEAN model and based on the profiling of people who choose to purchase a particular product.

According to Schoormans and van der Bel's analyses, product personality corresponds to a profile of human personality characteristics, and it can be used to describe a specific product and to differentiate it from others. A product's personality is relevant since it corresponds to an expression of customer preference, as consumers tend to prefer to select products with a personality like their own.

The models developed by these scholars also demonstrate a congruence between a person's psychological profile and the kind of products they prefer to purchase – goods or experiences, for example. In the case of Rolex watches, there is also an affinity between the customer and the product's personality.

Having applied the OCEAN model to a sample of consumers using an app, Schoormans has defined the five traits characterising the personality of Rolex timepieces and Rolex lovers. The results show that Rolex's first trait is, above all, that of being 'conventional', as the brand mainly expresses values linked to tradition. In regard to its second trait, concepts of 'accuracy' and 'precision' emerge, predictable considering the product category, and these concepts respond to expectations concerning the function of a Rolex. People expect a Rolex watch to do what it is meant to do: always tell the time, with high precision and quality. These characteristics, however, also combine with the values of 'simplicity and consistency'; these apply to the uniform messaging from the brand, the harmonious aesthetic through which it is expressed and its strong connection to the company's heritage that it has kept over time.

For its third trait under this model, Rolex receives a score that slides in somewhere between 'liveliness' and 'modesty'. This indicates that people perceive these watches as the preference of customers who are neither very extroverted nor especially self-centred. Rolex watches can be worn by sociable people who, however, also see the value of their personal space.

Another Rolex characteristic is that it could be defined as 'tolerant'. The brand's values maintain a balance between 'audacity' and 'attention to others' since, while it works to establish its own style, it is also committed to satisfying the needs of its user.

Lastly, Rolex can be defined as 'concrete' and 'confident'. The brand conveys an impression of calm and security and those who select it demonstrate a strong personality that is not easily swayed.

Remi Guillemin

PRECIOUS RARITIES

Rolex offers a wide range of watch models fit for any wrist, ranging from the more classic to the professional. All combine functionality and style, united by a design that has remained iconic over the years. Remi Guillemin, head of the Europe Watches and Wristwatches department at Christie's, explains what collectors are seeking.

Rolex has kept its success within a marketing strategy which the company has carried out over time through various channels and methods of action, all with the aim of positioning itself in the minds of consumers as the luxury good par excellence – not just in the context of watchmaking, but in a more general context. With a strong connection to sporting events, the Swiss company has attached its name to the most exclusive of sporting activities, limited to certain social classes – from Grand Prix races to golf tournaments, from tennis Grand Slam tournaments to horseback riding and sailing. The scheme features leading figures from sport, economic and social scenes, who have helped to define the product's exclusivity. Added to this is the increasingly frequent presence of prominent celebrities representing the brand. Ultimately, Rolex has managed to enter the collective imagination as a successful brand with a top position in the luxury sector.

'The Geneva *maison*'s watches have high prestige', comments Guillemin. 'Their value is dependent on their high quality, which is always guaranteed by their manufacturer. Success, however, is born of a set of intangible characteristics defining these products, which have been supported over time through messaging centred more on appealing to emotions than on providing information.'

Rolex timepieces are representative of cultural heritage, being suitable heirlooms for passing between generations within a family. This is also thanks to a design that, apart from being completely recognisable, maintains its identity and value over time.

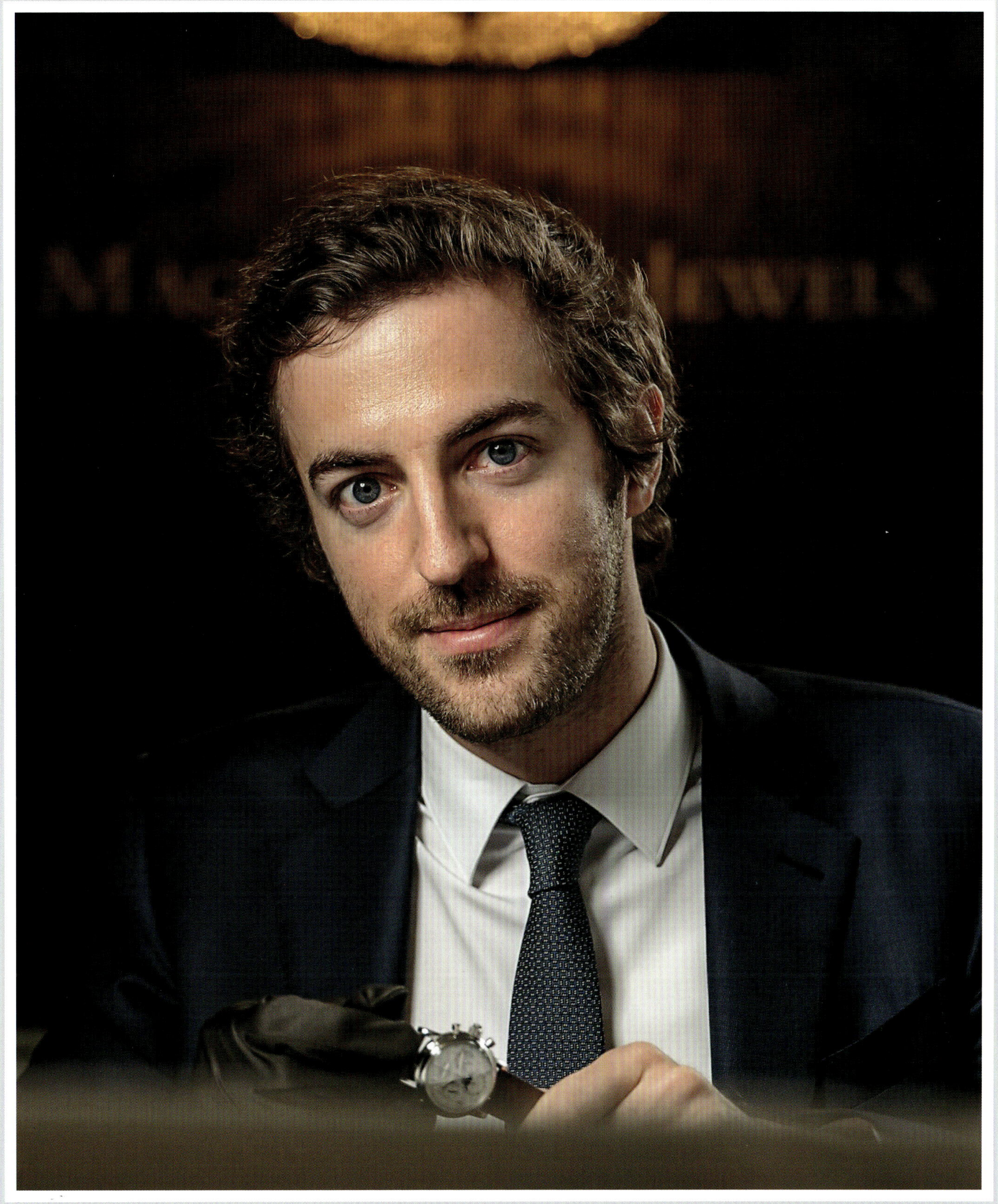

Remi Guillemin, Head of the Europe Watches & Wristwatches Department at Christie's.

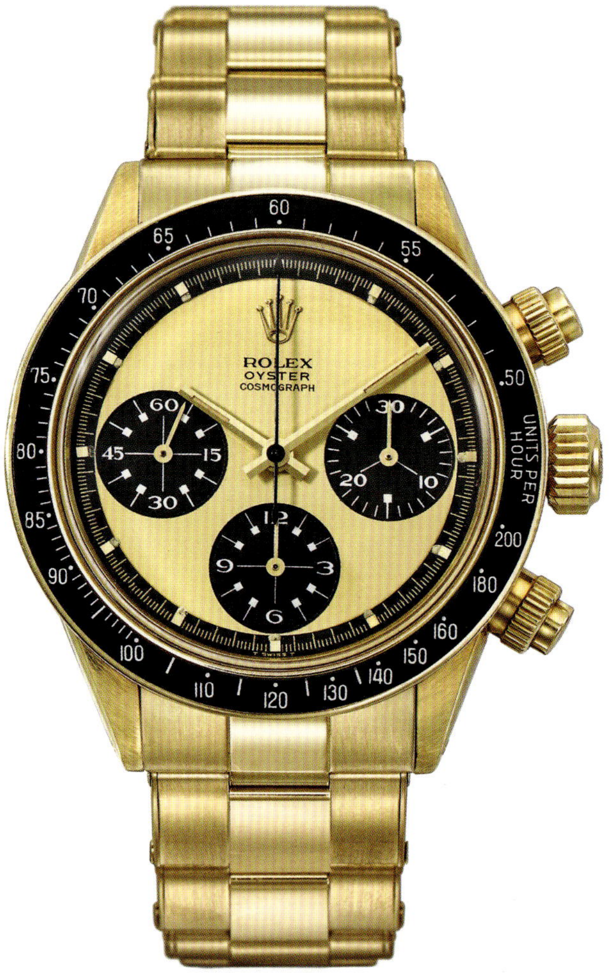

Rolex Daytona reference 6263, produced from 1963 – the first year the Daytona was produced – until about 1970.

'Rolex watches are always contemporary: with their unmistakable shape, they tell a story of ingenuity and sophisticated technique. They are comparable to the Kelly bag by Hermès or Porsche's Carrera model.'

Demand for Rolex watches remains high for both modern and 'vintage' models, without too many differences between requests from different areas of the market like Europe, Asia and the United States. The audience of collectors is very broad; it includes those who are seeking a very specific type of watch – divers, for example – and yet a Rolex Submariner is also counted among the possessions of Rolex lovers who hunt for and purchase various models.

'For everyone, a piece's distinctiveness and rarity define its value, which can reach very high figures', Guillemin comments. Among such watches is the reference 6263 Daytona Paul Newman, with an unusual light lemon-yellow dial, sold in 2022 at Christie's Geneva for 3,414,000 Swiss francs. It is one of three known examples of the 18-carat gold edition with a dial of this colour and nicknamed 'The Legend'. Also known as the Daytona Lemon, it owes its name to its dial's yellow colour as described just above, which is very different from the model's more common shades ranging from cream to champagne. It is also distinguished by the white lettering on its sub-dials. Rolex typically produced Paul Newman Cosmographs in yellow gold with champagne or gold numerals and indices on their sub-dials, which blended with the hue of the dial as a whole. In this watch, on the other hand, the white numerals contrast sharply with the dark background.

'The watch dates to 1969. In the 1960s, the *maison* often experimented, changing dials and colours, resulting in the creation of rare pieces that are now highly sought after', says Guillemin.

The Rolex Deepsea Special No. 1 sold at auction for the sum of 1,890,000 Swiss francs in 2021. It is a special watch, as one of its first prototypes: it was, in fact, the specimen that

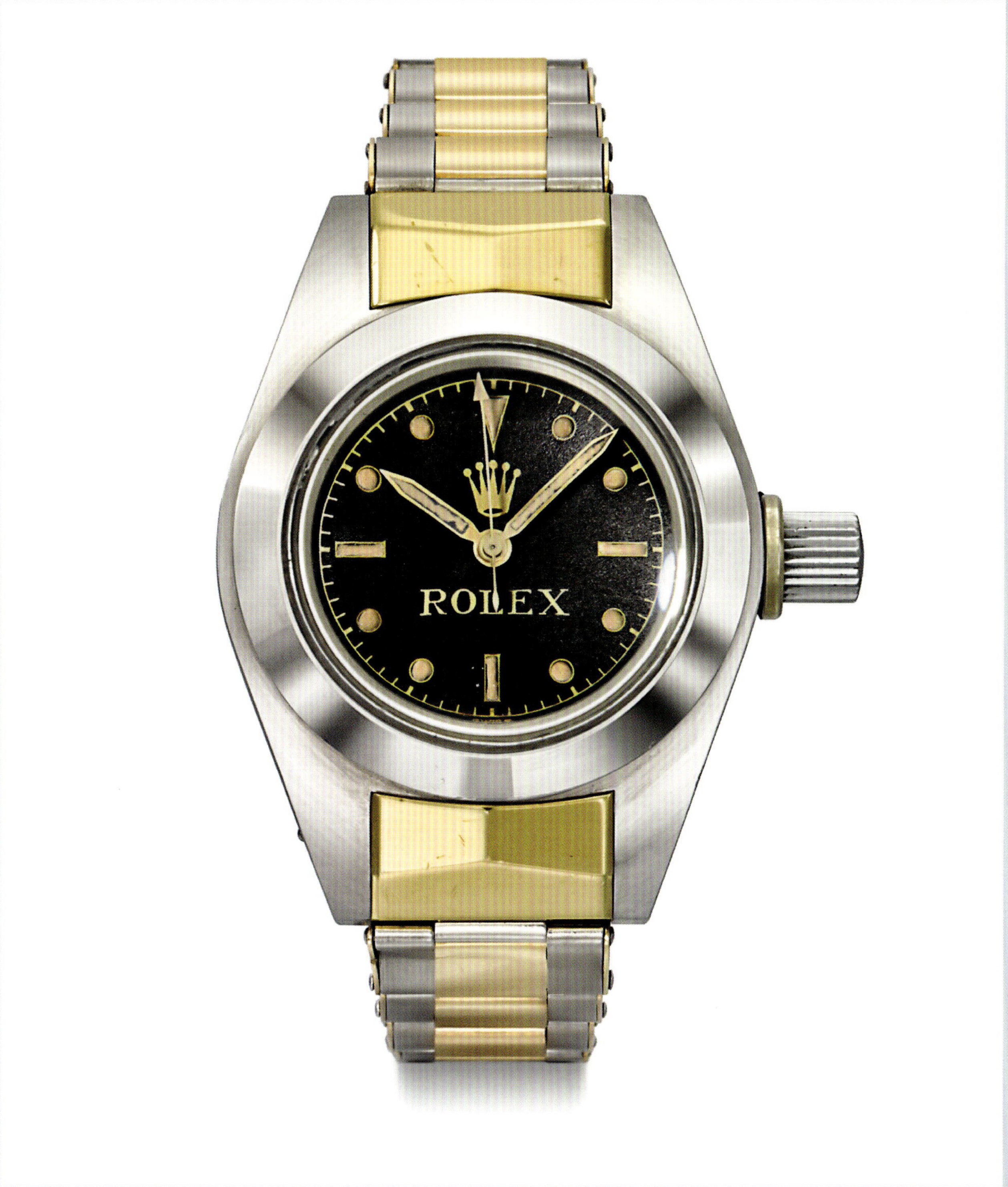

In 1960, Rolex embarked on a second mission with another prototype. This new 'Deepsea Special' was created to withstand the most extreme conditions. It actually plunged to over 10,000m below sea level in the Mariana Trench – the deepest known point on Earth – with Jacques Piccard and Don Walsh. The first Deepsea Special was, however, Rolex's point of departure for developing diving watches.

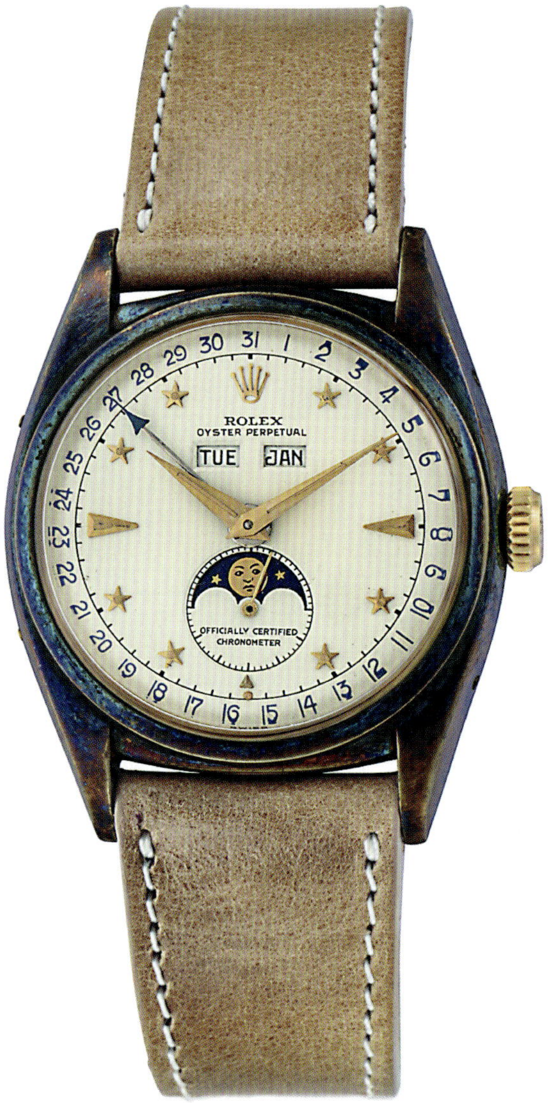

Reference 6062, with a silver dial. This exceptional 6062 'Stelline' or 'Star Dial' watch was introduced in 1950 and produced for just ten years. Nicknamed 'The Dark Star', a dark patina developed on its 18ct yellow-gold case over time and is a unique characteristic of this model.

on 30 September 1953 was attached to the hull of Auguste Piccard's bathyscaphe *Trieste* for its inaugural test in water at a depth of 3,150 metres. Another sale in the over one-million-euro range was that of one of the few complicated models made by Rolex: an automatic wristwatch with a triple calendar and moon phases, reference 6062, in 18-carat gold. The piece is extremely rare, as a dark patina covered the yellow-gold case over time, rendering it one of a kind. Nicknamed 'The Dark Star', it fetched $1,572,500 in New York in 2018.

Also in 2018, the auction house recorded another exceptional sale, that of a reference 6538 'Big Crown' Submariner, sold for $1,068,500. The feature that made this watch unique was its 'gilt' Explorer dial with Arabic numerals at 3, 6 and 9 o'clock, a true rarity for the reference. Another important detail was the inscription of its water resistance, rendered in red rather than white.

It is no secret that gem-set Daytonas are some of the 'hottest' watches on the market. Since the first automatic Daytona's very introduction, Rolex has offered its customers the opportunity to purchase the iconic model in highly precious editions, hence the emergence of models enriched with rubies, emeralds and sapphires. Although production of these was mostly from the 1990s onwards, there are two manual Daytona references in which Rolex had already demonstrated its mastery in the art of gem-setting, references so rare that both their production numbers added together amount to just fifty units. These are references 6269 and 6270, featuring a diamond-encrusted dial and bezel. Christie's specifically auctioned a reference 6269, which sold in Geneva, in 2022, for 1,614,000 Swiss francs.

Details make the difference, especially when it comes to Rolex, where the apparent imperfection of a dial can transform even the most common reference into a rare prize.

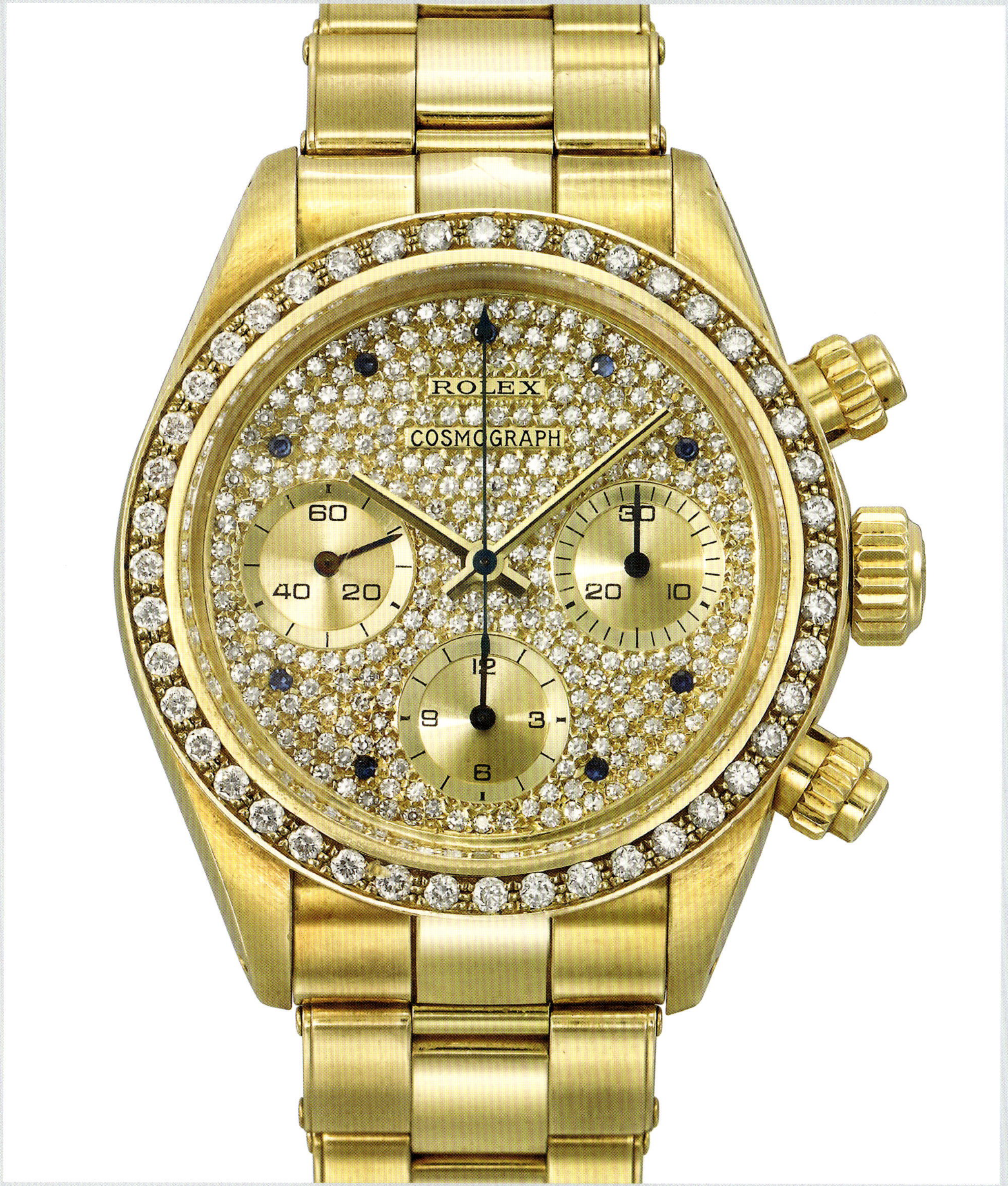

In the Daytona reference 6269, the brilliance of the diamonds meets that of its gold case. The dial is set with 240 brilliant-cut diamonds and nine sapphires for indices.

Beppe Vicenti

THE ARCHETYPE OF THE SOVEREIGN

The positioning of the Rolex brand is that of a leader, conveying authority, security and power. The following is a reading of these values via a selection of its historical advertisements.

In the early twentieth century, the Swiss psychologist and analyst Carl Jung developed his theory of archetypes. According to this theory, we each come into the world with a primitive model arising from the collective unconscious; this realm is derived from human experience over the course of the millennia and used as the basis for one's construction of a personal identity.

Jungian archetypes represent a typology of the human soul and are used to identify characteristic traits in a person's identity. They are allocated to four main quadrants, defined according to fundamental qualities. Archetypes have a very strong symbolic value, represented as characters in stories, myths and fables; in reality, these are schemes of the mind that bring together innate needs, motivations, objectives, fears and aspirations.

For a brand, it is useful to understand its dominant archetype, as it helps to construct a narrative that engages with the ideal customer, building an emotional affinity, and to define a strategy.

According to Beppe Vicenti, an expert on processes of signification in the world of brands and advertising, Rolex's dominant archetype is that of the 'sovereign', a king, queen or other ruler. The sovereign is a born leader, with a desire to govern and control. A sovereign aims to create something, whether a successful business, community or family; they show authority and define procedures. The sovereign archetype is one of seeking order and control, precision and order.

Vicenti explains 'It is important to emphasise that the sovereign archetype is symbolised in the royal crown, the exact mark of one's connection to the divine and of the power it lends the individual in the physical world.'

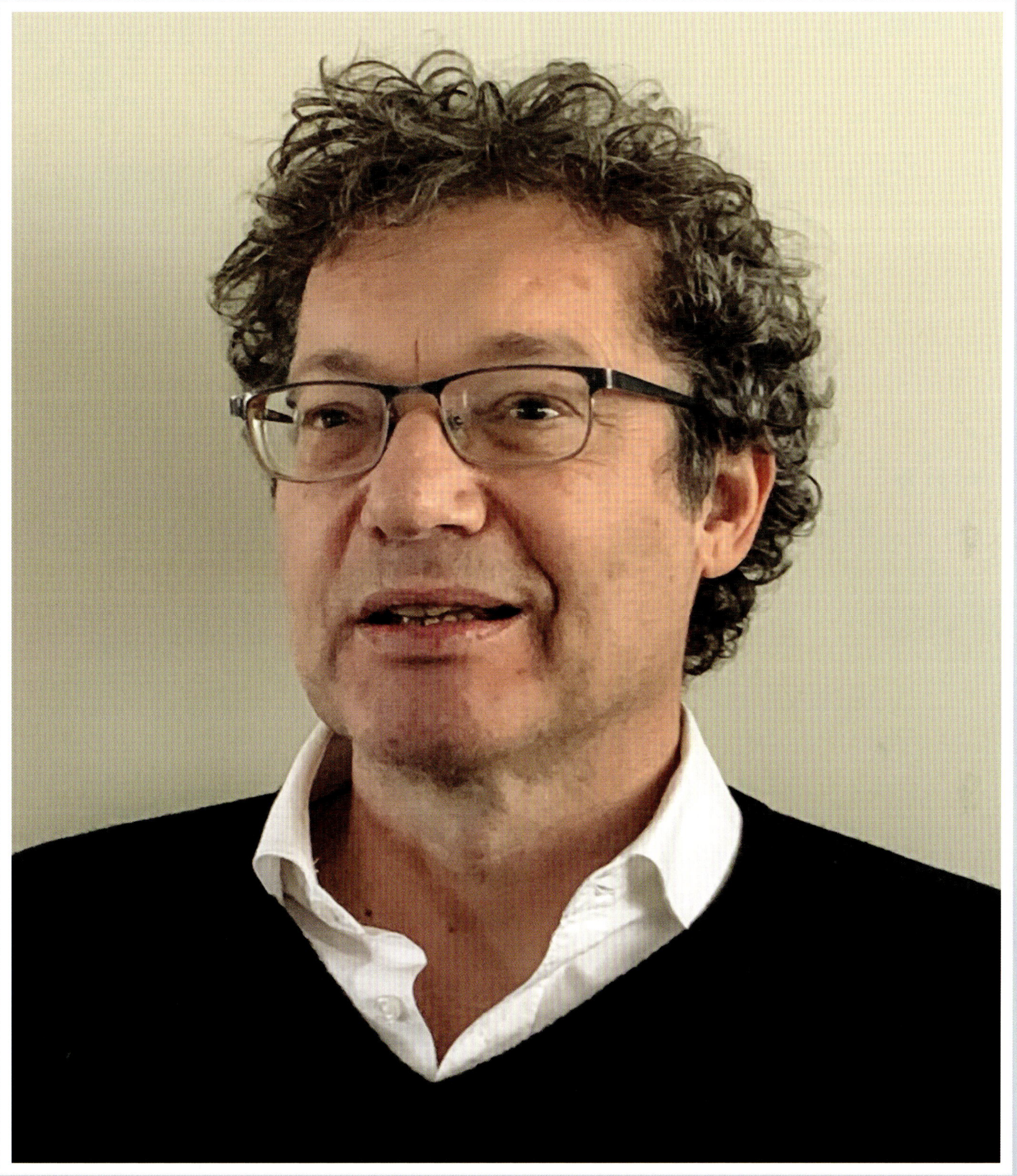

Having graduated in philosophy with a specialisation in psychology in Milan, Beppe Vicenti taught and conducted research for several years at the University of Milan's department of Dynamic Psychology. He was also responsible for annual courses on Meeting and Interviewing Techniques at the same department's Postgraduate School. For more than 25 years, he has dedicated himself to motivation research, collaborating with various institutes and even directing research in the qualitative department of IPSOS. In 2001, he founded NEXTEST, an organisation specialised in strategic research.

The high value of Rolex as a brand is the outcome of a journey that began long ago, dating to the company's founding years.

Advertising campaigns for the Oyster Perpetual from the 1950s underscore precisely this linkage of the brand with power.

'Men who guide the destinies of the world wear Rolex watches' was the *maison*'s claim. The phrase accompanied various images conveying authority: a captain plotting the route, a gentleman speaking to an audience, a man seated inside a car waving to a crowd, three figures shown from behind seated before a group of photographers, politicians in a context reminiscent of a UN meeting, someone preparing to record a speech for major international radio stations, a highly decorated soldier. These contexts are evocative of familiar situations, and while faces cannot be revealed, the text accompanying them intervenes to resolve this issue, stating: 'We cannot mention their names or show pictures of them. It would not be fitting to do so, for they include royalty, the heads of states, great service commanders. But we invite you to look carefully at the next pictures that you see of them, at their wrists as well as their faces and clothes. You will notice that in almost every case they wear a wrist-watch. That watch will most likely have been made by Rolex of Geneva.' The advertisements contain product information about the Oyster Perpetual, the 'culmination of three Rolex triumphs', the receipt of their first Official Timing Certificate for a wrist chronometer in 1910, the invention of the Oyster case in 1926, and the self-winding movement in 1931.

Over the years, advertising became more graphic and less discursive, yet the concept of power remained, suggested by a series of headlines announcing 'If you were flying the Concorde tomorrow you'd wear a Rolex', 'If you were speaking here tomorrow you'd wear a Rolex', 'If you were negotiating here tomorrow you'd wear a Rolex', 'If taming oil well fires were your job you'd wear a Rolex', or 'If you were sailing alone around Cape Horn tomorrow you'd wear a Rolex'. This series of ads, which appeared in the late sixties and the seventies, had in common an idea of adventure and courage, always in 'exclusive' situations. The text highlights the reliability of these timepieces, being faithful, valuable instruments: 'Because so much of the work is done by hand, it takes more than a year to make a Rolex. Which is why a Rolex will last you a lifetime.'

Linked directly to the power of its name, and a collective spirit, is a message

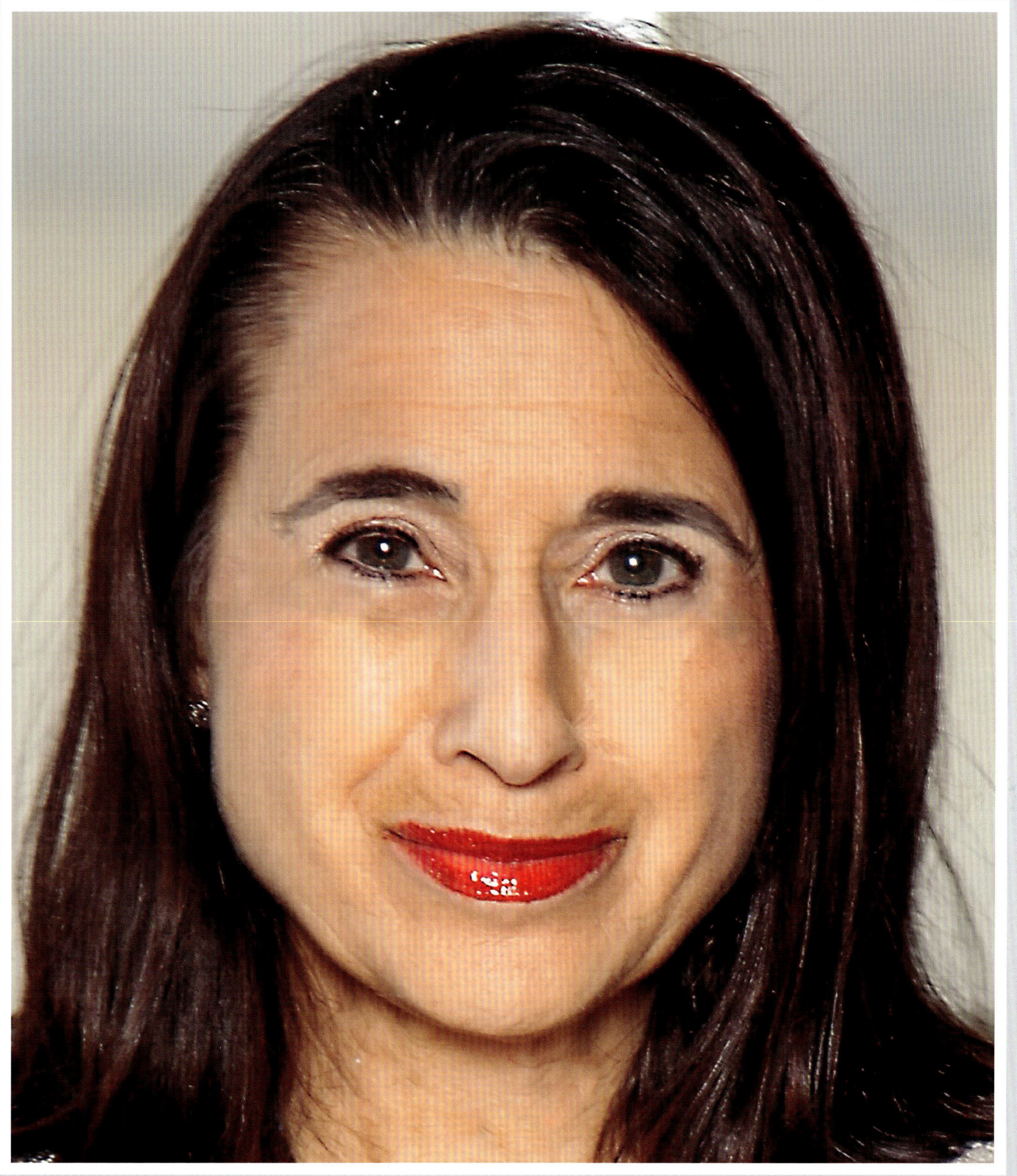

Daryn Schnipper

'When it introduced many of its premium models, like the Daytona, the Submariner, the GMT-Master or the Explorer, Rolex initially marketed them as "technical" watches, aimed at well-identified groups of professionals. There was then a knock-on effect, which was to create models touted as indispensable for racecar drivers, divers, pilots and explorers of all stripes. Over time, watches evolved into something more than just a time-telling device that a lucky few could afford, and Rolex watches became an object of desire for a large swath of people.'

The positive sentiment perceived toward Rolex, which is concretely demonstrated in purchase preferences, and the loyalty toward the brand and its products, are very high: What are the reasons for this attitude?

'Customers' loyalty to the Rolex brand lies in the company's fidelity to the models it offers on the market and in its stylistic consistency, which is expressed in the *maison*'s decision to update existing models both in terms of aesthetics and in terms of mechanics through subtle modifications that reinforce the brand's identity with potential customers. The dominance of Rolex is synonymous with success.'

According to Schnipper, customer preferences are directed toward models like the Cosmograph Daytona and the Submariner.

Sotheby's has in fact obtained record results for several Daytonas, including an example of reference 6264 with a 'John Player Special' version of the Paul Newman dial, which fetched $1.5 million at the auction house's British branch. The Rolex Daytona Paul Newman 'John Player Special' dates back to 1969 and its nickname derives from the John Player livery of the Lotus Formula 1 team's cars in the 1970s. It is one of the rarest configurations of the reference 6264, as most were produced in steel. Only ten examples were made in gold, and only a few of them feature the Paul Newman dial in black and gold, which was given the 'John Player Special' nickname.

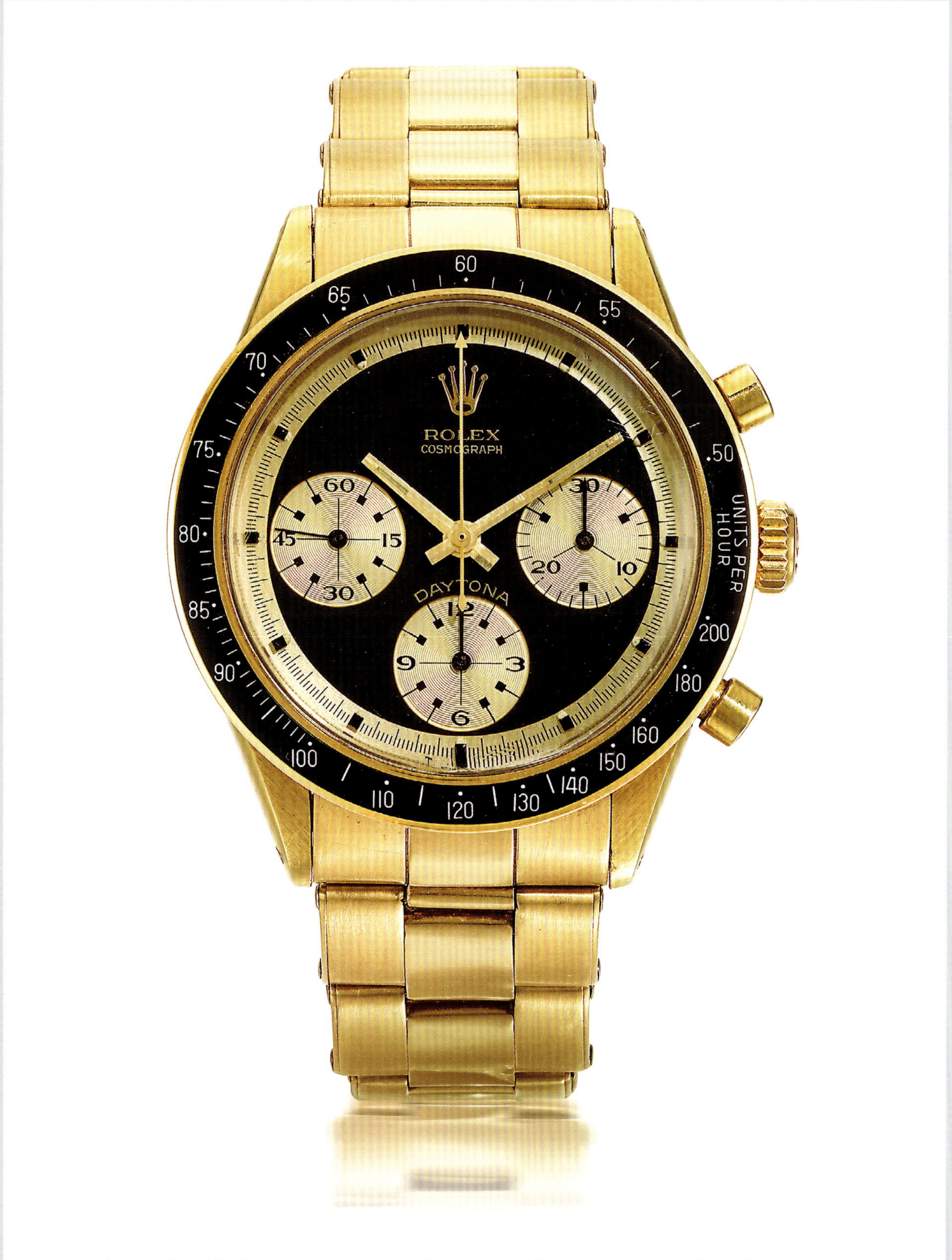

Daytona 'John Player Special': the black dial and 18ct gold bracelet make this watch a rare and unique piece.

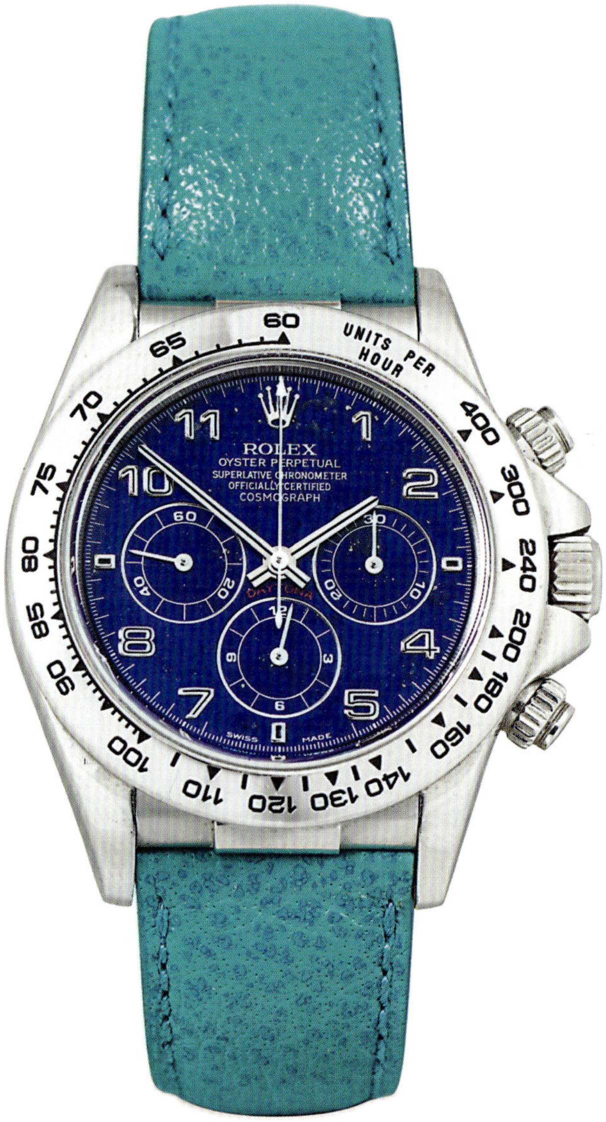

Daytona Lapis Lazuli: the platinum Daytona 16516 was crafted with a blue lapis lazuli stone dial.

Two more of the most recent Daytonas have achieved extraordinary results. These are two examples of the reference 16516 in platinum with hardstone dials. The first, with a lapis lazuli dial, sold for $3.27 million in July 2020. The second, meanwhile, sold for $3.1 million one year later at the Sotheby's Hong Kong Important Watches Auction; it had a turquoise stone dial. Both watches date to the 1990s and are animated by calibre 4030, with Arabic numerals on the dial.

This is certainly an unusual combination of materials for a sport model like the Daytona, which was predominantly made of steel, a metal better suited to meeting sporting challenges. Variants in yellow gold and white gold have been produced over the years, some versions set with diamonds and coloured stones. Platinum models are rare.

These Daytonas, having positioned themselves on the list of the world's most expensive watches, are special in that they are part of a group of five specimens that Rolex CEO Patrick Heiniger had had made and then given to friends and collaborators in the 1990s. Each limited-edition watch in the set features a 40-millimetre platinum case and hardstone dial. The models have been nicknamed 'Zenith' because their calibre is based on the Zenith El Primero movement. Three were auctioned by Sotheby's (a version with a black dial in Tahitian mother-of-pearl sold in 2018), and their configuration is now known; what remains secret, however, is that of the last two, said to each be unique. Perhaps one day the mystery will be revealed.

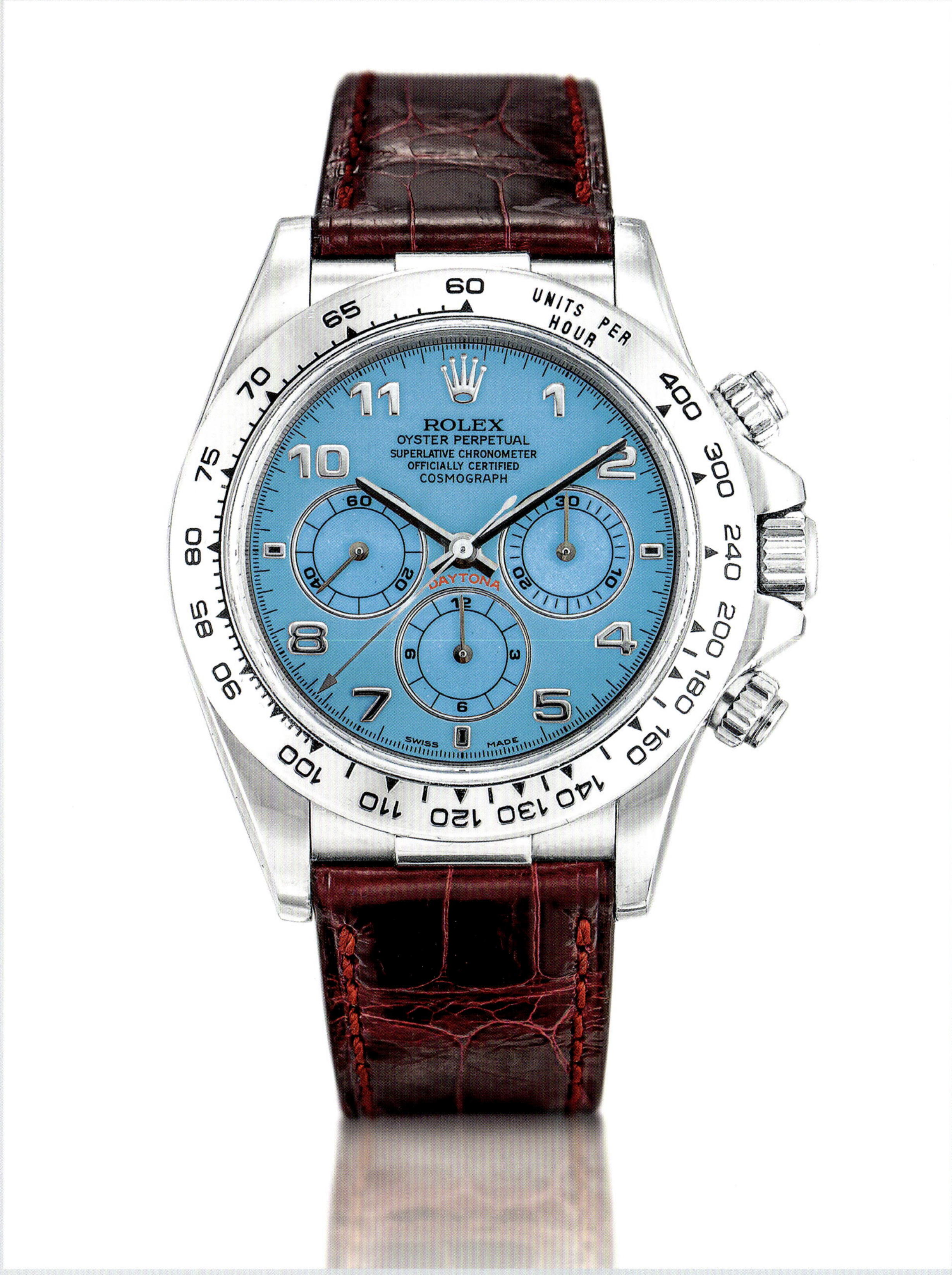

Daytona Turquoise: this particular model has a turquoise mineral stone dial so dazzling that it nearly sparkles. Its liveliness is offset by an alligator leather strap with a platinum buckle.

MARA CAPPELLETTI

A native of Milan, Mara Cappelletti graduated in foreign languages and literature from IULM University, Milan. She then studied the history of jewellery and the history of Eastern art at Sotheby's Institute of Art in London.

A freelance journalist, she is the author of several publications on the history of jewellery, ethnic jewellery and the *Maestri del tempo* ('Masters of Time') series published by 24 Ore Cultura as well as the *Investing in Wristwatches* book series, created in collaboration with Sotheby's and published by ACC Art Books.

Cappelletti designs and curates exhibitions on the themes of jewellery and fashion, as well as conducting research for their catalogues. Among these exhibitions was *Gioielli di Gusto e Stile Milano – Storie di Eleganza* (Milan's Tasteful and Stylish Jewellery – Stories of Elegance), held in February 2020 at Palazzo Morando – Milan's Museum of Costume, Fashion and Image.

She is president of the Associazione Culturale Stile e Storia, which deals with research and appreciation of historical heritage, highlighting the relationship between art, fashion and jewelry as part of the custom of their time.

Since 2019, she has been an adjunct professor in the history of jewellery at the University of Milan as part of the Master's program in Publishing and Communication, Fashion Cultures. Since 2020, she has been collaborating with the Raffles Milan Fashion and Design Institute, teaching the history of jewellery.

Cappelletti also holds conferences and workshops at museums, cultural associations and art galleries.